Advances in Sweeteners

Advances in Sweeteners

Edited by

T.H. GRENBY
Department of Oral Medicine & Pathology
UMDS
Guy's Hospital
London

BLACKIE ACADEMIC & PROFESSIONAL
An Imprint of Chapman & Hall
London · Glasgow · Weinheim · New York · Tokyo · Melbourne · Madras

Published by
Blackie Academic & Professional, an imprint of Chapman & Hall,
Wester Cleddens Road, Bishopbriggs, Glasgow G64 2NZ

Chapman & Hall, 2–6 Boundary Row, London SE1 8HN, UK

Blackie Academic & Professional, Wester Cleddens Road, Bishopbriggs, Glasgow G64 2NZ, UK

Chapman & Hall GmbH, Pappelallee 3, 69469 Weinheim, Germany

Chapman & Hall USA, Fourth Floor, 115 Fifth Avenue, New York NY 10003, USA

Chapman & Hall Japan, ITP-Japan, Kyowa Building, 3F, 2-2-1 Hirakawacho, Chiyoda-ku, Tokyo 102, Japan

DA Book (Aust.) Pty Ltd, 648 Whitehorse Road, Mitcham 3132, Victoria, Australia

Chapman & Hall India, R. Seshadri, 32 Second Main Road, CIT East, Madras 600 035, India

First edition 1996

Softcover reprint of the hardcover 1st edition 1996

Typeset in 10/12pt Times by Cambrian Typesetters, Frimley, Surrey

ISBN-13: 978-1-4612-8522-9 e-ISBN-13: 978-1-4613-1229-1
DOI: 10.1007/978-1-4613-1229-1

A catalogue record for this book is available from the British Library
Library of Congress Catalog Card Number: 95–80533

∞ Printed on permanent acid-free text paper, manufactured in accordance with ANSI/NISO Z39.48-1992 (Permanence of Paper).

Preface

The subject of sweeteners continues to advance and expand, but the progress that is being made may not be apparent for all to see, owing to changes that have been taking place in how research is funded and the locations where it is now mainly done. In former times scientific advancement was rated as a prized part of the output of academic research laboratories and institutions. Today, however, it is increasingly likely that major advances emanate chiefly from the research and development units of industrial and commercial enterprises and organisations. This means of course that the work becomes more focused on achieving specific marketing objectives, but because of the high level of commitment, cost and dedicated input required, publication of the findings tends to take a lower priority, and may actually be barred if there is any risk of loss of the commercial edge or advantage which has been one of the targets of the research.

Thus one of the objects of preparing this book has been to collect together information that might otherwise remain unpublished on advances in the field of sweeteners. Of the fifteen contributions which form the chapters, only 13% originate from academic departments, whereas in earlier books of reviews on similar topics, contributions from academic sources accounted for as much as 50% (*Developments in Sweeteners*, vols 2 and 3, 1987 and 1989) and 64% (*Progress in Sweeteners*, 1989). This is partly a reflection of reduced activity in the discovery and development of new types of sweeteners, and partly because of the growing realisation of the importance of technical matters such as the blending of sweeteners and their applications in calorie control, marketing issues and regulatory and legislative considerations, in determining which products arrive on the market and are available for the consumer or the manufacturer of finished food products to buy.

It may sound cynical, but it has been said that at present there is little incentive to seek and synthesise new sweeteners because the range that now exists is adequate for most purposes, the cost of developing new ones would be prohibitive, and the process of evaluating all their properties and completing a full programme of all necessary toxicological testing has become impossibly demanding. Emphasis has therefore shifted to improving the ways of using the limited number of sweeteners now approved and guiding other promising compounds through the regulatory minefields of the EU, the USA and other countries. It is a source of great satisfaction that detailed chapters on regulatory requirements and procedures on both sides of the Atlantic are included in this book, and it is hoped that they will provide useful information and guidelines for any individual or any organisation with the task of piloting a new sweetener through the legislatory processes of the EU or the USA. The authors had the brief of laying out the steps required as clearly as possible, and it is up to the reader to judge to what extent this has been achieved, but a hint of frustration at the length and uncertainty of outcome of some of the procedures may be discerned in places.

One point of difference from earlier books on similar topics is the reduced emphasis on the development of new intense sweeteners and the increasing interest in materials that can provide bulk as well as sweetness. Three chapters in this book deal with specific intense or high-potency products, while it contains five on various bulk sweeteners and their applications, with a sixth on polyols in general. This may be a reflection of three factors: (i) that it is considered that there is now an adequate range of intense or high-potency sweeteners available for table-top and most other applications; (ii) the difficulty and expense of steering a new synthetic sweetening agent through the stiff regulatory hurdles; and (iii) the perception that a particular need now is for an improvement in quality in confectionery and other products requiring bulk sweeteners, along with the support that re-formulation of products in this way can give to claims for non-cariogenicity, suitability for diabetics and reduced caloric value.

The exploitation of the properties of the polyols and various hydrogenated carbohydrate derivatives for these purposes is a relatively new and interesting field of development. One advantage is that most of the materials themselves are not new, so that the process of getting regulatory approval for them may not be as fraught with uncertainty as in the case of a completely new compound synthesised especially for its sweet taste intensity.

Progress in this area, the advances that have been made in relating molecular structure to sweet taste and the classification of sweet compounds according to their molecular configurations, are reviewed in the opening chapter. This is a fertile research topic on which much progress is now being made. A special feature is that it is multidisciplinary, attracting researchers from different backgrounds because it can be studied by a wide variety of techniques, including chemical, physical and physiological approaches. As advances continue to be made in this and other fields of research associated with sweeteners, we may concur with the words of Erasmus Darwin in 1792, that "a fool is a man who never tried an experiment in his life."

One matter that badly needs to be aired and improved is the public perception of sweeteners and their place in foods and drinks. The extent of unenlightenment and misconception is staggering. Assigning E-numbers to sweeteners is construed as a way of foisting additives on to unsuspecting consumers rather than protecting them against the use of objectionable food ingredients. Sweeteners are perceived as unnecessary 'artificial' or 'synthetic' chemicals, replacing healthier natural products. To combat these unbalanced views there is a growing need for sensible, reliable and easily understandable information to be made available, possibly by segments of the food and drinks industry with a particular interest in marketing new sweeteners or products that contain them. This topic of consumer education is one that may have to be explored, and would be well worth addressing in any future book on matters relating to the use of sweeteners.

To conclude, all the contributors to this book deserve sincere thanks for their efforts in preparing their chapters, adhering for the most part to tight deadlines, and the great pains they have taken to render their chapters into clear English. The tight deadlines were required because the majority of the chapters were based on presentations given at a Royal Society of Chemistry Update Symposium on Sweeteners on 17 February 1995, so it was important that they should be published with the least delay to ensure that they retained the maximum topical interest. Thanks are also due to the Publishers for smoothing the way to achieving this.

T.H.G.

Contributors

G.G. Birch	Department of Food Science & Technology, University of Reading, Whiteknights, PO Box 226, Reading RG6 2AP, UK
F.J. Broulik	McNeil Speciality Products Co, 501 George Street, PO Box 2400, New Brunswick, NJ 08903–2400, USA
M. Gonze	Eridania Béghin-Say, Vilvoorde Research & Development Centre, Havenstraat 84, B-1800 Vilvoorde, The Netherlands
J. Goossens	Eridania Béghin-Say, Vilvoorde Research & Development Centre, Havenstraat 84, B-1800 Vilvoorde, The Netherlands
I.R. Gordon	GIRACT, 24 Pre-Colomb, CH-1290 Versoix/ Geneva, Switzerland
M.E. Hendrick	Central Research–Food Science Group, Pfizer Inc., Groton, CT 06304, USA
M. Heume	Cerestar/Eridania Béghin-Say, Euro Food Centre, Havenstraat 84, B-1800 Vilvoorde, Belgium
M.R. Jenner	Tate & Lyle Speciality Sweeteners, PO Box 68, Whiteknights, Reading RG6 2BX, UK
M.G. Lindley	Lintech, Reading University Innovation Centre, PO Box 68, Whiteknights, Reading RG6 6BX, UK
T.H. Grenby	Department of Oral Medicine & Pathology, UMDS, Guy's Hospital, London SE1 9RT, UK
H.L. Mitchell	Pfizer Food Science, Pfizer Ltd., 10 Dover Road, Sandwich CT13 0BN, UK
P.R. Murray	Pfizer Food Science, Pfizer Ltd., 10 Dover Road, Sandwich CT13 0BN, UK

M.-O. Portmann	Leatherhead Food RA, Randalls Road, Leatherhead, KT22 7RY, UK
A. Rapaille	Cerestar/Eridania Béghin-Say, Euro Food Centre, Havenstraat 84, B-1800 Vilvoorde, Belgium
A.G. Renwick	Clinical Pharmacology Group, University of Southampton, Biomedical Sciences Building, Bassett Crescent East, Southampton SO16 7PX, UK
B.J. Richards	Additives & Novel Foods Division, Ministry of Agriculture, Fisheries & Food, Ergon House, 17 Smith Square, London SW1P 3JR, UK
H. Schiweck	University of Hohenheim, Institute of Food Technology, Garbenstraße 25, D-70599 Stuttgart, Germany
W.J. Spillane	Department of Chemistry, University College Galway, Ireland
G.-W. von Rymon Lipinski	Hoechst Aktiengesellschaft, D-6230 Frankfurt/Main, Germany
I. Willibald-Ettle	Palatinit Süßungsmittel GmbH, Gottlieb-Daimler Straße 12a, D-68165 Mannheim, Germany
S.C. Ziesenitz	Südzucker AG Mannheim/Ochsenfurt, ZAQ/ Nutrition Department, Marktbreiter Straße 74, D-97199 Ochsenfurt, Germany

Contents

1 Molecular structure and sweet taste

W.J. SPILLANE

1.1 Introduction

This subject is an extremly broad one and literally dozens of classes of sweet-tasting compounds are known. Within these classes many analogues and derivatives of the 'lead' or parent compound have usually been synthesised, so that in some cases hundreds of such compounds have been made.

With such a broad canvas, there will have to be a large number of omissions, and thus, for example, amino acids, various types of carbohydrates, sulfones and a number of other major classes have been excluded.

Attention is focused on the effect of molecular structure on taste, especially sweet taste, for about 12 well known sweetener classes. Lastly because of the special interest at University College, Galway in the sulfamate (cyclamate) group, these are discussed more extensively than the other groups. In Figure 1.1 the diversity of structural types giving rise to sweetness is well illustrated with the six examples given.

At the present time there are four main ways of finding new sweeteners:

(i) the serendipitous or accidental discovery;
(ii) screening of the literature for references to sweet taste;
(iii) development of molecular structure–taste relationships (SARs) for different classes;
(iv) 'purpose-built' design of sweeteners.

For most of the known classes of sweeteners the 'lead' or 'first discovered' sweetener was usually the result of an accidental discovery (method i). Alchemists and early chemists tended to include tasting in their scrutiny of compounds that they had made. Because of safety considerations and possible legal problems new compounds are, unfortunately, rarely tasted nowadays. Synthesis of further derivatives and analogues within a class or grouping might then be carried out, leading to new sweeteners and eventually, after a number had been made, tentative structure–taste conclusions could be drawn (method iii). In Figure 1.2 sucralose, a trichlorogalactosucrose developed by Hough and his collaborators through the accidental discovery of the sweetness of a tetrachlorogalactosucrose

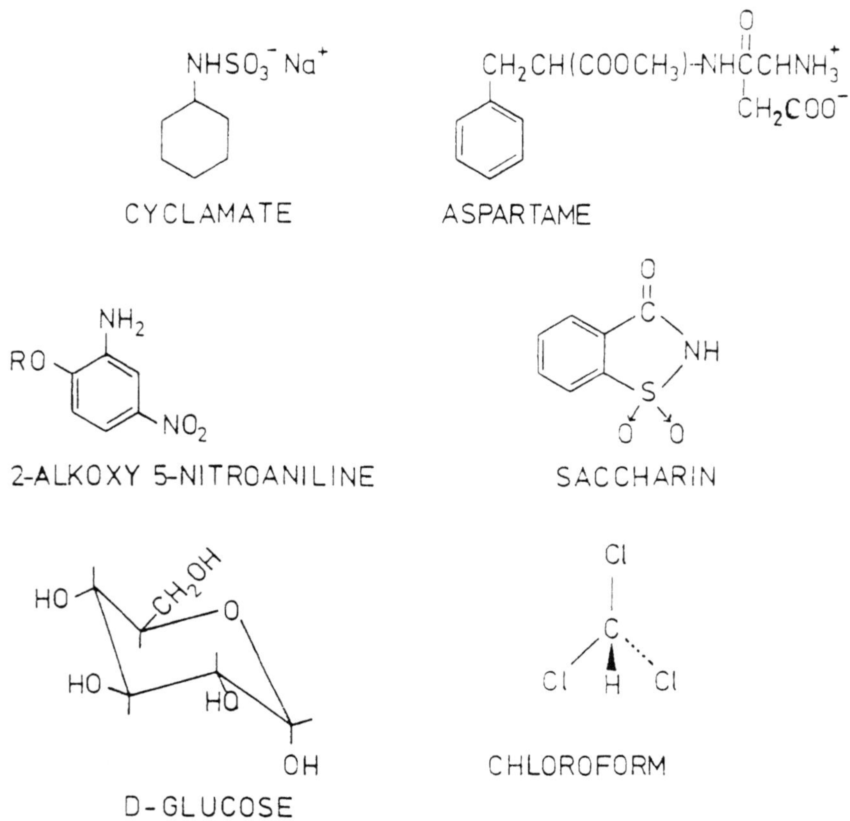

Figure 1.1 The diversity of sweetener structural types.

(Hough and Phadnis, 1976) is shown. The development of alitame (Figure 1.2) could be said to have involved a combination of methods i, iii and iv.

Method ii, while limited in its potential, is important. One of its more spectacular successes has been the rediscovery of the sweetener hernandulcin, a sesquiterpene known to the Aztecs (Figure 1.2). Kinghorn and his associates (Kinghorn and Soejarto, 1991), working from 16th-century Spanish literature, were able to trace back the sweet principle to a plant *Lippia dulcis*. This mode of discovery is of course just a part of a widely used strategy to find new, medicinally important compounds.

1.2 Development of structure–taste theories

Early structure–taste theories attempted to identify certain common structural features of tastants and to attribute a specific taste to these. One

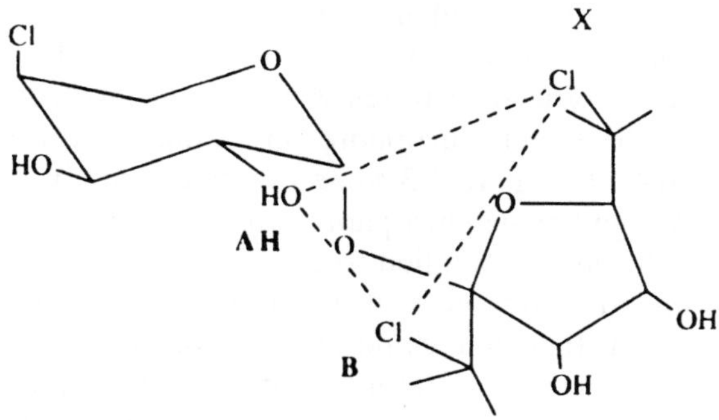

Sucralose

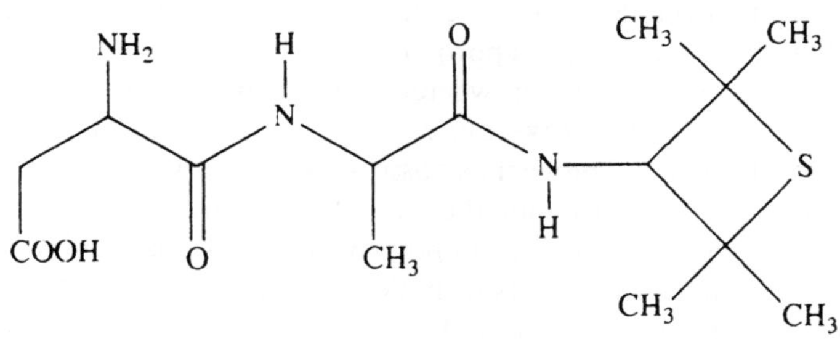

Alitame

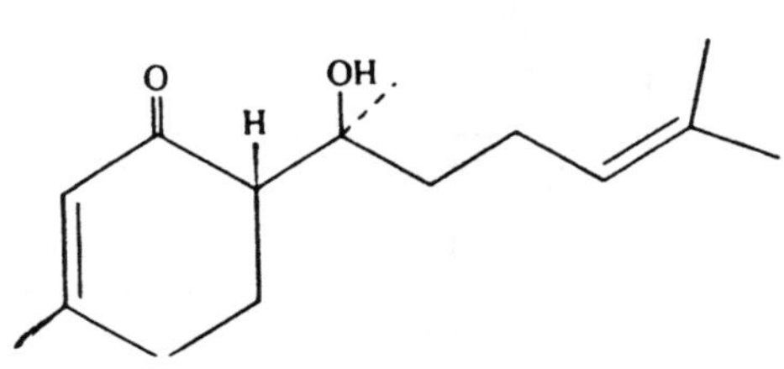

Hernandulcin

Figure 1.2 Some important sweeteners.

of the better known theories identified two distinct factors which should be present in a molecule for sweetness to be perceived: a glucophore (causing sweetness) and an otherwise tasteless entity, an auxogluc, which, when combined with the sweet glucophore, intensifies that sweetness (Oertly and Myers, 1919). In Figure 1.3 some examples are given for four sweet compounds. Altogether six glucophores and nine auxoglucs were defined. The theory could only be applied to explain the sweet tastes of aliphatic compounds, and the sweetness of a number of intense sweeteners e.g. saccharin known at that time could not be explained.

The importance of a migratory hydrogen atom for sweetness was proposed by Kodama in 1920. This 'vibratory' hydrogen could move from place to place, giving rise to different tautomeric forms. Loosely bound hydrogens are crucial to the present widely accepted theory of sweetness put forward by Shallenberger and Acree (1967) and illustrated in Figure 1.4(a). This AH/B theory suggests that two H-bonds are formed between the tastant and the receptor. In the tastant molecule an electronegative centre (B) and a centre (HA) capable of donating a hydrogen atom are sought. Two complementary sites are then required on the protein receptor so that the 'Shallenberger mechanism' can operate. The distance from AH to B should be approximately 3 Å. Likely AH/B sites are indicated in some well known sweeteners in Figure 1.4 and in Figure 1.2 for sucralose. The theory works quite well, but other factors must also be involved, otherwise all molecules possessing these 'sweet-inducing' entities would be sweet, which is not the case. Spatial, hydrophobic/hydrophilic and electronic effects are very important too, and this will be evident on examining some of the effects of molecular structure on taste for various classes of sweeteners. The strength of the Shallenberger–Acree theory and

Compound		**Glucophore**	**Auxogluc**
Glycol	OH OH H_2C-CH_2	OH OH H_2C-CH-	$-H$
Glycerol	OH OH OH $H_2C-CH-CH_2$	OH OH H_2C-CH-	$-CH_2OH$
Glycine	$HOOC-CH_2-NH_2$	$HOOC-CH-NH_2$	$-H$
Chloroform	$CHCl_3$	CCl_3	$-H$

Figure 1.3 Glucophores and auxoglucs for four sweet compounds.

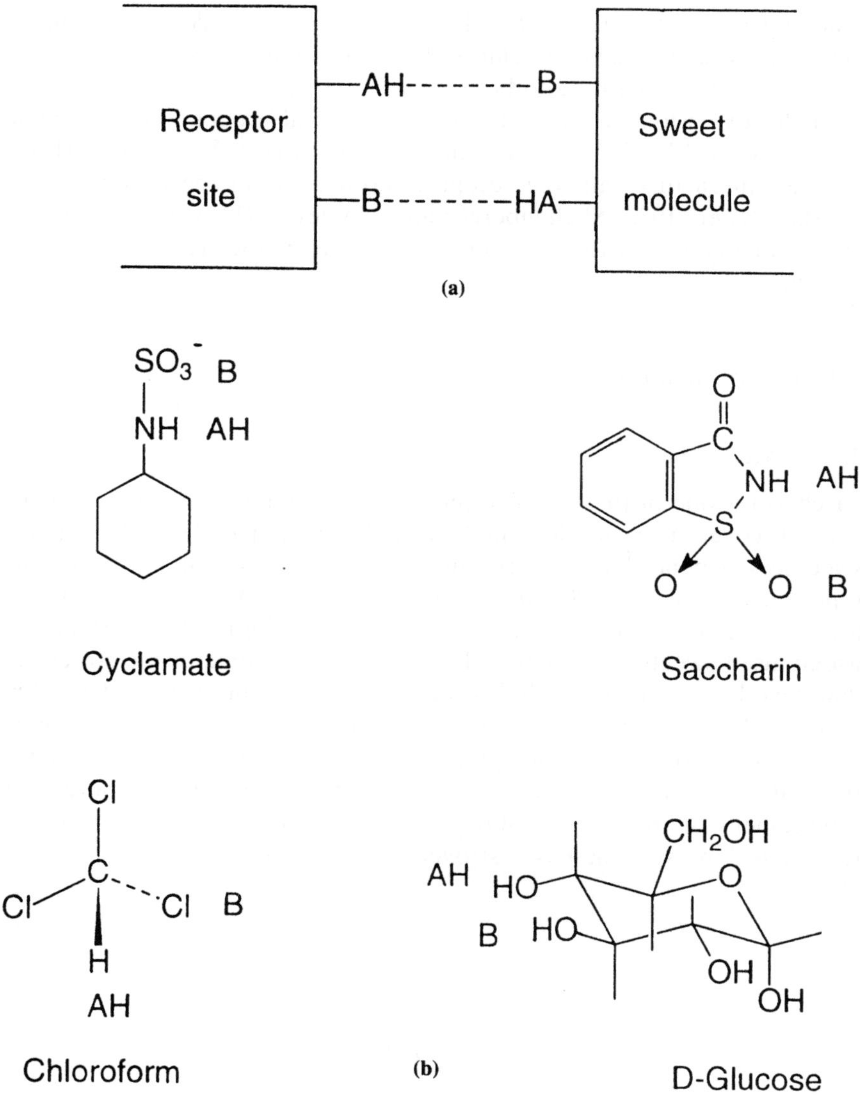

Figure 1.4 (a) AH/B theory. (b) AH/B sites in well known sweeteners.

its extensions (see last paragraph, p. 6) lies in its almost universal application. Other general theories of sweetness such as Belitz's e(electrophilic)/n(nucleophilic) theory, which strive to identify such centres in molecules, have not enjoyed the same wide usage. Thus, the AH/B theory has to be regarded as the principal interclass theory of sweetness. As will be seen many intraclass structure–taste relationships have been developed but, in general, a structure–activity relationship (SAR) that holds for one class of tastants will not hold for another.

In the extension of the AH/B theory, several groups have sought to identify a third binding site, variously labelled the *x*, δ or γ site. This is seen as interacting with the receptor through dispersion (hydrophobic) bonding (Kier, 1972; Shallenberger and Lindley, 1977; van der Heijden, 1978). In Figure 1.5 the optimum dimensions of the 'Kier triangle' are shown.

1.3 Classes of sweeteners

1.3.1 Saccharin

Saccharin is one of the oldest of the intense synthetic sweeteners, having been discovered by accident in 1878 and patented in 1885. Its relative sweetness depends on the concentration in solution but it is variously reported as being 200–700 times as sweet as sucrose. It has now been in use for over one hundred years and is still of the of the most important and widely used sweeteners. Figures 1.6 and 1.7 show some of the structures that have been studied for their taste. The AH/B centres are indicated in Figure 1.4 and the X centre (third site) is usually placed in the aromatic ring. Not surprisingly, *N*-methylsaccharin (Figure 1.6) is not sweet since the crucial H has been removed. It is also interesting to note that replacement of the CO by SO_2 group retains sweetness but when the reverse is done sweetness is destroyed (Figure 1.7).

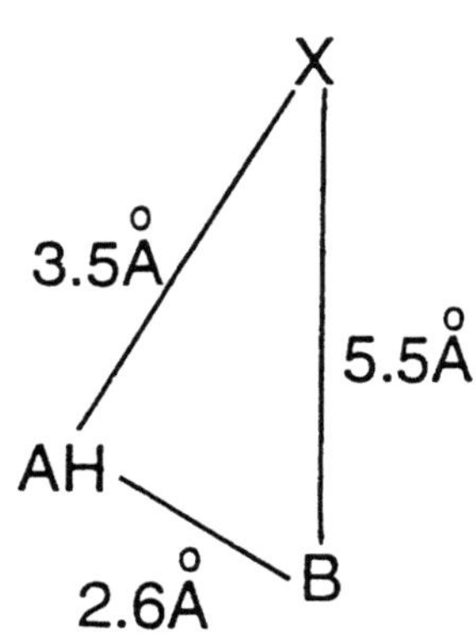

Figure 1.5 'Kier triangle'.

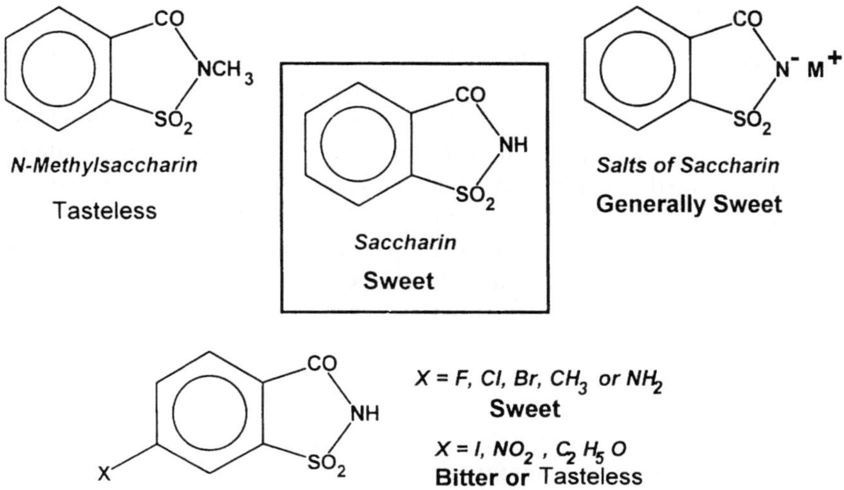

Figure 1.6 Modification of saccharin structure to alter taste.

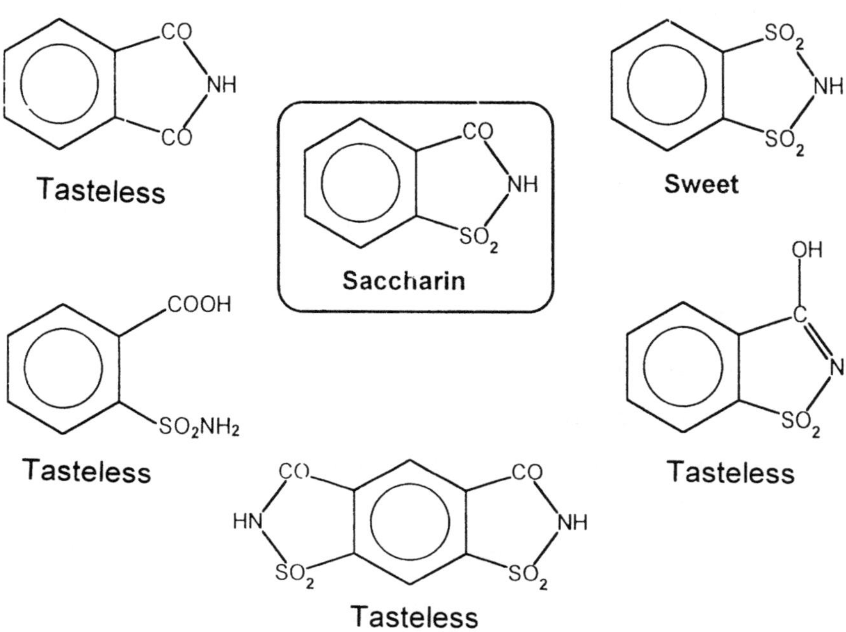

Figure 1.7 Further taste studies related to saccharin.

1.3.2 Acesulfame-K

Discovered in 1973, acesulfame-K is, like saccharin, now cleared for use in the EU. It is about 200 times as sweet as sucrose. Reaction of butyne and fluorosulfonylisocyanate led to the synthesis of the dimethyl compound (Figure 1.8, top right), which was found to be sweet, but subsequent synthesis of further dihydrooxathiazinone dioxides led to the other compounds shown in Figure 1.8, including acesulfame-K (top left), the most suitable sweetener of this class for general use. The AH/B/X sites are indicated for acesulfame-K, but it should be noted that the B and X sites will vary for the different derivatives shown.

1.3.3 Nitroanilines

The nitroaniline sweeteners, discovered in the 1940s, included what was at that time the sweetest compound known to man, P-4000, in which $R = O(CH_2)_2CH_3$ (Figure 1.9). Because of the carcinogenic properties of aromatic amines they are unsuitable for commercial use. They have however been widely used in quantitative structure–activity relationship (QSAR) studies, since relative sweetness data are available for a fairly representative set. One such QSAR is shown, and it indicates that π, the hydrophobic parameter, and σ, the Hammett electronic constant, show that hydrophilic and electronic effects are important in determining the relative sweetness of these compounds.

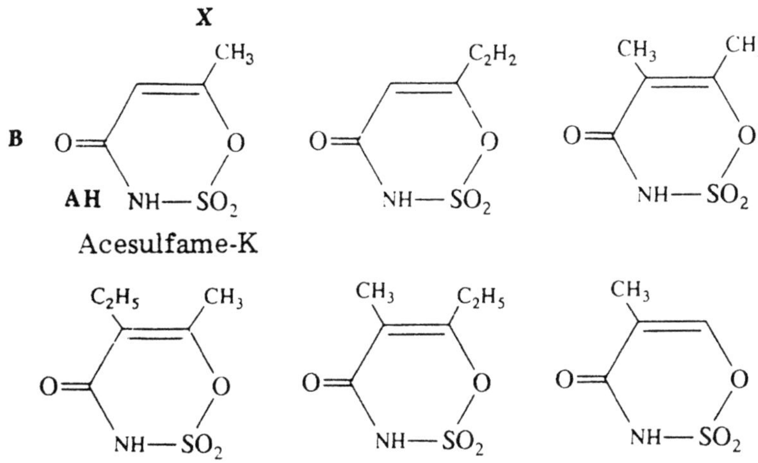

Figure 1.8 Compounds related to acesulfame-K.

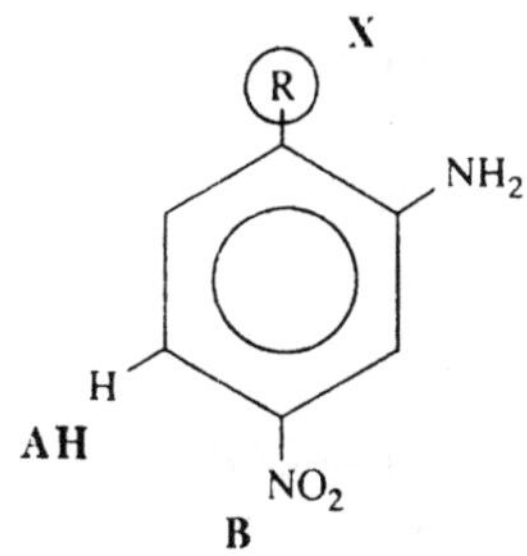

Nitroanilines

R	log RS
$O(CH_2)_2CH_3$	3.613
$OCH_2CH{=}CH_2$	3.301
I	3.097
OCH_2CH_3	2.978
$O(CH_2)_3CH_3$	3.000
Br	2.903
$OCH(CH_3)_2$	2.778
Cl	2.602
OCH_3	2.342
CH_3	–
F	1.502
H	–

Figure 1.9 Modification of sweetener properties of nitroanilines. $\log RS = 1.61\pi - 1.830\sigma + 1.73$. $r = 0.936$. $s = 0.28$.

1.3.4 Aspartame

Aspartame is one of the best known and most widely used of the new sweeteners. It was discovered by chance in 1969, though it had been made previously (1966) but had not been tasted. L-Asparyl-L-phenylalanine methyl ester (aspartame) is the top compound in Figure 1.10. The results of some structure–taste studies are also shown in Figure 1.10. Aspartame is the first entry and the last column gives the relative sweetness compared to sucrose. Hundreds of derivatives/analogues of aspartame have been synthesised and tasted.

1.3.5 'Designer' supersweeteners

In a major development in the sweetener field, Tinti and Nofre (1991) 'combined' two known sweeteners i.e. aspartame and cyanosuosan (Figure 1.11) to design a new supersweetener, called superaspartame, which is about 8000 times as sweet as sucrose and much sweeter than the components from which it was made. Superaspartame combines not only

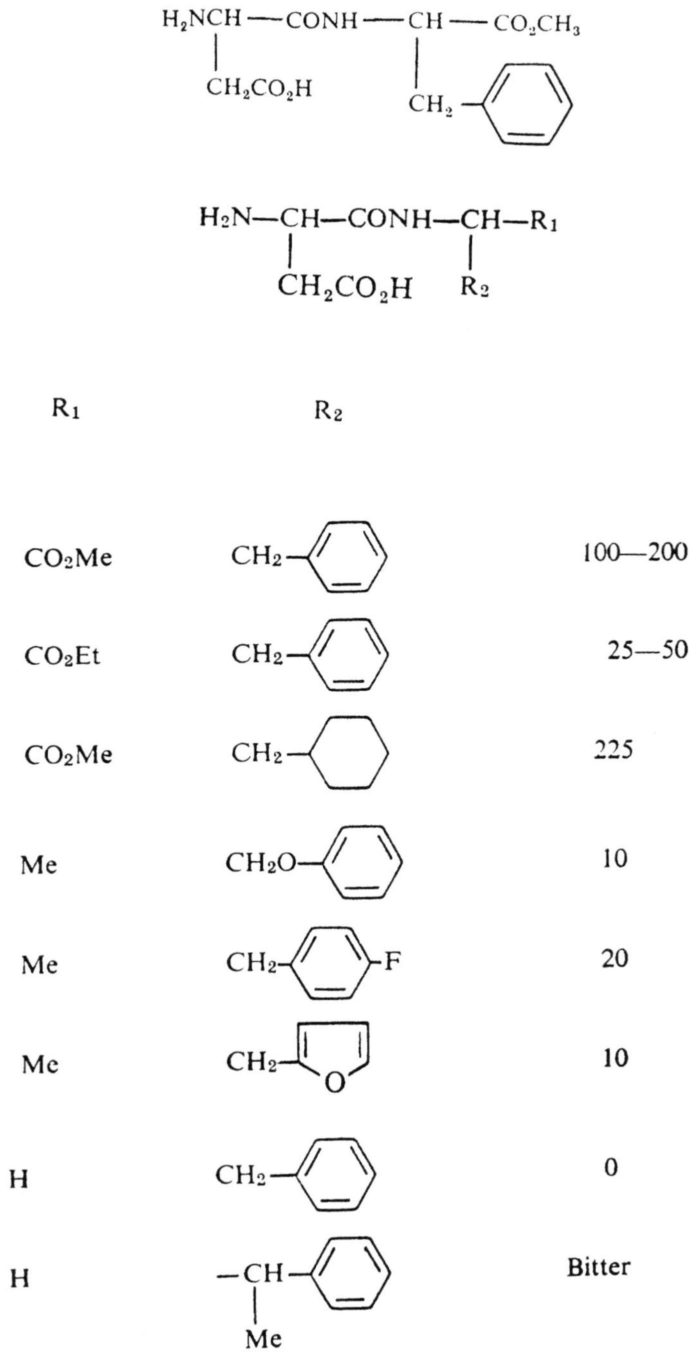

Figure 1.10 Structure–taste studies related to aspartame.

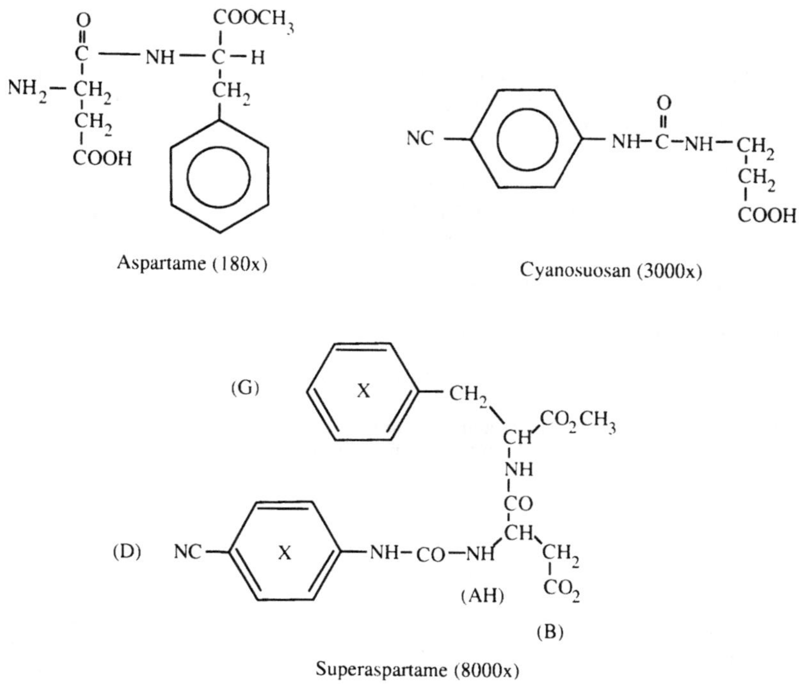

Figure 1.11 Design of superaspartame.

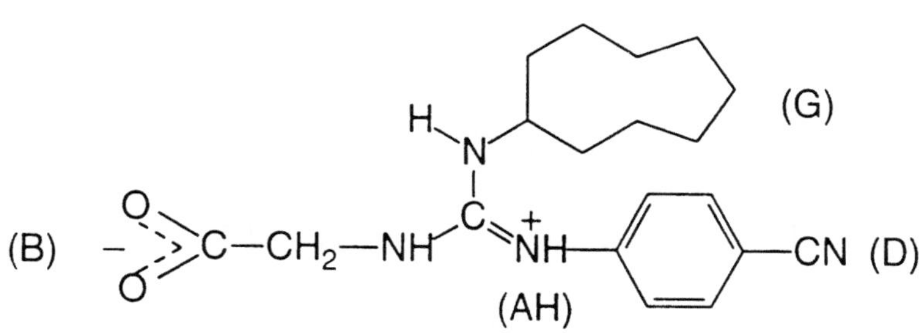

Figure 1.12 Sucrononic acid.

the common 2-aminomalonyl function of aspartame and cyanosuosan, but also the other molecular features of the individual 'parent' sweeteners. These workers have identified AH/B/X sites and additionally D and G sites. Further more recent work has led to the development of sucrononic acid, said to be approximately 200 000 times sweeter than sucrose (Figure 1.12).

Monellin (found in 'serendipity berries' – West African plant). Polypeptide

Subunit A
Arg-Glu-Ile-Lys-Gly-Tyr-Glu-Tyr-Gln-Leu-Tyr-Val-Tyr-Ala-Ser-Asp-Lys-Leu-Phe-Arg-Ala-Asn-Ile-Ser-Gln-Asn-Tyr-Lys-Thr-Arg-Gly-Arg-Lys-Leu-Leu-Arg-Phe-Asp-Gly-Pro-Val-Pro-Pro-Pro.

Subunit B
Gly-Glu-Trp-Glu-Ile-Ile-Asp-Ile-Gly-Pro-Phe-Thr-Gln-Asn-Leu-Gly-Lys-Phe-Ala-Val-Asp-Glu-Glu-Asn-Lys-Ile-Gly-Gln-Tyr-Gly-Arg-Leu-Thr-Phe-Asn-Lys-Val-Ile-Arg-Pro-Cys-Met-Lys-Lys-Thr-Ile-Tyr-Glu-Glu-Asn.

Thaumatin (West African shrub) (Sweetness comparable to monellin). Polypeptide.

Miraculin ('Miracle fruit'). Glycoprotein.

Curculin (from fruits of *Curculigo latifolia*). Polypeptide.

Figure 1.13 Naturally occurring protein sweeteners, showing composition of the two amino acid subunits (A and B) of monellin (mol. wt. ~10 700).

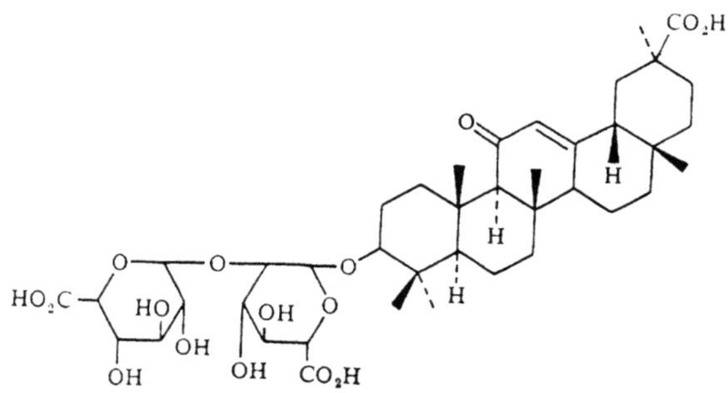

Glycyrrhizic acid

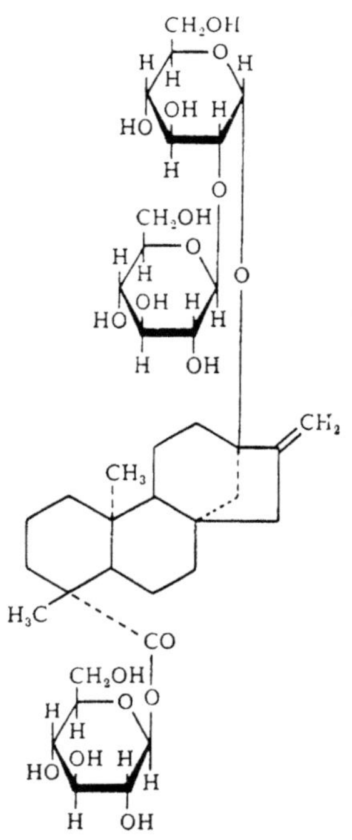

Stevioside

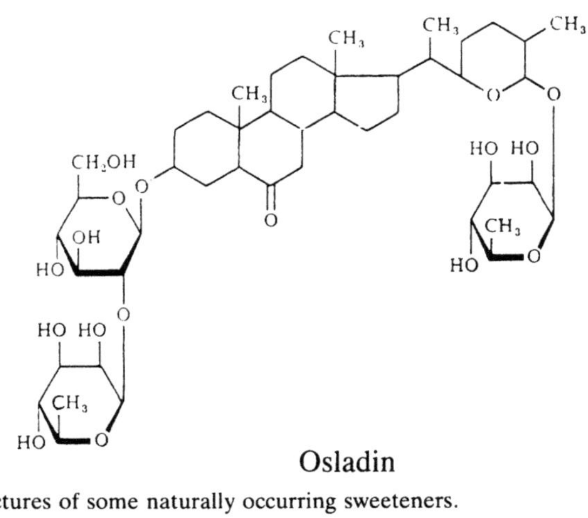

Osladin

Figure 1.14 Structures of some naturally occurring sweeteners.

1.3.6 Some naturally occurring sweeteners

Figures 1.13–1.15 deal with some remarkable naturally occurring sweeteners. The composition of the two amino acid subunits (A and B) of monellin (mol. wt. ~10 700) are shown in Figure 1.13. Monellin is about 3000 times as sweet as sucrose. Thaumatin (mol. wt. ~14 000) has a slight liquorice aftertaste and is 750 times as sweet as sucrose. Micraculin is a taste modifier and causes substances to taste sweet though it is not sweet itself. Curculin (1990) is sweet and is also a taste modifier.

Stevioside makes up 6% by weight of the leaves of the Paraguayan plant *Stevia rebaudiana* and is about 300 times as sweet as sucrose. Osladin is a steroidal saponin about 3000 times as sweet as sucrose. Glycyrrhizic acid is about 50 times as sweet as sucrose. It has been known for about 4000 years and is extracted from liquorice and the roots of various plants. Periandrin, a terpene glycoside, has been isolated from the roots of the Japanese plant, *Periandra dulcis*. Its relative sweetness is similar to that of glycyrrhizin.

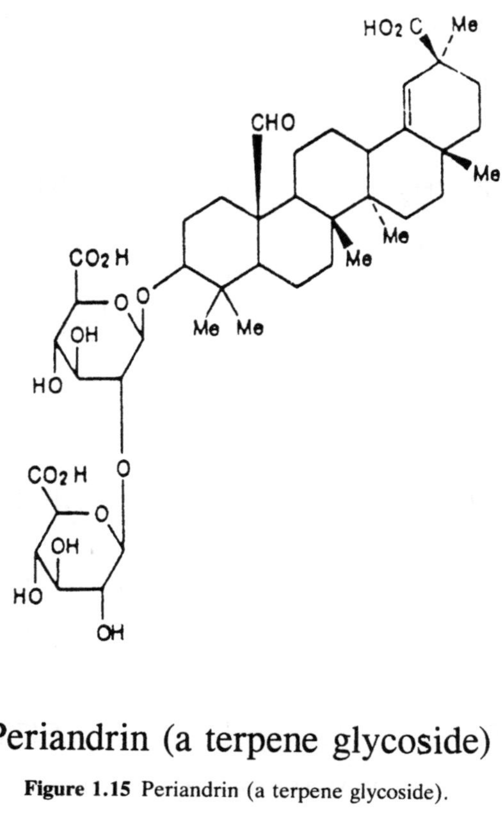

Figure 1.15 Periandrin (a terpene glycoside).

1.3.7 Ureas

Dulcin (Figure 1.16), about 250 times as sweet as sucrose, is the best known of the urea sweeteners. Figure 1.16 illustrates the limitations in the positioning of the $-OC_2H_5$ and $-CH_3$ groups on the aromatic ring if sweetness is to be retained. Clearly shorter AH–X and B–X distances are unfavourable for the retention of sweetness.

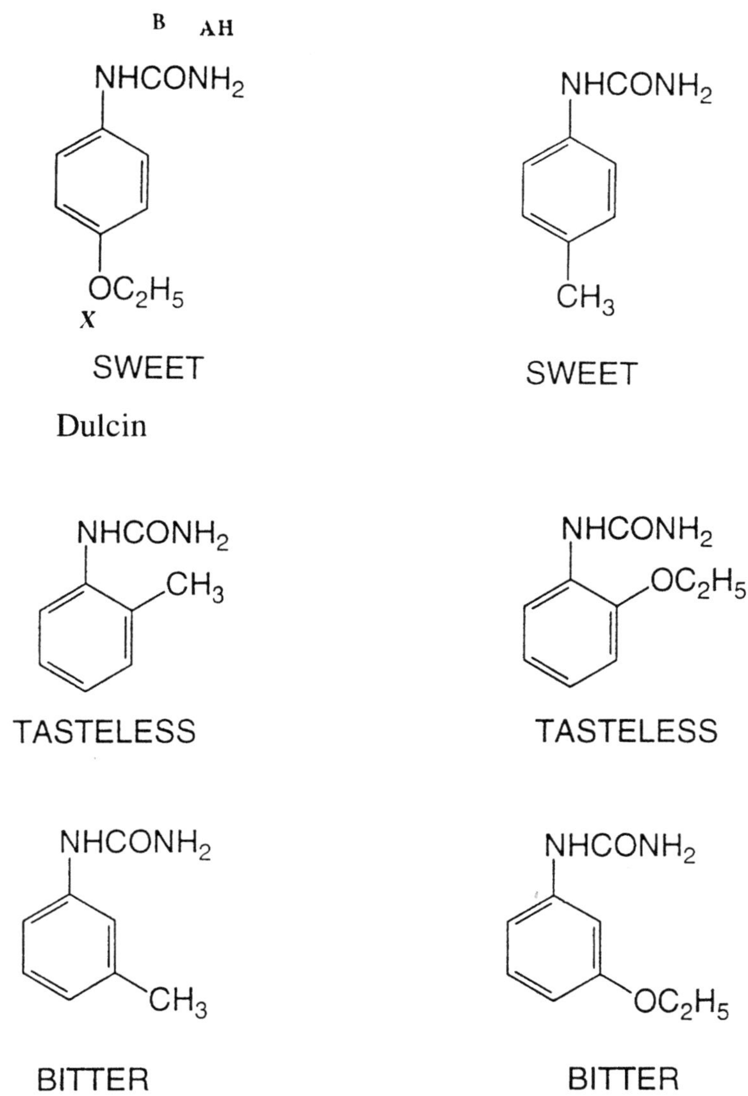

Figure 1.16 Modification of dulcin sweeteners.

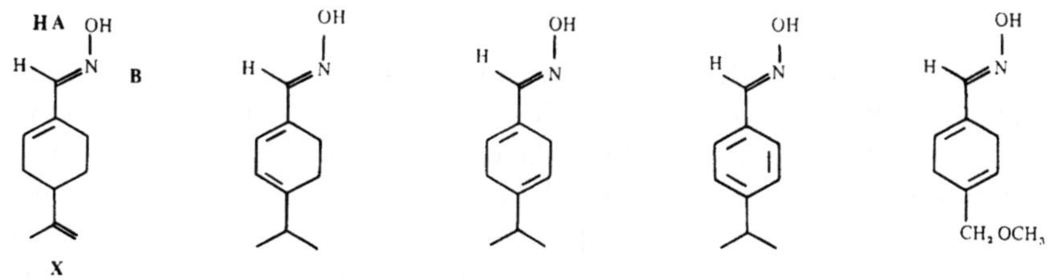

SRI Oxime V

Figure 1.17 Compounds based on SRI Oxime V.

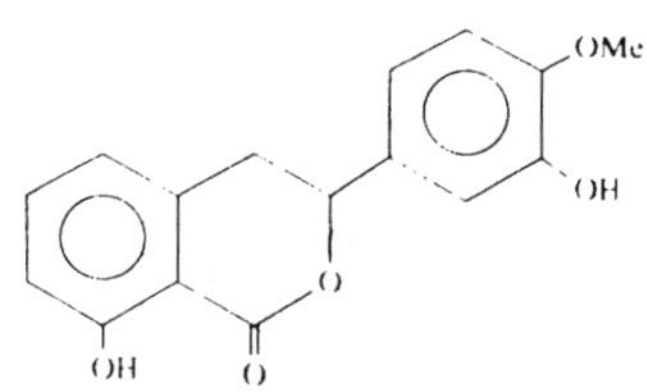

Phyllodulcin (400x)

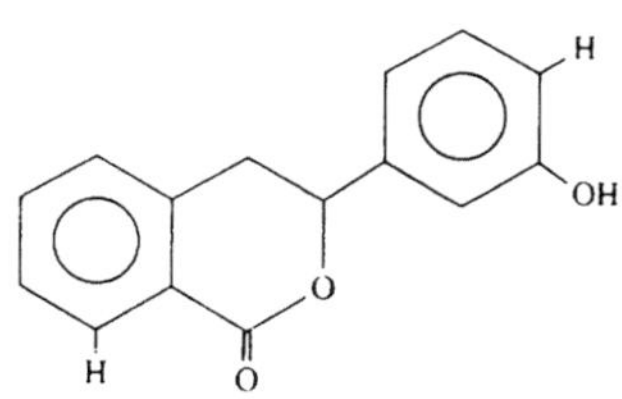

Tasteless

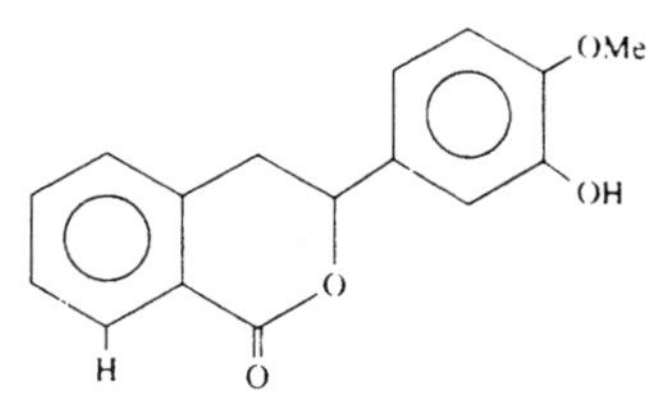

Sweet (100x)

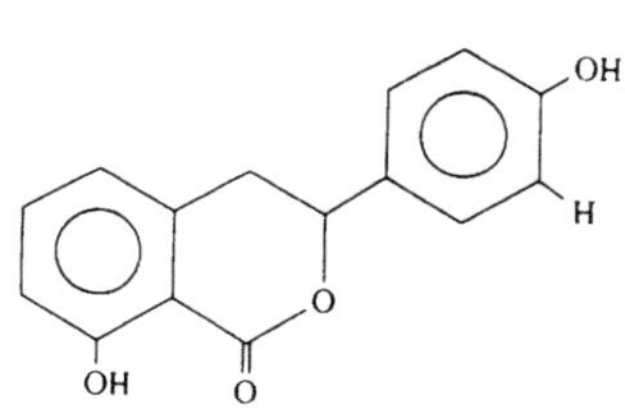

Tasteless

Figure 1.18 Modification of phyllodulcin sweetness.

1.3.8 Oximes

Perillartine (perillaldehyde oxime) (1920) is the purified oil from the Japanese plant *Perilla namkinensis*. Some of the extensive synthetic work carried out by Acton and Stone (Acton *et al.*, 1970) is shown. The parent compound is on the left of Figure 1.17. SRI Oxime V, which is 450 times as sweet as sucrose, was specially prepared by Acton and Stone to overcome the low water solubility of the other oximes, which restricts their usage. A C=C conjugated to the oxime is an essential requirement for sweetness.

1.3.9 Isocoumarins

Phyllodulcin, isolated from the leaves of *Hydrangea macrophylla seringe*, is about 200–300 times as sweet as sucrose. It is a 3,4-dihydroisocoumarin

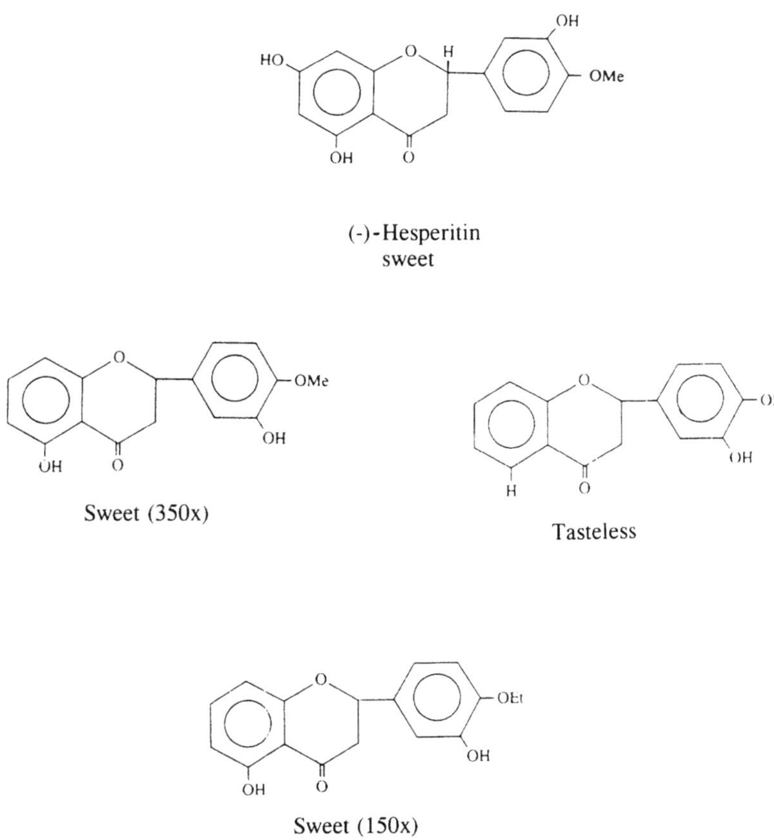

Figure 1.19 Modification of (−)-hesperitin sweetness.

derivative. Two AH/B locations are easily recognisable in these structures. Some limited structure–taste data are given in Figure 1.18.

1.3.10 Flavones

Flavones, which are structurally very similar to the phyllodulcins, are widely distributed in nature and many are sweet. Some are shown in Figure 1.19.

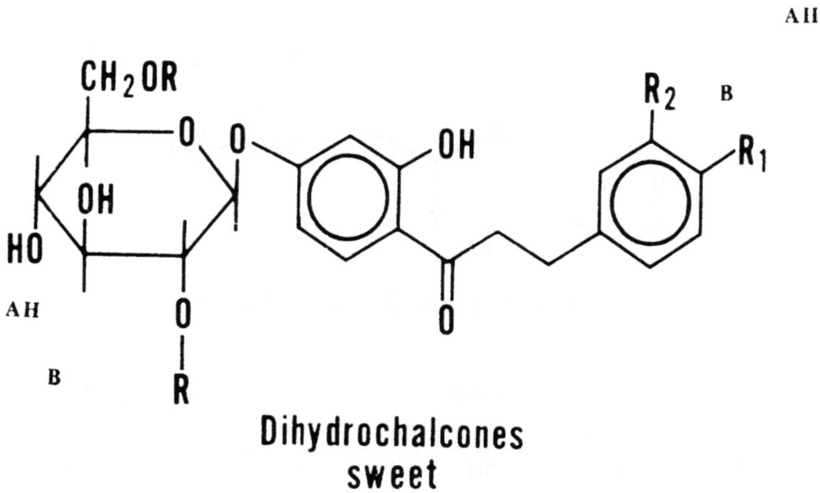

Naringin DHC, R_1 =OH; R_2 = H; R = α-L-Rhamnosyl

Neohesperidin DHC, R_1 = OCH_3; R_2 = OH; R = α-L-Rhamnosyl

Hesperetin DHC, R_1 = OCH_3; R_2 = OH; R = H

α-L-Rhamnosyl =

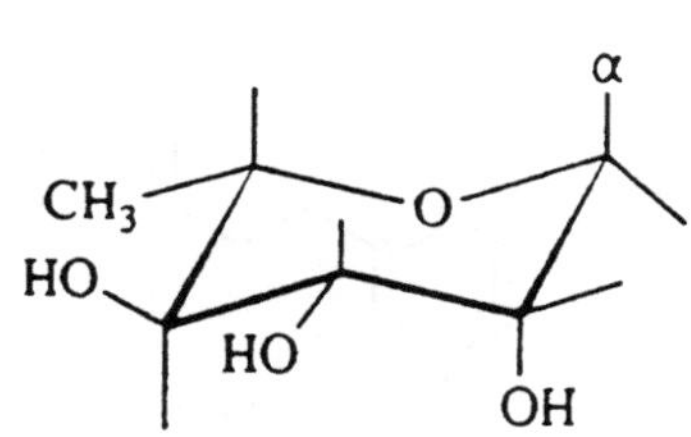

Figure 1.20 Dihydrochalcones.

1.3.11 Dihydrochalcones

The DHC sweeteners are prepared from naturally-occurring bitter-tasting flavones from citrus fruits, and are thus examples of semisynthetic sweeteners. The three best known DHCs are illustrated in Figure 1.20, and are prepared by chemical reaction from naringin, neohesperidin and hesperetin. They are approximately 250 times as sweet as sucrose and have a menthol-like aftertaste.

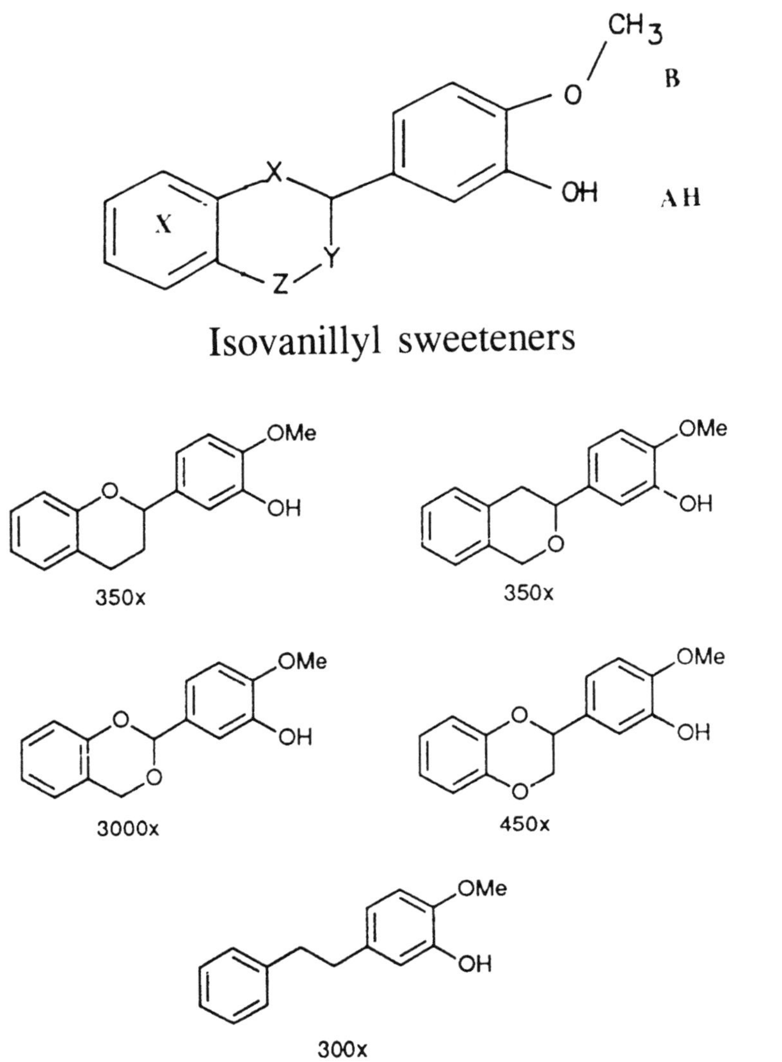

Figure 1.21 Isovanillyl sweeteners.

1.3.12 *Isovanillyls*

The isovanillyl sweeteners, some of which are shown in Figure 1.21, are the subject of active study by an Italian group. The AH/B/X sites are indicated, and the numbers written under the structures show the sweetness compared to sucrose.

1.3.13 *Sulfamates*

The 'first-discovered' compound was *N*-cyclohexylsulfamate (1937) shown at the top of Figure 1.22. The cation does not affect the relative sweetness

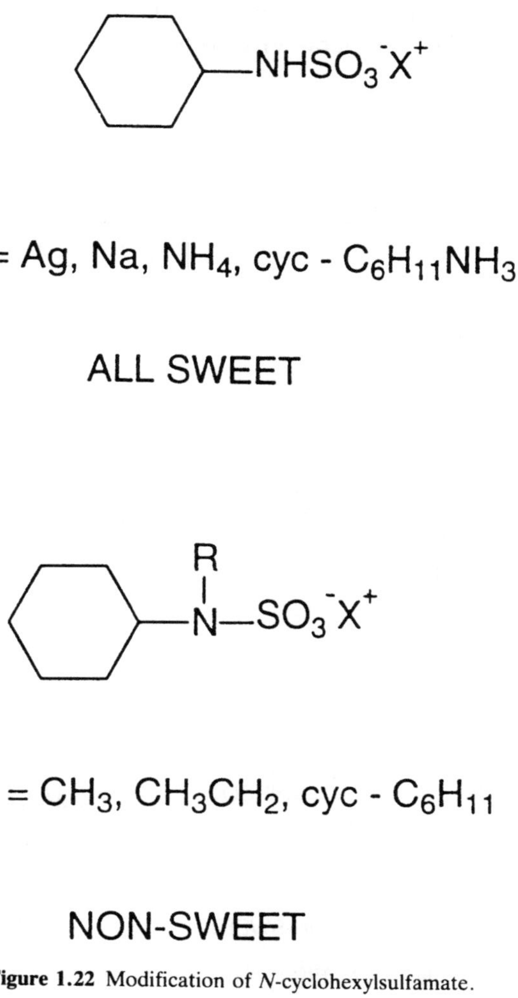

Figure 1.22 Modification of *N*-cyclohexylsulfamate.

(RS), and all of those shown have RS~40 (based on a 3% sucrose solution). Replacement of the amino hydrogen results in loss of sweetness. Several groups have been active in structure-taste studies and some results are shown in the figures. In Figure 1.23 the effect of ring size has been examined and in Figure 1.24 the structures of some sweet aliphatic sulfames are shown. It has been found that in plots of the length of the R group (x) in $RNHSO_3^-Na^+$ versus the volume of the R group (V) the sweet compounds tend to lie in a reasonably well defined area (A, Figure 1.25). Compounds falling in areas B, C and D are too large, too large and long, and too long respectively, to conveniently fit the receptor site, and thus they do not elicit a sweet taste. Compounds below area A may be considered to give too loose a fit at the receptor site.

These x/V plots can also be used to explain the sweetness of certain meta-substituted aromatic sulfamates and the absence of sweetness in the corresponding ortho- and para-derivatives. They have a certain predictive

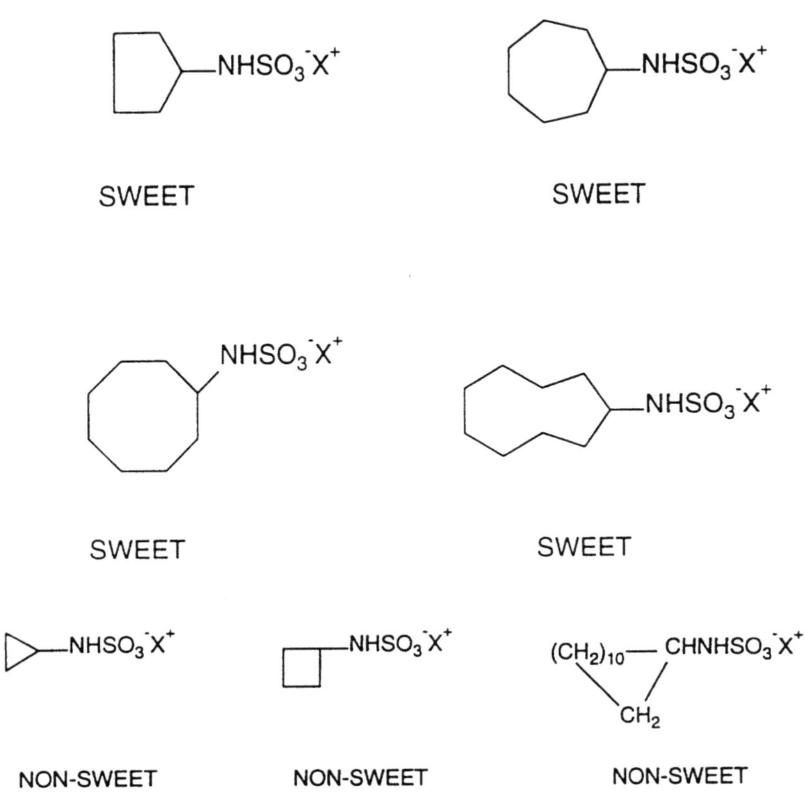

Figure 1.23 Effect of ring size on sweetness.

$CH_3CH_2CH_2NHSO_3^-X^+$

n-Propylsulfamate

SWEET

$CH_3CH_2CH_2CH_2NHSO_3^-X^+$

n-Butylsulfamate

SWEET

$(CH_3)_2CHCH_2NHSO_3^-X^+$

Isobutylsulfamate

SWEET

$CH_3CH_2CH(CH_3)CH_2NHSO_3^-X^+$

2-Methylbutylsulfamate

SWEET

$(CH_3)_2CH(CH_2)_2NHSO_3^-X^+$

Isoamylsulfamate

SWEET

$(CH_3)_3CCH_2NHSO_3^-X^+$

Neopentylsulfamate

SWEET

Figure 1.24 Sweet aliphatic sulfamates.

power, so that in Figure 1.26 it can be predicted that cyclobutylmethylsulfamate will be sweet, since measurements show that it falls within area A. 2-Methyl- and 3-methylcyclobutylsulfamates are predicted to be sweet on the same basis.

The x and V (x, y, z, i.e. length, width and height of R) measurements were made with Corey–Pauling–Koltun models. Volumes were also calculated using a suitable computer programme, and they correlated well with the V_{CPK} volumes. Figure 1.27 shows one such plot with V_{CPK} plotted against van der Waals volumes (V_W). For heterosulfamates (Figure 1.28) and disubstituted aromatic sulfamates (Figure 1.29) it is more difficult to develop semi-quantitative SARs such as those represented in the x/V plots.

Attempts are currently being made to develop a QSAR for about 20 structurally diverse sulfamates for which RS (relative sweetness) data exist, in the Department of Chemistry at University College, Galway. This is

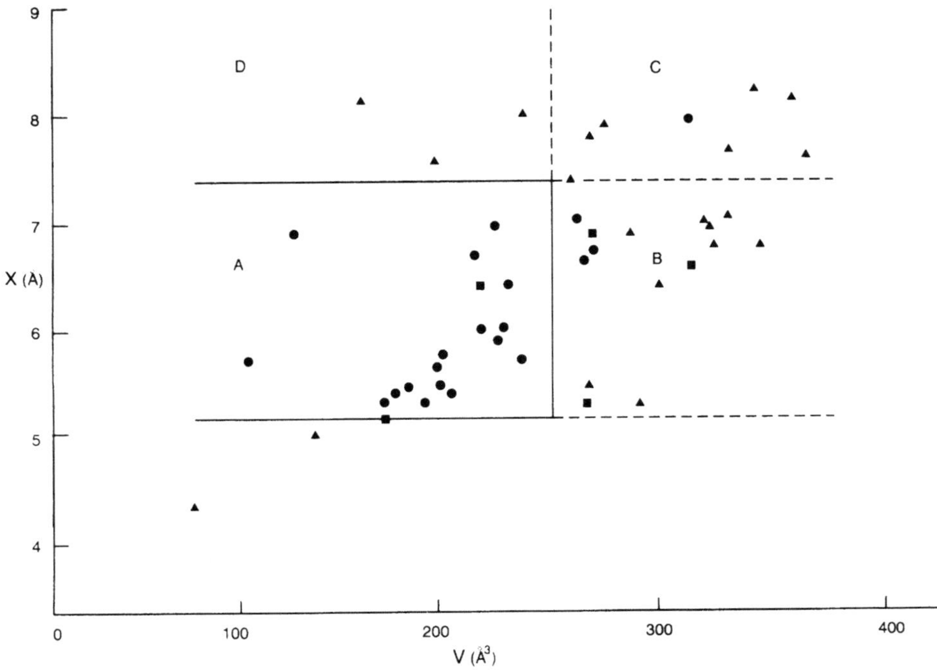

Figure 1.25 Effect of aliphatic group (R), length (x) and volume (V) on sweetness. ●, sweet; ▲, non-sweet; ■, bitter.

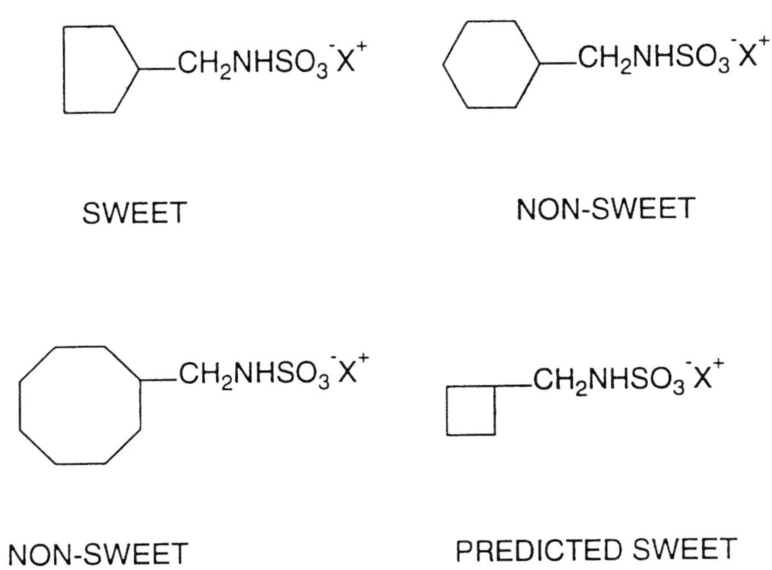

Figure 1.26 Prediction of sweetness from x/V plots.

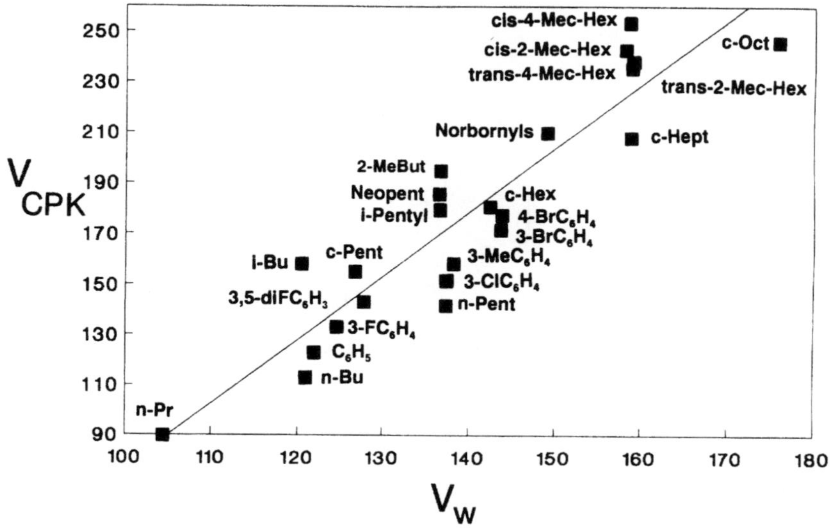

Figure 1.27 Plot of V_{CPK} against V_W for sulfamates.

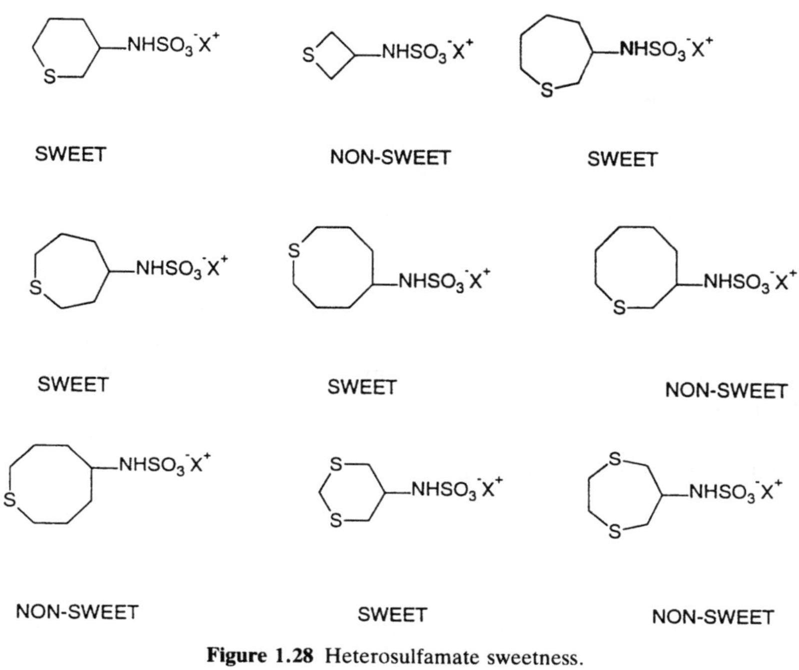

Figure 1.28 Heterosulfamate sweetness.

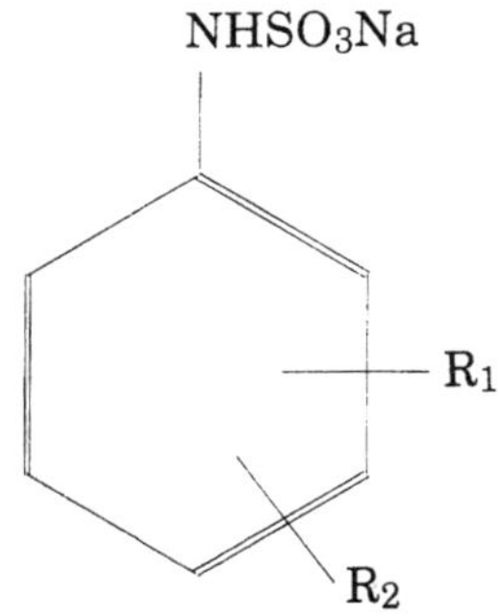

Figure 1.29 Disubstituted aromatic sulfamate.

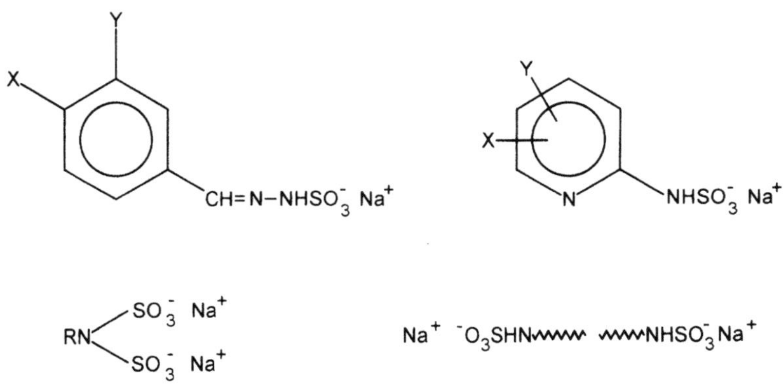

$ArSO_2NHSO_3^-\ Na^+$ $ArN(R)SO_2NHSO_3^-\ Na^+$

$ArCONHSO_3^-\ Na^+$

Figure 1.30 Representative sulfamates used in QSAR studies.

expected to be on the lines of that for the nitroanilines (see Figure 1.9). In other work an extended sulfamation programme is in hand starting from a diverse array of compounds (including nitroaniline and phenylurea tastants) and synthesising their mono- and disulfamates. Some representative sulfamates are shown in Figure 1.30. Results and developments in these areas will be reported in due course.

References

Acton, E.M., Leaffer, M.A., Oliver, S.M. and Stone, H. (1970) Structure–taste relationships in oximes related to perillartine. *J. Agric. Food Chem.* **18**, 1061–1068.

van der Heijden, A., Brussel, L.B.P. and Peer H.G. (1978) Chemoreception of sweet-tasting dipeptide esters: a third binding site. *Food Chem.* **3**, 207–211.
Hough, L. and Phadnis, S.P. (1976) Enhancement in the sweetness of sucrose. *Nature* **263**, 800.
Kier, L.B. (1972) A molecular theory of sweet taste. *J. Pharm. Sci.* **61**, 1394–1397.
Kinghorn, A.D. and Soejarto, D.D. (1991) New highly sweet compounds from natural sources, in *Sweeteners Discovery, Molecular Design and Chemoreception* (eds D.E. Walters, F.T. Orthoefer and G.E. DuBois), ACS Symp. Series 450, American Chemical Society, Washington DC. pp. 14–27.
Kodama, S. (1920) Taste. *J. Chem. Soc. (Japan)* **41**, 495–534.
Oertly, E. and Myers, R.G. (1919) Simple relations between the constitution of aliphatic compounds and their sweet taste. *J. Amer. Chem. Soc.* **41**, 855–867.
Shallenberger, R.S. and Acree, T.E. (1967) Molecular theory of sweet taste. *Nature* **216**, 480–482.
Shallenberger, R.S. and Lindley, M.G. (1977) A lipophilic-hydrophobic attribute and component in the stereochemistry of sweetness. *Food Chem.* **2**, 145–153.
Tinti, J.-M. and Nofre, C. (1991) Design of sweeteners. A rational approach, in *Sweeteners, Discovery, Molecular Design and Chemoreception* (eds D.E. Walters, F.T. Orthoefer and G.E. DuBois), ACS Symp. Series 450, American Chemical Society, Washington DC. pp. 88–99.

Further reading

Mathlouthi, M., Kanters, J.A. and Birch, G.G. (eds) (1993) *Sweet-Taste Chemoreception*. Elsevier Applied Science, London.
Shallenberger, R.S. (1993) *Taste Chemistry*. Blackie Academic & Professional, Glasgow.
Walters, D.E., Orthoefer, F.T. and DuBois, G.E. (eds) (1991) *Sweeteners, Discovery Molecular Design and Chemoreception*, ACS Symp. Series 450. American Chemical Society, Washington, DC.

2 Essential information on regulatory and legislative matters pertaining to sweeteners in the United Kingdom and the European Union*

B.J. RICHARDS

2.1 Current legislative position of sweeteners

The use in the United Kingdom of artificial sweeteners in foodstuffs is controlled by the Sweeteners in Food Regulations 1983 (S.I. 1983 No. 1211), as amended by the Sweeteners in Food (Amendment) Regulations 1988 (S.I. 1988 No. 2112). These provide for the use of the sweeteners listed in Table 2.1 in most foods. Thus, although restricted to the use of those sweeteners on the permitted list, food manufacturers have considerable scope for using them in a wide range of foodstuffs. There is, of course, a history of the use of sweeteners in products for diabetics but more recently manufacturers have not been slow to make use of this relatively light regulatory regime, particularly in relation to the production of diet and non-cariogenic foods.

Manufacturers in other EU member states have been doing the same to varying degrees but in compliance with their own national legislation, as until now there has been no specific EU legislation on sweeteners. This has inevitably led to barriers to trade. In the mid 1980s when the decision on the creation and implementation of a European Single Market were taken, it was clear that food law throughout the member states would have to be harmonised. One of the major areas to be covered was that of food additives, including sweeteners.

The first stage in the harmonisation of EU food additives legislation was the adoption at the end of 1988 of a Council Directive 'on the approximation of the laws of the Member States concerning food additives authorised for use in foodstuffs intended for human consumption' (89/107/EEC– *O.J.* No. L 40, 11.2.89, p. 27). This is commonly known as the 'food additives framework directive'. It set out general criteria for the use of food additives and provided for the later adoption of specific provisions in respect of the various classes of additives listed.

Table 2.1 Sweeteners permitted for use in the United Kingdom under the Sweeteners in Food Regulations 1983 (as amended)

Acesulfame potassium
Aspartame
Hydrogenated glucose syrup
Isomalt
Lactitol
Mannitol
Saccharin
Sodium saccharin
Calcium saccharin
Sorbitol
Sorbitol syrup
Thaumatin
Xylitol

Sweeteners were the first class of additives to be covered by a proposal for a more specific directive to be made under the framework directive. The reasons they came first were that there was no existing EU legislation controlling them (as there was for some other additives), and there were relatively few substances to be covered.

2.2 The EU Sweeteners Directive

Negotiation of the directive in Brussels took nearly four years from the time the formal proposal was made by the European Commission until adoption and publication of the directive in the *Official Journal*. All the member states were starting from different positions and each was trying to protect not only its own food manufacturers but also its consumers. Eventually, on 30 June 1994, a directive 'on sweeteners for use in foodstuffs' was adopted (94/35/EC – *O.J.* No. L 237, 10.9.94, p. 3).

The directive defines and controls the use of sweeteners and specifically excludes the use of foodstuffs with sweetening properties (such as sugar and honey) from its scope. It lists the foodstuff categories in which the various sweeteners may be used and at what levels, makes certain additional requirements for the labelling of table-top sweeteners, and sets out a timetable for the establishment of a system of monitoring of sweeteners consumption.

It is important to note that the directive controls the substances listed *only when they are performing a sweetening function*. This means that uses which are controlled in other legislation, such as sorbitol being used as a humectant, or some of the intense sweeteners being used as flavour enhancers, are not covered. Sweeteners are thus defined as food additives "which are used:

- to impart a sweet taste to foodstuffs;
- as table-top sweeteners."

The directive provides for the use of the sweeteners listed in Table 2.2. Those which are not currently permitted for use in the United Kingdom are maltitol (E965(i)), cyclamic acid and its Na and Ca salts (E952), potassium saccharin (E954) and neohesperidine DC (E959). As far as the continued inclusion of cyclamates is concerned, the European Commission has stated that it is essential that the results of the EU Scientific Committee for Food (SCF)'s current study on the conversion of cyclamates into cyclohexylamine must be available before implementation of the directive. Its statement continued that "if the SCF's examination of these results were to show that the status of cyclamates needed to be changed, the Commission would immediately take appropriate measures." At the time of writing the SCF's examination is still underway.

The permitted sweeteners to be used in foodstuffs and table-top sweeteners must comply with purity criteria which are being drawn up and which will shortly be included in a Commission directive to be adopted by the Standing Committee for Foodstuffs under the provisions of the food additives framework directive.

The Commission's original draft of these purity criteria was based largely on specifications already laid down by the Joint FAO/WHO Expert Committee on Food Additives (JECFA) but it has been revised following

Table 2.2 Sweeteners permitted for use under European Parliament and Council Directive 94/35/EC on Sweeteners for Use in Foodstuffs

E420	Sorbitol
	(i) Sorbitol
	(ii) Sorbitol syrup
E421	Mannitol
E953	Isomalt
E965	Maltitol
	(i) Maltitol
	(ii) Maltitol syrup
E966	Lactitol
E967	Xylitol
E950	Acesulfame K
E951	Aspartame
E952	Cyclamic acid and its Na and Ca salts
E954	Saccharin and its Na, K and Ca salts
E957	Thaumatin
E959	Neohesperidine DC

discussions with the EU member states. There has been some concern that the proposed Commission directive does not include methods of analysis to be used in assessing adherence to the purity criteria. However, one of the introductory paragraphs to the directive gives some guidance, stating:

> "Whereas it is necessary to take account of the specifications and analytical techniques for sweeteners as set out in the Codex Alimentarius and the Joint FAO/WHO Expert Committee on Food Additives (JECFA)."

Sweeteners may only be used in foodstuffs which fall within the categories listed in the annex to the directive. Their use in foods especially prepared for infants and young children will be prohibited unless specifically provided for in other EU legislation. Such legislation has yet to be proposed. For this purpose 'infants' are taken to be under the age of twelve months and 'young children' between the ages of one and three years.

Many of the foodstuffs categories listed in the annex to the directive are described as 'with no added sugar' or 'energy reduced'. The effect of this is to limit further the range of foods in which permitted sweeteners may be used as products must comply with the definitions of these terms. These are:

- 'with no added sugar': without any added mono- or disaccharides or any other foodstuff used for its sweetening properties;
- 'energy reduced': with an energy value reduced by at least 30% compared with the original foodstuff or a similar product.

The annex to the directive specifies a maximum usable dose for each sweetener in relation to each foodstuff category. These maximum levels relate to ready-to-eat foodstuffs prepared according to any instructions for use. For instance, the amount of sweetener in a bottle of concentrated squash may be higher than that listed in the annex provided that there are instructions for dilution which would result in the ready-to-drink squash containing a level of sweetener no higher than that specified in the annex. The use of two or more sweeteners in combination in a single foodstuff will be permitted provided that each sweetener is separately permitted to be used in that foodstuff and the maximum level of use for each individual sweetener is respected. The polyols are permitted for use in a range of foodstuffs at a level of *quantum satis*. This term is not defined in the directive as it is in other food additives directives. It is defined in these directives as meaning that

> "[the additives] shall be used in accordance with good manufacturing practice, at a level not higher than is necessary to achieve the intended purpose and provided they do not mislead the consumer".

It seems sensible to use this same interpretation in respect of the sweeteners. At the time of writing it is not known whether the directive is likely to be amended to include such a definition.

There is also no provision in the directive, as there is in the other additives directives, for what is known as 'carry-over'. Such a provision would have permitted the presence of a sweetener in a compound foodstuff to the level that it would be permitted in one of the ingredients of that foodstuff. The converse ('reverse carry-over') would also have permitted a sweetener to be used in an ingredient (in which it would not normally be permitted) provided that the ingredient was to be used solely in the preparation of a compound food which was allowed to contain the sweetener. Again, at the time of writing it is not known whether such a provision will be included in the directive in the future.

The foodstuffs categories listed in the annex to the directive are described in relatively general terms. This is to allow a necessary degree of flexibility to accommodate variations in the types of foodstuffs produced in the different member states. If each category required had been closely defined the length of the annex would have increased considerably! It is possible that some products might meet the description of more than one of the categories listed. A few of these categories cover particular national speciality products which have therefore not been translated into the other EU languages.

The directive contains some specific labelling requirements in relation to table-top sweeteners. In addition to existing labelling requirements, the sales description of table-top sweeteners will have to include the term '. . . -based table-top sweetener', using the name(s) of the sweetening substance(s) used in its composition. There are also extra labelling requirements for table-top sweeteners containing polyols and/or aspartame. The labels must bear these warnings:

- polyols: "excessive consumption may induce laxative effects";
- aspartame: "contains a source of phenylalanine".

Some manufacturers are already doing this on a voluntary basis in the United Kingdom. The directive also requires provisions to be adopted concerning the details to appear on labels of foodstuffs which contain sweeteners "in order to make their presence clear" and warnings concerning the presence of certain sweeteners in foodstuffs. At the time of writing it is not clear how detailed these provisions will be.

Another important aspect of the directive is the requirement for the establishment within three years of adoption of the directive, of a system of consumer surveys to monitor sweetener consumption. The United Kingdom already has experience of such surveys which could be built upon. The extension of such surveys to cover the whole European Union is to be welcomed. However, the scope and methodology of the survey system have still to be agreed in Brussels. Within a further two years the Commission will have to submit to the European Parliament a report based on the information obtained from the monitoring system, on changes in the sweeteners market, levels of use, and whether conditions of use need to be further restricted to ensure that use does not exceed the acceptable daily intake. If necessary, proposals for amendment of the directive to take account of such changes will have to accompany the report.

2.3 Implementation of the directive in United Kingdom law

The directive requires member states to bring into force the necessary legislation or administrative provisions for its implementation by 31 December 1995. Such legislation has to permit by the same date trade in and use of products conforming to the directive. It also has to prohibit by 30 June 1996 trade in and use of products which do not conform. There is, however, an important provision for the marketing until stocks are exhausted of non-complying products which were put on the market or labelled before that date.

In the United Kingdom implementation will be effected by means of regulations made under the Food Safety Act 1990. These will supersede the Sweeteners in Food Regulations 1983 (as amended) and will also replace any provisions for the use of sweeteners contained in food compositional regulations such as those covering soft drinks, ice cream and jam and similar products. It is hoped that the legislation will not need to reiterate the detailed provisions of the directive but will simply contain reference to them along with necessary provisions relating to penalties, enforcement and amendment and revocation of other legislation. The legislation will also implement the Commission Directive on the purity criteria for permitted sweeteners referred to above.

Ministers are required by the Food Safety Act 1990 not to make such regulations until they have consulted organisations representative of interests likely to be substantially affected by the legislation. There will thus be consultation on the draft sweeteners regulations with food manufacturers, manufacturers of sweeteners, consumer bodies, food retailers and food law enforcement authorities. It is the custom for the responses to, and the results of, any such consultation to be made publicly available, which will be the case here. All this must be done in time for the regulations to be introduced and in force by 31 December 1995 at the latest.

2.4 Amendment of the Sweeteners Directive

Given the length of time devoted to the negotiation of the directive in the Council of Ministers and the European Parliament, it is not surprising that various scientific and technical developments in the use of sweeteners took place during that period. This has meant that some new products containing sweeteners are now on the market but are not included in the directive. Others have been developed and are almost ready to be launched. In view of this the European Commission and the Council of Ministers agreed that an amendment to the directive taking account of such developments should be proposed with a view to its adoption by the time trade in products not conforming with the directive has to be prohibited – 30 June 1996. The aim is to try to ensure that products which are not covered by the directive but which are currenty legally marketed in the member states do not have to be removed from the shelves after that date. It is hoped that such an amendment would also cover issues such as those referred to above on 'carry-over' and the definition of *quantum satis*. At the time of writing a proposal for such an amendment is still awaited. Once it is negotiated and adopted it will be implemented in United Kingdom law by means of an amendment to the regulations which will implement the main directive.

2.5 Procedures for the addition of new sweeteners to the permitted list

Before any food additive is considered for inclusion in permitted lists in EU legislation it must undergo a safety evaluation by or on behalf of the SCF. The Committee considers the results of toxicological studies carried out for the manufacturer and, if appropriate, establishes an Acceptable Daily Intake (ADI) for the additive. The Commission, in consultation with industry and consumer bodies and representatives of the member states, may then draw up conditions of use which will ensure that consumption remains within the ADI. It then proposes appropriate legislation to the Council of Ministers and the European Parliament. If, therefore, a manufacturer wishes to market a new additive or food product containing that additive they may request the SCF to assess its safety and the Commission subsequently to draft the appropriate legislation.

In the case of sweeteners there are already some new substances which have been developed (sucralose, alitame) but which were not far enough through the safety evaluation process to be included in the sweeteners directive. Indeed it is possible that the safety assessment will not be complete in time for them to be included in the amendment to the directive already foreshadowed. Once evaluation is complete the manufacturers may request a further amendment to make provision for the use of the new sweeteners. In such a case the proposed legislation will have to undergo the full procedures laid down in Article 100a of the Treaty of Rome and to be agreed by both the Council of Ministers and the European Parliament.

There is also another route open to those who wish the legislation to be amended. The food additives framework directive provides for national authorisation of additives which may have been developed since the adoption of the permitted list. Such authorisation is subject to certain conditions. It is limited to a maximum period of two years, and foodstuffs containing the additive must be officially monitored. The foodstuffs in question may also be required to be specially labelled. Before expiry of the two-year period the member state concerned may request the Commission to include the additive in the permitted list. It must provide evidence supporting such an inclusion and indicate how the additive is to be used. If the request is considered by the Commission to be justified it must propose an amendment to the relevant directive. The amendment will then go through the Treaty procedure described above and must be considered by the Council of Ministers and the European Parliament within 18 months of the Commission's proposal being made. If either the Commission fails to make a proposal within the initial two years or the Council and Parliament fail to act within the subsequent 18 months the national authorisation must be cancelled. No new authorisation for the same additive may be granted unless it is justified by scientific or technical development made since the cancellation of the original national authorisation.

For such a national authorisation to be made in the United Kingdom the first step would be for the independent Food Advisory Committee (FAC) to consider a case of need for the additive. This would involve the company concerned providing the Committee with sufficient data to demonstrate that the additive performs a new function in the food, or performs an existing function better with clear benefits to the consumer. In assessing such benefits the Committee takes into account, amongst other things, the need for food to be presented in a palatable and attractive manner, the extension of dietary choice, the need for nutritional supplementation and any economic advantage. This consideration of need is necessary because under the provisions of the Food Safety Act 1990 ministers are required to consider the desirability of restricting as far as possible the use of substances of no nutritional value as foods or ingredients of food. If the FAC were to conclude that the case had been justified it would refer the additive to the Committee on the Toxicity of Chemicals in Food, Consumer Products and the Environment (COT), whose task is to assess the safety in use of the additive. The COT would consider all the toxicity data available on the new additive from animal and other laboratory studies, together with any information arising from human exposure. These data would be used to establish the maximum dose of the additive which causes no adverse effects in the animal – the no 'effect level'. The COT would then apply appropriate safety margins to take account of differences between the animals and humans and between individual humans. It would finally express its advice in the form of an acceptable daily intake (ADI) for the additive. If necessary other expert committees would also be consulted on issues such as genetic modification or nutrition. Once the safety assessment was complete the FAC would use the COT's advice in considering whether any conditions of use for the additive would be required, taking into account the types of food which might contain the additive and likely consumption patterns. Purity criteria for the additive would also have to be drawn up and agreed. If, at the end of this process, ministers accepted a recommendation from the FAC that use of the additive should be permitted, the United Kingdom regulations would be amended accordingly and the European Commission duly notified under the provisions of the food additives framework directive.

As well as being under consideration by the SCF, the new sweeteners sucralose and alitame are currently going through this United Kingdom approval process. The FAC has agreed a case of need for both sweeteners, and the safety assessments are in train. In cases such as these there is nothing to prevent a company simultaneously requesting approval from both the EU and a single member state. It is for the company concerned to decide which route is preferable. One advantage of the 'national' route is that a definite timetable for European action is laid down once national authorisation has been achieved. The length of time needed for achieving that authorisation will inevitably vary from one member state to another, depending on the complexity and extent of their approval procedures.

2.6 Conclusions

Although the EU directive is more restrictive than the United Kingdom regulations it replaces, it is understood to cover all the important current uses of sweeteners in the United Kingdom. New developments should be covered by the forthcoming amendment to the directive. There will be significant advantages for trade in foodstuffs between EU member states with the dismantling of trade barriers caused by the varying legislation and conditions of use which have existed until now. Above all, consumer safety continues to be protected by means of the rigorous evaluation of each individual sweetener which takes place before its use is permitted in food at all.

3 Regulatory processes for new sweeteners in the United States of America

F.J. BROULIK

Obtaining United States Food and Drug Administration (FDA) approval of a food additive petition for a new low-calorie sweetener is an unpredictable process that can generate over 30 000 pages of data, take 10–20 years, cost hundreds of millions of dollars, and exhaust the patience, endurance and resources of even the most determined and assiduous company. FDA's statutory requirements for a food additive petition appear to be comprehensive, straightforward and objective, but there are many points in the data generation process where judgment and interpretation are necessary and where consulting the FDA is critical. Despite the petitioner's best efforts to submit a petition that is well-organized, scientifically balanced and complete, the FDA is almost certain to ask for more information – sometimes simple background or explanatory material, but at other times extensive new data that can add years to the process.

The FDA's review process is painstakingly slow, due primarily to its ultra-conservative safety evaluation, especially with regard to sweeteners, as they will be widely consumed and attract controversy. Even though the FDA has taken steps to streamline the process, its food staff and budget continue to decline while Congress continues to add to its mandated responsibilities. Unless or until there is a substantial improvement in FDA's resources and in its food additive review and approval process, there is little reason to expect that companies will pursue the development of new sweetening compounds.

3.1 FDA's responsibilities versus resources

The FDA regulates all food products in the USA except for meat and poultry products and frozen and dried eggs, which are under the authority of the US Department of Agriculture, and the labeling of alcoholic beverages and tobacco, which are regulated by the US Department of the Treasury. Within the FDA, the Center for Food Safety and Applied Nutrition (CFSAN) carries out the agency's food responsibilities with a staff of 912, which is now back to the same level as in 1984, after dipping even lower in the last 11 years. Its budget of $222 million has remained the same for the last four years, with no adjustment for inflation. While the CFSAN's resources have remained basically unchanged over the last four years, Congress has continued to pass legislation imposing added responsibilities on CFSAN, such as the Nutrition Labeling and Education Act of 1990 and the Dietary Supplement Health and Education Act of 1994. As a consequence, CFSAN's priorities are not additive approvals but more immediate food safety concerns such as dealing with outbreaks of microbiological contamination and establishing regulations for HACCP (Hazard Analysis and Critical Control Point) systems to prevent them, as well as meeting Congressionally imposed deadlines for new labeling and supplement regulations.

3.1.1 FDA's regulation of sweeteners as additives

While a new low-calorie sweetener could theoretically fall under the less rigorous regulatory classification of Generally Recognized as Safe (GRAS), the focus on this chapter is solely on the process by which a sweetener can become a regulated food additive. Generally, low-calorie sweeteners would not meet the requirements for GRAS status because they are man-made (new molecular entities) rather than naturally occurring substances. GRAS substances are considered safe because their past extensive use has not shown any harmful effects.

The *Code of Federal Regulations* (1994), or CFR, (section 170.3 of title 21) defines food additives as:

> "substances . . . the intended use of which results or may reasonably be expected to result, directly or indirectly, either in their becoming a component of food or otherwise affecting the characteristics of food."

3.1.2 Requirements of food additives

The FDA has authority over food additives under the Federal Food, Drug, and Cosmetic Act (FD&C Act) of 1938 and the Act's Food Additives Amendment of 1958. Before 1958, the FDA had the burden of proving that the addition of an ingredient to a food rendered it injurious to health. The 1958 Amendment placed the burden of proof of safety on the additive's sponsor by requiring a premarket safety evaluation process. Based on the content of the sponsor's food additive petition, the FDA must be able to conclude that the additive is safe for its intended conditions of use. Safe, as defined by the CFR:

> "means that there is a reasonable certainty in the minds of competent scientists that the substance is not harmful under the intended conditions of use. It is impossible in the present state of scientific knowledge to establish with complete certainty the absolute harmlessness of the use of any substance."

Besides providing the FDA with evidence of safety, the sponsor of a proposed new food additive also must document its technological function. In the case of a sweetener, the sponsor would submit sensory data demonstrating the sweetening effect in prototype foods from each of the proposed categories of use. In addition, some sweeteners may qualify as flavor enhancers if the appropriate documentation is provided.

Unlike the requirements in the United Kingdom, Canada, Australia and other countries, the FDA's requirements do not call for a demonstration of need. This absense of a 'case of need' preprequisite means that the sponsor does not have to garner food industry interest in, or support for, potential use of the new sweetener; but it also means that the FDA does not enter into its decision-making equation the potential for technological innovation or advantage or consumer benefit.

3.1.3 Controversial nature of sweeteners

Intense sweeteners have been a regulatory lightning rod in the USA since Theodore Roosevelt was President. During his term in office (1901–1910), he personally intervened to keep saccharin publicly available, unwilling to tolerate the withdrawal of his favorite diet aid:

> "Anyone who says that saccharin is injurious is an idiot." (Lawler, 1986)

Even without Roosevelt to quash a renewed saccharin debate in 1910, Dr Henry Washington Wiley, first head of the FDA, failed in his attempt to remove saccharin from the market, presaging further conflicts between the regulators, the regulated and the calorie-conscious public. The FDA's subsequent attempt to ban saccharin in 1977 provoked a groundswell of public outcry that propelled Congress to override the FDA. The main legacy of these decades of regulatory thrusts and parries is a warning label on saccharin-containing products stating:

> "Use of this product may be hazardous to your health. This product contains saccharin which has been determined to cause cancer in laboratory animals."

Cyclamates went through a shorter, but bitter rise and fall in the 1950s and 1960s with their safety and reputation still in contention today. Similarly, in the 1970s and 1980s, aspartame withstood a rough ride of regulatory controversy that 24 years of consumer use still have not quelled.

The history of these three sweeteners underscores a fact that the FDA knows only too well: no other category of food additives has ever been so swept up in controversy as sweeteners, with the FDA always at the center of the turmoil. The unquenchable demand for sweetness without calories has resulted in numerous Congressional hearings, a Congressional moratorium on an FDA decision to ban, protracted administrative hearings, headline media coverage, consumer outcry over actual or attempted bans, and even a consumer group of sweetener 'victims'. FDA Commissioners have come and gone (with one having lost his job over a cyclamate dispute), but the FDA has become admittedly more sensitive in its dealings with sweeteners.

3.2 Regulatory strategy for a food additive petition

Seeking FDA approval of a petition for a new sweetener is a daunting undertaking, probably a once-per-career experience, which requires an unflagging commitment of time, money and manpower. Fledgling or resource-poor companies need not apply. For the company that does decide to embark on this course, its management must clearly understand and accept the magnitude of the commitment, the risks, and the potential for failure.

Some of the obstacles to success will be beyond the control of the petitioner:

- insufficient FDA resources;
- reorganization of FDA offices;
- revamping of FDA procedures;
- turnover in FDA staff;
- changes in FDA's data requirements or interpretations;
- inconsistent or equivocal or worse safety test results;
- negative comments submitted by third parties.

Other obstacles will be within the petitioner's power to avoid:

- failure to follow FDA's guidelines or advice;
- poorly designed or executed safety studies;
- insufficient or inadequate petition data;
- failure to fully illuminate potential problem areas in the data.

Two tactical regulatory considerations require early attention because they dictate the details of the petition's content and may have a bearing on the overall time to achieve approval. First, should the petition request a narrow or broad range of food uses? Second, should applications for regulatory approval be submitted concurrently in other countries?

There are two schools of thought on both these questions. The narrow, or limited category, approach means that there will be less information to generate (food formulations, sensory and stability data, and analytical methodologies) and a lower projected level of consumption. Theoretically, the FDA would have a lower hurdle to reach approval because of the more limited human exposure to the new compound, and it would be easy to add further food uses later. This has been the approach for aspartame and acesulfame-K, but the initial narrow approvals were neither quick nor simple and the subsequent category extensions have been excruciatingly slow.

On the other hand, the FDA has chosen to give the same level of scrutiny to the safety data in a petition for one food category as for ten or twenty. Also, as part of its safety evaluation, the FDA fully expects and plans for higher human intakes than projected for the initially petitioned categories. The FDA knows that any sweetener worth its salt will need to be in fairly broad use to be commercially viable. This was the approach taken in the 1986 and 1987 petitions for alitame and sucralose, which, as of March 1995, are still making their way through the exacting FDA review process.

As for the question of concurrent submissions in other countries, again there are conflicting considerations. There are very few deep-pocketed companies that would opt to submit petitions only in the USA and await FDA approval before submitting elsewhere. This approach is certainly the least difficult from a regulatory management perspective but the least practical from a business point of view. Needless to say, business needs invariably take priority.

In the ideal world, the petitioner would submit regulatory applications in the USA and several other countries at the same time (e.g. Canada, Australia, New Zealand and, in the era before the European Union's Sweeteners Directive, key European countries such as the United Kingdom and Germany) and would receive approval of his new sweetener in one or more countries while awaiting FDA approval. The petitioner could then ask the approving country or countries to recommend the sweetener for review by the Joint FAO/WHO Expert Committee on Food Additives (JECFA). JECFA would allocate an Acceptable Daily Intake (ADI) for the sweetener, which would satisfy the prerequisite for regulatory applications in some of the less-developed countries that look to JECFA for safety determinations. Then the FDA, with the added assurance and confidence engendered by these national and JECFA endorsements, would follow along with the grand prize in the sweetener race, the US market.

This chain of events would be a dream come true, but more likely it will be only a dream. It is wishful thinking that approvals in other countries or an ADI from JECFA will have a salutary effect on the FDA's decision-making process. The FDA is adamant that this is not the case, and that its review is *sui generis*. Even if every major regulatory body in the world plus every worldwide scientific expert agree that the sweetener is safe, the FDA will not approve until its scientists complete their review and say it is safe.

When this sequence of events can transmogrify from dream to nightmare, however, is when a regulatory authority identifies a safety question or issue that it believes only new data can resolve. Other agencies may not raise the issue or see the need for new data, but they are obliged to review such data while they still have the sweetener under review. The dilemma the petitioner faces is whether (a) to undertake the requested study for one authority while risking delays or even a standstill at the other agencies; or (b) to hold off doing the study while awaiting approval from the other agencies. When the FDA is the requesting agency, the former course makes sense because the USA is likely to be the petitioner's top priority as a business opportunity. In recent years, the sponsers of alitame and sucralose both took this course. When the FDA is not the requesting agency, course (b) may be the right one. At any rate, managing multi-country submissions can be challenging.

One further impediment with concurrent regulatory applications is the lack of harmony across countries in the basic safety data requirements. If possible the petitioner should plan and conduct a safety evaluation program that will meet all the targeted countries' varying needs. At the outset, this all-encompassing approach may be difficult to delineate and also may seem unnecessarily expensive, time-consuming or wasteful of test animals, but it will prove to be less risky and more efficient in the long run.

3.3 Generating data for a food additive petition

The FDA group responsible for reviewing food additive petitions and drafting the regulations permitting the use of new additives is the Office of Premarket Approval (OPA) in the Center for Food Safety and Applied Nutrition. The OPA's staff of 120 also have responsibility for evaluating petitions for GRAS food ingredients, color additives, indirect additives, and veterinary food additives, reviewing data on foods derived from new plant varieties, and providing advisory opinions on the regulatory status of food ingredients, which means answering 2500–3000 letters a year. Furthermore, OPA staff are involved in research on additive migration from packaging and in analytical methods development, as well as other food safety research.

As part of its January 1993 reorganization, the CFSAN established the OPA to serve as a centralized, coordinated focal point for petition review and approval in an effort to improve review quality and management. Despite the OPA's initiatives to simplify its handling of the large number of indirect additive petitions and to establish a team approach to direct additive reviews, it has not succeeded in reducing the backlog of pending petitions. At the end of 1994, there were 250 petitions pending, 42 of which were for direct additives, including the 1986 petition for alitame and the 1987 petition for sucralose.

3.3.1 Pre-petition reviews

It is critical to involve the OPA in the initial stages of petition development: just like the old saw about voting, the contacts should be early and often. The OPA has a delegated statutory requirement under section 170.20(b) of the CFR to provide guidance to prospective petitioners:

> Upon written request describing the proposed use of an additive and the proposed experiments to determine its safety, the Commissioner will advise a person who wishes to establish the safety of a food additive whether he believes the experiments planned will yield data adequate for an evaluation of the safety of the additive.

Because of the OPA's current backlog of pending petitions, it is trying to limit technical review of pre-petition submissions to protocols for unperformed studies. In the past, the prospective petitioner could submit data and summary information at key junctures in the collection process, and the OPA (or its predecessor group at CFSAN) would provide detailed feedback both in meetings and letters. In October 1991, however, the CFSAN stated that it would determine on a case-by-case basis whether to undertake technical reviews of specific information in a pre-petition submission (*Food Chemical News*, 1991). As of March 1995, the OPA is telling companies there will be a three month wait to receive comments on a protocol for a proposed study.

Nevertheless, a would-be petitioner should persevere with his efforts to elicit the OPA's reaction to his overall plans for the sweetener's safety evaluation program. Certainly, the OPA will have to review some amount of summary information in order to carry out its statutory duty responsibly. Such a summary would need to cover the sweetener's structure, its method of synthesis, and key physico-chemical characteristics such as stability and solubility.

The two most important operating principles in preparing a pre-petition submission are to understand and follow the FDA's written guidelines for safety studies and to highlight any problematic areas in the data already amassed. Trying to deviate from the FDA's regulations or guidelines or to gloss over potential issues in the data will not save anyone's time and will only tarnish what will be a long and challenging relationship. Moreover, the petitioner would benefit by having an outside expert inspect and prepare a critique of the draft submission to ensure that it is comprehensive, comprehensible, and, most importantly, objective.

When the prospective petitioner makes the submission, his covering letter should ask the OPA to establish a master file, which will be the repository for all pre-petition documents pertaining to the sweetener. All submitted materials plus the FDA's review notes and meeting minutes will be treated and held as confidential under the Freedom of Information Act

(FOIA). However, if a third party submits an FOIA request for information about the sweetener, the FDA is obliged to release all information that is publicly known. For example, if the petitioner has made public mention of a meeting with FDA concerning the sweetener, but has not disclosed the substance of the contact, the FDA would have to release only those portions of the materials that it knows are public, which could be just a copy of its meeting minutes with all the text blacked out except the list of participants. In cases where the FDA receives a FOIA request and is uncertain what information is public knowledge, the FDA will send a copy of its planned release to the petitioner for verification.

As part of his pre-petition submission, the petitioner should ask the OPA to schedule a meeting for three to four weeks later to allow the OPA time to review the material and prepare to discuss it. Many experts have written extensively on the subject of successful meetings with the FDA. While there are many permutations and subtleties, the basics are common sense:

1. An explicit agenda provided to the OPA in advance with the names and positions of the participants.
2. As few participants as possible, with the scientists who know the data leading the discussion.
3. A candid focus on potential issues or areas of uncertainty.
4. No lawyers, which not only affects the tone of the meeting, but also can inhibit discussion and detract from the science.
5. No surprises.
6. Painstaking preparation.

Following the meeting, the petitioner should document the discussion and decisions by preparing minutes and submitting them to the master file. The FDA will write back if it does not agree with the petitioner's account. The FDA will also prepare minutes for the master file, which the petitioner should request, although they tend to be sketchy and less than definitive. Even though the FDA can later change its views based on new learning or perspectives, it is well worth while to establish a paper trail (record of all contacts) that is as explicit and detailed as possible.

3.3.2 Elements of a food additive petition

The *Code of Federal Regulations* [21 CFR 171.1(c)] sets forth the FDA's six basic requirements for the contents of a food additive petition:

1. The additive's chemical identity and composition.
2. Its proposed uses and labeling.
3. Its intended technical effect.
4. Methods of analysis for the additive in food.

5. Full reports of all safety investigations.
6. An environmental assessment.

The Office of Pre-Market Approval has prepared expansive descriptions and recommendations for these requirements to help guide prospective petitioners in their preparations. Since these documents are readily available from the OPA, there is no reason to recapitulate the details here. Rather, what follows is a brief overview with emphasis on areas where difficulties can arise or caution is warranted.

3.3.3 Chemical identity and composition

The Chemistry Review Branch of the OPA has prepared a document covering all chemistry-related areas needed in a petition: *Recommendations for Submission of Chemical and Technological Data for Direct Food Additive and GRAS Food Ingredient Petitions*, May 1993 (hereinafter *Recommendations for Submission.*) These recommendations do not address administrative, toxicological, microbiological, nutritional, environmental assessment or labeling requirements.

The recommended identity and composition information includes chemical name, common name and trade name of the sweetener, Chemical Abstracts Service Registry Number when possible, empirical and structural formulae, molecular weight, characteristic physico-chemical properties such as melting point, specific gravity, solubility, etc., manufacturing process and specifications.

The OPA needs to see a detailed description of the manufacturing process, including a list of reagents, solvents, purification aids, etc., as well as reaction conditions and production controls, in order to identify and assess any potential or actual impurities in the sweetener. The level of detail to provide can be problematic.

During the petition preparation stage, the petitioner is probably still working to improve and refine the manufacturing process, so the description needs to avoid being too specific or restrictive. It is best to use either/or lists of chemicals to be employed, ranges of conditions for temperatures, times, pressures, pH, etc., and general descriptions of process controls. The key is to describe the basic chemistry of the process in such a way that the OPA will be assured on the impurity profile and process control while the petitioner will not be bound to a process that may turn out to be chemically or economically unfeasible.

Even though manufacturing methods and quality control procedures are exempt from disclosure under the FOIA, the petitioner should clearly mark each page of this section of the petition as exempt from disclosure. There also should be a shorter 'sanitized' version with all proprietary or confidential information deleted, which the OPA can use to respond to

FOIA requests. This will satisfy the statutory requirement that all information pertinent to the sweetener's safety must be released.

The proposed specifications for the sweetener require even more careful delineation than the process description. This is because the specifications prescribe the actual impurities and their levels that will be permitted in the commercial product, that is, the legal standard by which the FDA will regulate. Typically the purity of a new sweetener should be about 98%. Determining the exact composition of the remaining 2% requires considerable attention. The limits on the types and levels of impurities need to be tight enough to demonstrate a controlled process and a safe product, yet loose enough to allow for scale-up from pilot plant to commercial production and to permit batch-to-batch variability. Moreover, the specifications must prescribe a commercial product that is at least as clean as the material used in the core toxicology studies. Even though the types and levels of impurities generally decrease as the process is refined, it is wise not to be too optimistic in projecting what is achievable.

3.3.4 Usage and labeling

This section covers the proposed food product categories in which the sweetener will be used, the proposed maximal and typical amounts of the sweetener that will be added to those foods, the projected consumer exposure to the sweetener (known as the 'estimated daily intake', or EDI), its stability and fate in the petitioned foods, including the identity and amount of any degradation products, and any recommendations and directions for use, including specimens of labeling.

The projections of intake need to be calculated with great care to avoid being unduly conservative, that is, overestimated. That is because the EDI usually must not exceed "1/100th of the maximum amount demonstrated to be without harm to experimental animals" (21 CFR 170.22), which amount is termed the "acceptable daily intake" or ADI. If the EDI does exceed the ADI, the OPA will either have to limit the categories of use or not approve the sweetener, depending on the extent of excess of the EDI over the ADI.

The overall aim of regulatory agencies in both determining and evaluating EDIs is to take into proper consideration the potential for certain subgroups of the population to ingest much more of the additive than others. The OPA wants to see EDIs for the 'average' consumer and the 'high' consumer, which are generally represented by the consumers of a given food at the 50th and the 90th percentile, respectively, in an 'eaters-only' population. The OPA's use of the 90th percentile for the 'high' consumer contrasts with the UK MAFF's use of the 97.5th percentile, which results in a much more conservative regulatory risk assessment.

There are several approaches to determining the EDIs, but the critical part of the process is the construction of consumption models or assumptions. In the past the OPA wanted to see EDI projections based on a scenario in which the proposed sweetener would replace all the sugar in all the petitioned food categories. More recently, the OPA has agreed to more realistic assumptions, such as replacing sugar in selected categories while replacing the current low-calorie sweeteners in the remaining categories. Even these circumstances will result in a conservative, that is, inflated, set of values.

Furthermore, if the petitioner is submitting a narrow petition (requesting a few categories or limited use levels), the OPA is unlikely to accept a correspondingly narrow EDI as the basis for regulating the sweetener. The OPA cautions (*Recommendations for Submission*, 1993) that "the EDI must be based on reasonable consumption scenarios . . .; it cannot be based solely on the petitioner's current marketing plans." In other words, if the petition covers only chewing gum, the EDI cannot be based only on chewing gum consumption.

As for the required specimens of labeling, the petitioner needs to provide sample labels only for products he will distribute or market, such as table-top sachets, tablets, or liquids, and not for other companies' foods or beverages that will contain his sweetener. Such sample labels should provide a mock-up of the principal display panel, the ingredient statement, the nutrition label and any special use labeling. Currently, the only special use labeling the FDA has required is that for aspartame: (i) the statement "PHENYLKETONURICS: CONTAINS PHENYLALANINE"; and (ii) instructions not to use the table-top product in cooking or baking.

While low-calorie sweeteners have always been required to bear nutrition labeling (21 CFR 105.66), under the Nutrition Labeling and Education Act (NLEA), the new 'Nutrition Facts' panel for a table-top product requires more information in a rigidly prescribed format, with a now-standardized serving size (equivalent to the sweetness of two teaspoons of sugar). Even for such a relatively simple product, compliance with NLEA will entail careful interpretation of the complex regulation.

3.3.5 Intended technical effect

For a low-calorie sweetener petition, this section should present sensory data demonstrating the minimal, optimal and maximal concentrations of the sweetener needed to achieve a desired level of sweetness in the proposed food products. It is important to demonstrate that the sweetener has a maximal concentration in food, above which the food is unpalatable, for, without this technologically self-limiting use level, the OPA might need to impose tolerances (i.e. specified maximal use levels product by product). Sweeteners generally prove to have self-limiting use levels and therefore are approved for use according to Good Manufacturing Practices. The optimal concentration levels, of course, are the values the petitioner uses to project the Estimated Daily Intakes for the use section of the petition.

Naturally, developing the required sensory data means exposing human volunteers to the unapproved sweetener. Such a taste test program needs to incorporate a cautious, medically sanctioned escalation in exposure that is directly linked to the growing knowledge about the safety of the compound.

3.3.6 Methodology for analysis of the additive in food

The OPA's *Recommendations for Submission* give useful guidance on developing methods of analysis for a sweetener in food products. Such methods are necessary to demonstrate the degree of stability of the sweetener across the range of petitioned food products. Also, if the OPA must impose a limitation for safety reasons on the amount of the sweetener (or an impurity or degradation product) in particular foods, reliable methods are needed to quantify the substance in order to enforce the limit. Developing such methods often can be extremely difficult and time-consuming, particularly if there is a potential for interference or interactions in the food matrices.

3.3.7 Safety evaluation

This section comprises the bulk of a food additive petition, including sometimes more than 100 studies in seven species. The CFSAN has published extensive guidelines for the design and conduct of a safety evaluation program that will satisfy its toxicological requirements for food additives: *Toxicological Principles for the Safety Assessment of Direct Food Additives and Color Additives Used in Food* (1982), commonly known as the *Redbook*, plus the revised draft of *Redbook II* (1993).

There have been volumes of commentary about the two Redbooks, some highly critical, especially of the more far-reaching, less scientifically established testing areas added in *Redbook II*. At any rate, the petitioner should strive to adhere to the Redbook principles, certainly for the core studies, and should discuss any planned or desired deviations with the OPA toxicologists beforehand. While the OPA has been known to relent on some points, it has also been known to renege on them later. One point about *Redbook II* that the OPA fully concedes is that it has no decision elements, no road map posted with measurements and milestones that leads to the ultimate destination – approval.

One crucially important element in the conduct of safety studies is compliance with Good Laboratory Practices, or GLPs (21 CFR 58). Each study report should certify that the testing laboratory followed GLPs, or if not, where and why deviations occurred. As part of its safety evaluation process, the OPA will send a team of toxicologists and inspectors to the laboratory or laboratories where the key core studies were conducted to do a multi-day GLP audit of those studies. Even with clean toxicology data, GLP irregularities can jeopardize the validity of a study, or at least slow down the OPA's sign-off of the study. Furthermore, the FDA's inspection report is a publicly available document that cites all the identified major and minor shortcomings of the study, the report and the laboratory. To ensure compliance with GLPs, the petitioner should inspect the laboratory before signing a contract to place the study there and should conduct GLP audits throughout the duration of the study and preparation of the study report. Also at the outset the petitioner should inspect and approve the laboratory for its security measures in this unpredictable age of animal rights activism.

Another area requiring caution is studies in humans. Even though the FDA does not require clinical studies for food additives, they can provide appropriate confirmatory data correlating with the animal data. In accordance with the provisions in Parts 50 and 56 of the CFR, such studies must adhere to Good Clinical Practices and incorporate controls for the protection of the participants: (i) a detailed written protocol, including limits on intake, criteria for exclusion, and inclusion of a control group; (ii) informed consent of participants; (iii) review and approval by an Institutional Review Board; and (iv) mechanisms for fielding and resolving any health-related inquiries. In addition, the CFSAN's draft *Redbook II* provides recommended guidelines and considerations for the clinical evaluation of food additives.

3.3.8 Environmental assessment

The last section of a food additive petition evaluates the environmental impact of the additive's production as well as its use and disposal, and includes biodegradation and ecotoxicity studies, plus emissions quantification, control, and treatment at the manufacturing site, as well as lists of all applicable federal, state and local laws and permits. The assessment can be quite extensive and involved, especially for a sweetener that is not metabolized by the body and therefore enters intact into sewage and septic disposal systems.

Part 25 of 21 CFR sets forth the requirements and format for the assessment, and the CFSAN has published additional guidance and recommendations in *Preparing Environmental Assessments: General Suggestions* (1990) and *Environmental Assessment Technical Handbook* (1987). In the first document the CFSAN reminds the petitioner that the assessment will be publicly available and cautions that any confidential information, such as that pertaining to production capacity, should not be in the assessment but in a separate section of the petition.

3.4 Mechanics of a food additive petition

In 1987, the CFSAN issued *Guidelines for the Preparation of Petition Submissions*, which contains suggestions for paginating, binding and labeling a petition, in an effort to improve the management and maintenance of its voluminous records. It is well worth while following these guidelines because the OPA will repaginate, rebind, and relabel any petition not in this suggested form, which introduces opportunities for error when there are thousands of pages. Also, having the same pagination on the petitioner's copies as on the FDA's copies is especially helpful when the two parties are holding meetings or exchanging correspondence during the petition review.

There are other features the petitioner can incorporate into his multithousand page petition to make it as user- or reviewer-friendly as possible. The three most important are a detailed table of contents, an executive summary and a list of the safety studies with their volume/page location. Other helpful features include summaries of the required six major sections of the petition, a guide to the chemical structure and nomenclature of pertinent compounds (e.g. processing intermediates, impurities, degradation products), copies of the literature references, and a fiche copy of the petition.

Furthermore, while the FDA's regulations require submission of only three copies of the petition (which go to the toxicology review group, the chemistry review group and the Consumer Safety Officer who has day-to-day responsibility for the petition), the OPA can make good use of a fourth copy. Indeed, the OPA has even welcomed a fifth copy that has been purged of all confidential or proprietary information, which can be used as the master from which to make copies in response to FOIA requests.

3.5 The petition review and approval process

Once the petitioner submits a petition for approval of a new sweetener, the relationship between the regulator and the would-be regulated becomes more formal and less forthright. The Food Additives Amendment of 1958 plus the CFR stipulate certain key rules that the FDA must follow in reviewing and deciding on a petition's approvability, but most of the detailed workings of the process consist of non-statutory procedures, policies and guidelines that have evolved over the past 37 years.

The OPA has described all the elements of the process in its step-by-step Staff Manual Guide (*Management Porgram Policies and Procedures*, 1993). Figure 3.1 shows the OPA's simplified depiction of the movement of a petition through the initial review, the expert technical reviews, the decision points, the legal review, the FDA management reviews, and finally publication of a regulation in the *Federal Register*.

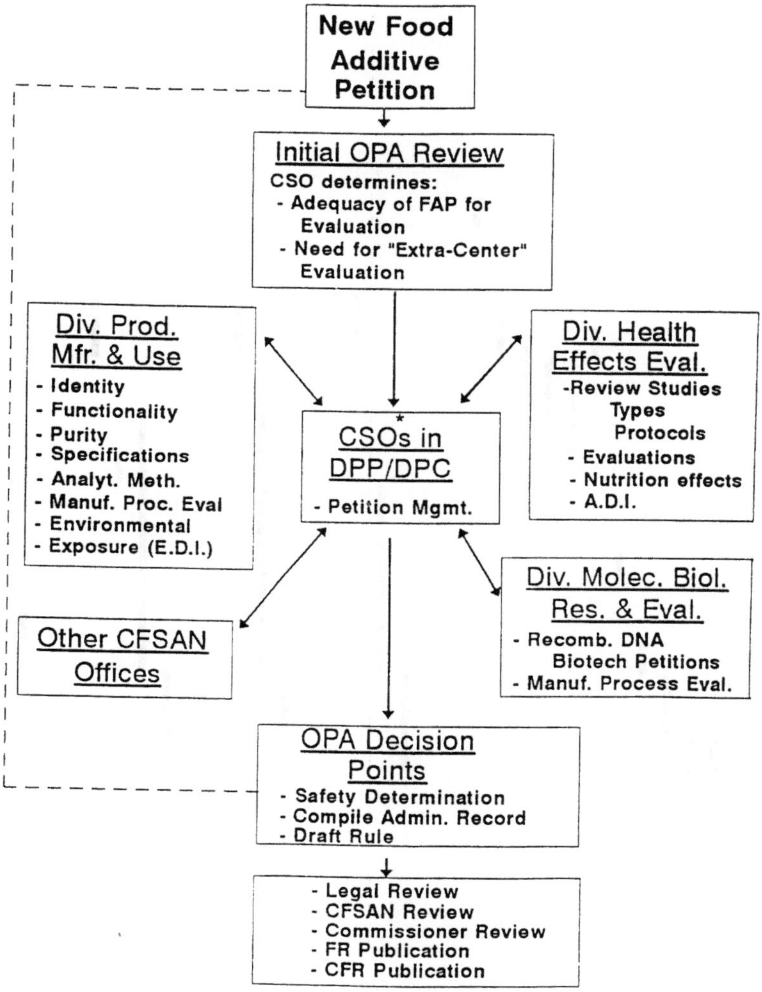

Figure 3.1 Food additive petition review in the office of premarket approval.* Consumer Safety Officers in Division of Product Policy or in Division of Petition Control.

What this figure does not show are the statutory times for acting on a petition. Section 409(c)(2) of the FD&C Act requires FDA to grant or deny a petition within 90 days or, before the 90th day, to extend the statutory time for an additional 90 days. Figure 3.2 (Rulis and Tarantino, 1993) shows the steps and times in this 'ideal' 180 day petition review cycle.

Clearly, this is not adhered to in practice: the OPA deals with this tight mandate by making time stand still, that is, allowing the clock to run only when it has all the information it requires, stopping the clock when it does

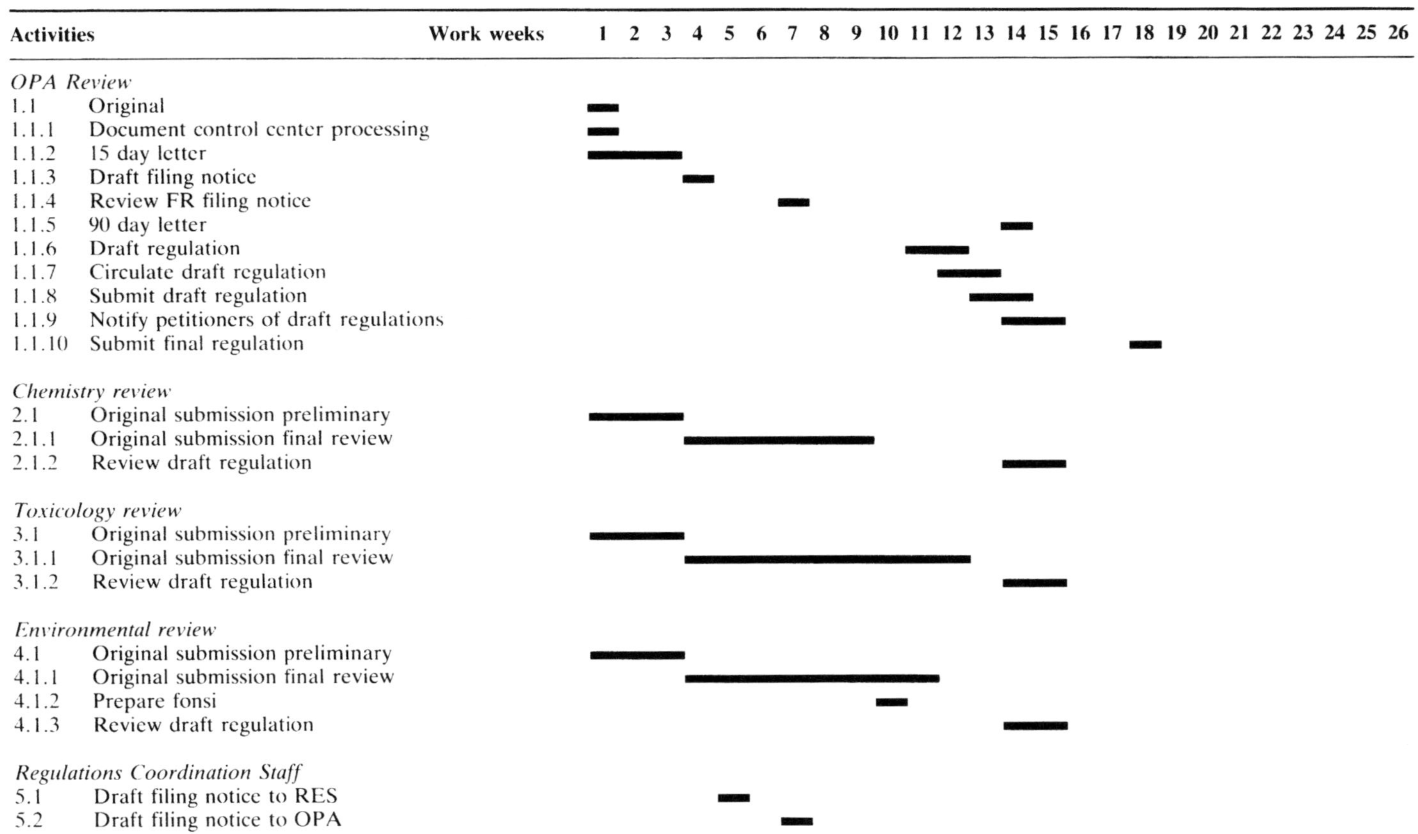
Activities
Work weeks
1 2 3 4 5 6 7 8 9 10 11 12 13 14 15 16 17 18 19 20 21 22 23 24 25 26
OPA Review
1.1 Original
1.1.1 Document control center processing
1.1.2 15 day letter
1.1.3 Draft filing notice
1.1.4 Review FR filing notice
1.1.5 90 day letter
1.1.6 Draft regulation
1.1.7 Circulate draft regulation
1.1.8 Submit draft regulation
1.1.9 Notify petitioners of draft regulations
1.1.10 Submit final regulation
Chemistry review
2.1 Original submission preliminary
2.1.1 Original submission final review
2.1.2 Review draft regulation
Toxicology review
3.1 Original submission preliminary
3.1.1 Original submission final review
3.1.2 Review draft regulation
Environmental review
4.1 Original submission preliminary
4.1.1 Original submission final review
4.1.2 Prepare fonsi
4.1.3 Review draft regulation
Regulations Coordination Staff
5.1 Draft filing notice to RES
5.2 Draft filing notice to OPA

5.3	Final filing notice submitted	
5.4	Draft regulation to RES	
5.5	Draft regulation to OPA	
5.6	Final regulation submitted	
General Counsel Review		
CFSAN/Dep. Comm'r/Commissioner		
7.1	Filing notice signed	
7.1.1	Office of Premarket Approval	
7.1.2	OPPSI Center Director	
7.2	Filing notice published	
7.3	Regulation signed	
7.3.1	Office of Premarket Approval	
7.3.2	OPPSI Center director	
7.3.3	Dep. Comm'r/Comissioner	
7.4	Regulation published	
8	CAC/QRAC	Concurrent with 2–4
9	Cosmetics and colors	Concurrent with 2–4
10	Mathematics	Concurrent with 2–4
11	Microbiology/Molecular biology	Concurrent with 2–4
12	Nutrition	Concurrent with 2–4
13	Pathology	Concurrent with 2–4
14	Other centers/agencies	Concurrent with 2–4

Figure 3.2 Food additive petition review. Cycle showing steps and times over 'ideal' 180 days.

not, and usually resetting the clock to zero when it receives the requested information from the petitioner.

What these two figures also cannot capture are the elements that characterize the review, evaluation and decision-making processes:

(a) The Consumer Safety Officer (CSO) in the OPA manages and coordinates the technical reviews, determines the need for any special review expertise outside the OPA or even outside of the CFSAN, compiles the administrative record that reflects each step of the process and that leads to a decision, and drafts the regulation. The CSO also serves as the 'gatekeeper' between the petitioner and the OPA.

(b) The review and evaluation process proceeds in a concurrent, decentralized fashion among the various technical experts, progresses to consensus on definitive endpoints for each separate scientific discipline, culminating in a consensus on the overall safety decision. The OPA's experts in chemistry evaluate information on the sweetener's physicochemical characteristics, purity, manufacturing process and exposure, then determine the specifications and EDI. The toxicology experts evaluate the entire safety data base and then determine the ADI. The experts in environmental science evaluate the environmental data and then reach a Finding of No Significant Impact, or FONSI. These consensus conclusions by discipline finally combine to create the basis for a consensus decision on safety and approvability.

(c) The evaluation process is 'recursive' (Rulis and Tarantino, 1993), or iterative. That is, the original data in the petition are the starting point from which flow any complexities and issues, the reviewers' requests for clarification or further data, the petitioner's responses, and the reviewers' interactions among themselves and with the petitioner to resolve their questions.

(d) The entire process is open to public scrutiny and challenge. Therefore, the administrative record must carefully document every step of the process and absolutely support the FDA's decision, in addition to serving as a deterrent or even the ultimate defence against any challenges or objections.

Once the OPA reaches a consensus on safety and determines that the sweetener is approvable, the process can still take a considerable time or take an unexpected turn. Much time is devoted to completing the administrative record, particularly the preamble that is published with the regulation in the *Federal Register*. The preamble is a critically important, often lengthy document that summarizes the salient features and conclusions of the technical review, including any issues or problematic areas and their resolution. Arriving at a final preamble requires extensive and time-consuming input and approval from all the technical reviewers, the regulatory experts and legal counsel, with sign-off by layers of the OPA, CFSAN, and FDA management.

Furthermore, during the preamble preparation stage, there may be developments that can slow or even stop the entire process. Toxicology reviewers may leave the FDA and the new toxicologists may have different perspectives. The reviewing lawyer may be uncomfortable with the strength or defensibility of the scientists' conclusions. Third-party groups may submit challenges on the safety of the sweetener, which the OPA must consider and resolve, as well as incorporate into the preamble. Even at this late stage, uncertainty and unpredictability reign, but the stakes are higher than earlier in the process because of the time and money invested. As the Director of the OPA has acknowledged, "The agency is aware that an evaluation delayed can be an evaluation denied because of the competitive pressure of the marketplace." (Rulis, 1990)

Even though the reviewers and the petitioner may have discussed and worked together on many elements of the data both before the petition and during the review stages, and even though the OPA has told the petitioner his sweetener is approvable, the petitioner does not have any role in the preparation of the preamble and will have little knowledge of its progress. The only knowledge he will have at this stage will be the content of the draft regulation, which the OPA will ask him to review to ensure that it accurately reflects the petitioned action. Reviewing the draft regulation is the only official glimpse the petitioner gets of the final stage of the process.

One new area for late-stage interaction is pre-approval inspection of the manufacturing plant. The FDA has recently inaugurated such inspections for new drugs but has yet to initiate a program for new food additives. Certainly adding this step will introduce more potential for delay and uncertainty.

With the publication of the regulation in the *Federal Register*, there is still one major remaining hurdle. Although the regulation is effective upon publication, under 21 CFR 12 there is a thirty-day period for submitting objections and a request for a formal evidentiary hearing. Each objection for which a hearing is requested must present "a detailed description and analysis of the factual information to be presented in support of the objection." The FDA reviews all objections and requests for hearings "as soon as possible" and determines whether a hearing is justified and whether the final regulation should be modified or revoked. The FDA grants a request for a hearing if the objecting party's submission shows that there is a "genuine and substantial issue of fact" that "can be resolved by available and specifically identified reliable evidence." It will not grant a hearing on issues of policy or law or "on the basis of mere allegations or denials or general descriptions of positions and contentions." If the FDA denies the hearing, it will publish an explanatory notice in the *Federal Register*. If it grants the hearing, the burden of proof is on the petitioner.

3.6 Conclusion

Recently, many public and private sector groups have roundly criticized the FDA's dismal record on petition reviews and approvals and have begun initiatives to effect change. Food companies, trade associations, law firms, public interest groups, foundations, former FDA officials, former members of Congress and, of course, current members of Congress have all joined the fray.

It is unclear whether these efforts will have any success or whether the FDA will or can chart a better course on its own or in concert with others. What is clear is that innovative new food additives cannot be brought into use for some time to come, and that in the meantime the companies already involved in the petition process have a long, expensive, frustrating and unpredictable road ahead.

> The race is not to the swift, nor the battle to the strong, neither yet bread to the wise, nor yet riches to men of understanding, nor yet favor to men of skill; but time and chance happeneth to them all.
>
> *Ecclesiastes 9:11*

References

Code of Federal Regulations (1994) US Government Printing Office, Washington, DC.

Environmental Assessment Technical Handbook (March 1987) Prepared by the Environmental Staffs of the Center for Food Safety and Applied Nutrition and the Center for Veterinary Medicine, US Food and Drug Administration, Washington, DC.

Federal Food, Drug, and Cosmetic Act, as amended, and related laws (1986) US Government Printing Office, Washington, DC.

Food Chemical News (Oct. 21, 1991) 61–62.

Guidelines for the Preparation of Petition Submissions (1987) Center for Food Safety and Applied Nutrition, US Food and Drug Administration, Washington, DC.

Lawler, P.F. (1986) *Sweet Talk*, The Media Institute, Washington, DC, p. 8.

Management Program Policies and Procedures; Policies, Authority, and Procedures for Food and Color Additive Petitions and GRAS Affirmation Petitions; Staff Manual Guide (April 15, 1993) Center for Food Safety and Applied Nutrition, US Food and Drug Administration, Washington, DC.

Preparing Environmental Assessments: General Suggestions (August 1990) Prepared by the Environmental Impact Section, Center for Food Safety and Applied Nutrition, US Food and Drug Administration, Washington, DC.

Recommendations for Submission of Chemical and Technological Data for Direct Food Additive and GRAS Food Ingredient Petitions (May 1993) Chemistry Review Branch,

Office of Premarket Approval, Center for Food Safety and Applied Nutrition, US Food and Drug Administration, Washington, DC.
Rulis, A.M. (1990) The Food and Drug Administration's Food Additive Petition Review Process. *Food Drug Cosmetic Law Journal* **45**, 533–544.
Rulis, A.M. and Tarantino, L.M. (1993) The Food Additive Petition Process: Recent Data. *Food and Drug Law Journal* **48**, 137–151.
Toxicological Principles for the Safety Assessment of Direct Food Additives and Color Additives Used in Food (1982) US Food and Drug Administration, Washington, DC.
Toxicological Principles for the Safety Assessment of Direct Food Additives and Color Additives Used in Food, Redbook II (Draft, 1993) US Food and Drug Administration, Washington, DC.

4 Physiological properties of polyols in comparison with easily metabolisable saccharides

H. SCHIWECK and S.C. ZIESENITZ

Summary

The main features of the metabolism of the polyols erythritol, xylitol, sorbitol, mannitol, lactitol, isomalt, maltitol and hydrogenated starch hydrolysates (HSH), plus those of L-sorbose, are described, including their metabolism, absorption, glycemic effects, energetic utilisation, gastro-intestinal tolerance and dental properties. Although similarities exist between these polyols, the data show that the metabolic characteristics of each polyol, especially their caloric values, have to be considered individually. For consumer acceptance the sensoric and technological properties of the respective polyols are also important. Xylitol is to be preferred for the production of toothfriendly sweets with menthol and peppermint tastes, which enhance the cooling effect. Isomalt, on account of its sensoric and physical properties, is especially suited for the production of sweets and food products with taste profiles similar to that of sucrose. Hydrogenated starch hydrolysates (HSH) can replace glucose syrups in conventional food products and are an important component in confectionery.

4.1 Introduction

Up to 1950 dietary carbohydrates, including mono- and disaccharides, oligo- and polysaccharides and starches were generally assumed to be completely digested and their energy was assumed to be totally bioavailable. It was known that carbohydrates digested in the small intestine are glucogenic and hyperglycemic and therefore their use needed to be controlled in the diabetic diet. Dietary fibre on the contrary was believed not to provide available energy for the human body. Recent progress in nutrition research has changed opinion on the calorific availability of dietary fibres, non-starch polysaccharides (NSP) and resistant starches (RS) formed during heat treatment of starchy foods during the cooking procedure (Siljeström *et al.*, 1989), which used to be regarded as non-nutrient components of the diet.

Dietary fibre is now regarded as a source of some energy for humans, as it has been established that dietary fibre and non starch polysaccharides make some energetic contribution via their bacterial fermentation products, i.e. short chain volatile fatty acids (SCFA), which are absorbed from the colon. Thus dietary fibre and NSP are now classified as partially available in terms of energy, but not in terms of glucose availability, as short chain fatty acids can be an important source of carbon and energy not only for ruminants but also for nonruminant mammals (Bugaut and Bentéjac, 1993).

Also the concept of a common physiological caloric value for monosaccharides changed when differences in postprandial oxidation rates of glucose and fructose were observed (Macdonald, 1984). Fructose, given as pure substance or as part of a meal, induced 10% greater thermogenesis than an equivalent amount of glucose (Tappy *et al.*, 1986; Schwarz *et al.*, 1989). When it was infused parenterally the thermic effect of fructose was higher than that of glucose (Schwarz *et al.*, 1992). The large thermic response to fructose is probably due to stimulation of gluconeogenesis and glycogen synthesis in the liver. This can be explained by the hydrolysis of 3.5–4.5 mol ATP/mol fructose stored as glycogen, compared with 2.5 mol ATP/mol glucose stored (Tappy and Jéquier, 1993). Although fructose and glucose are both hexose monosaccharides, they are metabolised quite differently. A large proportion of either intravenous or orally administered glucose escapes liver uptake and is disposed of by peripheral tissues, while fructose is essentially metabolised in the liver.

The main impetus for the expansion of use of dietary polyols was the demand for sweeteners suitable for diabetics which have no impact on blood glucose concentrations. The ideal sweeteners for diabetics were believed to be those substances not increasing postprandial blood glucose concentrations. The very first substances suggested as sweeteners for diabetics were fructose (Külz, 1874; Leuthardt and Stuhlfauth, 1960; Mehnert and Förster, 1979), sorbitol (Thannhauser and Meyer, 1929), and L-sorbose (Grieshaber, 1936a,b; Huchette and Leroy, 1973), followed by xylitol (Mellinghoff, 1961).

4.1.1 Sorbitol and fructose

Sorbitol has to be oxidised by sorbitol dehydrogenase to fructose, which is metabolised in the liver more rapidly than glucose. Fructose is an established sweetener for diabetics and is used in several countries. Metabolic and nutritional aspects of fructose had been reviewed comprehensively by Henry *et al.* (1991) and in a monograph by Forbes and Bowman (1993). The metabolism of fructose differs from that of glucose in several respects (Heinz, 1973): The liver enzyme responsible for the phosphorylation of glucose, liver hexokinase, has a higher affinity for glucose than for fructose, whereas glucokinase is inactive on fructose

(Vinuela *et al.*, 1963). Fructose is phosphorylated to fructose-1-phosphate by fructokinase (Leuthardt and Testa, 1951; Cori *et al.*, 1951; Hers, 1952), which can also catalyse the phosphorylation of L-sorbose and D-tagatose (Adelman *et al.*, 1967).

4.1.2 L-Sorbose

Since its discovery in 1852, L-sorbose, an isomer to D-fructose, has attracted interest due to its structural similarity to D-fructose. L-Sorbose was also regarded as an alternative diabetic sugar, as it did not induce hyperglycemia or insulinemia. Grieshaber reported in 1936 on the poor absorption and the high urinary excretion (12–14%) of L-sorbose by man (Grieshaber, 1936a,b).

Metabolic studies with L-sorbose in rats were reported in the 1950s (Burns *et al.*, 1955) and in the 1970s (Würsch *et al.*, 1979), at the same time as the low cariogenic potential of L-sorbose was recognised in rat caries studies (Mühlemann, 1976; Mühlemann and Schneider, 1976; Lohmann *et al.*, 1981). In the mid-1970s, about forty years after the antiketogenic effect of L-sorbose with diabetics, its poor absorption and partial excretion were described, it was suggested as a calorie-reduced sweetener for diabetics with a low cariogenic potential (Huchette and Leroy, 1973). However, as Zürrer *et al.* (1978) and Keller and Kistler (1978) had observed adverse hemolytic effects on erythrocytes in beagle dogs, the safety of L-sorbose came to be regarded as questionable. Siebert *et al.* (1980) reported from 90-day growth experiments with rats fed either 10% L-sorbose or glucose, that liver weights, plasma lactate dehydrogenase levels, and the percentage of eosinophilic cells amongst white blood corpuscles were highly significantly increased after feeding L-sorbose. These increases in liver weights and disturbances in blood cell composition called for some caution in claiming that L-sorbose could be used as a dietetic sweetener.

More than ten years later interest in L-sorbose returned, and it was again proposed as a sweetener for diabetics, based on studies in diabetic mice and rats (Furuse *et al.*, 1991, 1993; Tamura *et al.*, 1991). These recent experiments in mice and rats did not however eliminate the early concerns on adverse health effects induced by L-sorbose, such as the data reported on increased organ weights (liver, kidney, heart). The authors commented that metabolic organs seemed to be influenced by dietary sorbose, but data on the controversial issue of sorbose affecting blood composition were missing in their publications. Furthermore any word of caution on restricting the use of L-sorbose for dietary purposes on the basis of increased weights of metabolic organs was lacking. The studies were not designed to solve these controversial issues. The increase in liver weights and other organs could be a cause of serious concern, and require further experimental effort.

In summary, metabolic studies with L-sorbose demonstrated its defects at a very early stage of experimentation, deterring further research on its use for nutritional purposes (Zürrer *et al.*, 1978; Keller and Kistler, 1978; Siebert *et al.*, 1980; Goto *et al.*, 1994).

4.1.3 L-Sugars

The basis for studies with L-sugars was presumably that the human body cannot utilise L-sugars. This theoretical assumption that they might be non-caloric attracted a great deal of interest and much effort was put into their synthesis or production. Meanwhile the claim of the non-utilisation of L-sugars as calorically inert substances has been questioned by a series of animal growth and metabolic experiments with different L-sugars, such as L-glucose, L-fructose and L-sucrose. Not all the results have been published, but the most interesting challenge of the L-sugar concept came from digestibility experiments with rats fed either D- or L-fructose, in which L-fructose had a physiological caloric value which was not all negligible (Klein *et al.*, 1993). In fact the digestibility of L-fructose in rats was about 10% lower than that of D-fructose. This was contradictory to the widely propagated claim that L-sugars are noncaloric and might therefore be ideal as bulk sweeteners. The experimental data did not support the concept that L-sugars have no nutritive contribution. It must concluded that L-sugars, especially L-isomers of regular dietary carbohydrates like D-glucose and D-fructose, are bioavailable to approximately the same extent as their D-sugar counterparts. Comparative experiments also demonstrated that L-sugars differ in their sensoric properties (sweetness, sensoric profile) from their corresponding D-sugar isomers (Schiweck, unpublished results).

4.1.4 Mannitol

Mannitol, known since 1806, is obtained as a byproduct from the production of sorbitol using fructose-based raw materials. Mannitol has always been of minor importance as a sweetener for diabetics and is predominantly used for special applications such as in chewing gum and bodying for pharmaceutical products. The relative low interest in mannitol as a sweetener is generally ascribed to its technological and physiological properties, as its sweetness and solubility are limited and is not well tolerated, with a relatively high laxative potential. Mannitol was one of the first carbohydrates recognised as having a reduced caloric value in man (Nasrallah and Iber, 1969).

4.1.5 Xylitol

Xylitol is a normal metabolite of the glucuronate-xylulose cycle. It is oxidised to D-xylulose, which is phosphorylated to D-xylulose-5-phosphate and subsequently metabolised to fructose-6-phosphate, and thus fed via the nonoxidative part of the pentose-phosphate pathway into the glycolytic chain, or it may be transformed to glucose-1-phosphate, a precursor of glycogen (Bässler *et al.*, 1962, 1966; Lang, 1968; Touster, 1974; Froesch and Jakob, 1974).

At the beginning of the 1950s the low fermentability of sorbitol and mannitol by caries-inducing microorganisms and dental plaque were recognised (Fosdick *et al.*, £957). These properties received more attention once xylitol was produced on a large scale, and its noncariogenicity was established by the Turku sugar studies in the 1970s (Scheinin and Mäkinen, 1975).

This brief review of the development of monosaccharide-derived polyols and their applications gives the three main physiological characteristics in which they differ from regular nutritive carbohydrates. These polyol-specific properties are:

(i) suitability for diabetics;
(ii) noncariogenicity;
(iii) reduced physiological caloric value.

Attention then transferred to the development of disaccharide alcohols like isomalt, maltitol, lactitol and HSH (hydrogenated starch hydrolysates), starting in the 1970s. Since then these disaccharide alcohols have been approved for use in the European Union (94/35/EC) and elsewhere in recent years. In the USA substances are individually regulated as published in the *Code of Federal Regulations*. The following polyols are now available on the market: sorbitol, mannitol, xylitol (monosaccharide alcohols) and isomalt, lactitol, maltitol and maltitol/sorbitol-containing syrups (disaccharide alcohols).

4.2 Digestion

The digestion of disaccharide alcohols is different in principle from the digestion of monosaccharide alcohols, as disaccharide alcohols have to be cleaved enzymatically into their monomers by the digestive carbohydrases of the intestinal mucosa, in an analogous way to the digestive cleavage of disaccharides. The absorption of disaccharide alcohols is practically negligible, as only traces of the intact substances may be absorbed in a concentration-dependent manner, after which they are excreted almost quantitatively unchanged in the urine.

4.2.1 Cleavage of disaccharides, disaccharide alcohols and oligosaccharide alcohols

The same principle that applies to polymeric carbohydrates also applies to disaccharide alcohols. They have to be cleaved first, to make their building blocks available for absorption. The carbohydrate digestive enzymes reside in the intestinal mucosa, which has maltase as well as sucrase–isomaltase activity. Isomalt, maltitol and HSH (hydrogenated starch hydrolysates) are substrates of these digestive enzymes to varying extents, depending on the affinities of the disaccharide alcohols to these enzymes. α-Glucosylhexitols like maltitol and isomalt are thus hydrolysed to their monomers at lower rates than are dietary disaccharides such as maltose, sucrose or isomaltose and isomaltulose. A low cleavage rate *in vitro* may reflect incomplete hydrolysis *in vivo*. This is a generalisation but it is corroborated by recovery studies with fistulated animals (van Weerden and Huisman, 1993a,b) and with colectomised man (Kroneberg *et al.*, 1979; Langkilde *et al.*, 1994).

Table 4.1 demonstrates the different affinities of disaccharides and disaccharide alcohols to human digestive enzyme preparations, and the influence of the type of glycosidic linkage as well as the effect of hydrogenation of disaccharides on k_m and relative cleavage rates compared to the reference substrate maltose (Ziesenitz and Siebert, 1987). *In vitro* incubations with human intestinal biopsy samples demonstrated in addition that maltitol is relatively easily hydrolysed whereas isomalt and lactitol are

Table 4.1 Cleavage of disaccharides and hydrogenated disaccharides by a pooled sample ($n = 8$) of human jejunal mucosa

Substrate		k_m (mM)	v_{max} (nmol/min per mg protein)	(%)
glc α(1→1) fru	Glucosylfructose/Trehalulose	21	136	15
glc α(1→β2) fru	Saccharose	22	280	31
glc α(1→3) fru	Turanose	20	120	13
glc α(1→5) fru	Leucrose	11	315	35
glc α(1→6) fru	Isomaltulose (Palatinose®)	9	73	8
glc α(1→4) glc	Maltose	9	900	100
glc α(1→6) glc	Isomaltose	7	234	26
glc α(1→1) mtl	Glucosylmannitol	11	32	4
glc α(1→1) mtl and glc α(1→1) sorb	Hydrogenated trehalulose	3	35	4
glc α(1→3) mtl and glc α(1→3) sorb	Hydrogenated turanose	27	22	2
glc α(1→4) sorb	Maltitol, cristalline	13	112	12
glc α(1→4) sorb	Malbit®	50–70	250–310	28–34
glc α(1→6) sorb	Glucosylsorbitol	8	67	7
glc α(1→1) mtl and glc α(1→6) sorb	Palatinit®/Isomalt	5	35	4

Source: Ziesenitz and Siebert (1987)

poor substrates for human intestinal disaccharidases (Nilsson and Jägerstad, 1987). From maltitol-based syrups (HSH) which contain hydrogenated oligosaccharides in $\alpha(1\rightarrow4)$ glycosidic linkages, e.g. maltotriitol, glucose is liberated rapidly, as $\alpha(1\rightarrow4)$-linked glucose-oligosaccharides or maltose are the natural substrates for sucrase–isomaltase complex in the human mucosa. Lactitol, a galactosylhexitol, is not a substrate for the sucrase/isomaltase complex and was reported to have only 10% of the hydrolytic cleavage rate of lactose (van Velthuijsen and Blankers, 1991).

4.3 Absorption of polyols

The monosaccharide-derived polyols like mannitol, sorbitol, xylitol or erythritol are absorbed by a concentration-dependent diffusion process, in contrast to the active uptake mechanism of glucose. The permeability of these polyols across membranes is theoretically in favour of substances of lower molecular weight, but it is well established that the permeability of the small intestine differs widely along the intestinal tract. Erythritol as well as mannitol – while not significantly metabolised – are known as fluid phase markers and are eventually used as probe molecules to measure intestinal membrane permeability (Yuasa *et al.*, 1989; Elia *et al.*, 1987). The absorption rate of erythritol measured *in vitro* is correlated to the intestinal permeability as estimated from perfusion experiments in Wistar rats (Yuasa *et al.*, 1989). Data on the absorption of polyols in humans are diverse. The data for polyol absorption presented here concentrate as far as possible on direct quantitative measurements of small bowel absorption/disappearance in humans obtained either by ileal aspiration techniques or from ileostomised persons.

The pros and cons of these methods are well known. The major advantage lies in the fact that the resulting data are directly related to man and that extrapolation from test species is not mandatory. The limitations of the data are that ileal intubation may affect intestinal transport processes, or might be different under ileostomised conditions, and that the disappearance of the substance under study from the intestinal lumen might be overestimated due to bacterial utilisation under ileostomised conditions.

In general, glucose absorption in the upper intestine is termed complete due to the active uptake mechanism for glucose. The absorption of polyols on their intestinal passage however, because of the concentration-dependent uptake mechanism, is slower and incomplete, and substantial portions may be fermented in the colon.

Mehnert *et al.* (1959, 1975) compared the relative absorption of 5 g each of sorbitol, xylitol, glucose, galactose, fructose and disaccharides in humans, using the intestinal aspiration technique. They generally observed slower absorption rates for sorbitol and xylitol compared to glucose. The absorption rate from xylitol was higher than from equal doses of sorbitol in humans, using the balloon aspiration method with a disappearance of 33% for xylitol compared with the complete disappearance of glucose in 30 min, compared with 27% for sorbitol (Mehnert *et al.*, 1975). The absorption capacities for sorbitol etc. were in the order of 10–20 g absorbed from the intestinal lumen per hour (Mehnert and Förster, 1979). Sorbitol absorption is not restricted to the upper gut, although the absorption rate in the terminal ileum is slower than in the upper part of the intestine (Lauwers *et al.*, 1985). Saunders and Wiggins (1981) reported that 26% of ingested mannitol had disappeared from the small intestine of ileostomy patients, whereas Langkilde *et al.* (1994) observed the disappearance of 73% of ingested sorbitol in ileostomy patients after consumption of 2 × 15 g given in the form of chocolate.

Data reported on the absorption capacity of xylitol in humans are also conflicting. Asano *et al.* (1973), measuring the ileal output by aspiration of lumen contents, reported that 50–95% of ingested xylitol had disappeared from the small intestine in humans who had consumed doses of 5–30 g/day over two to three weeks.

Hiele *et al.* (1993) concluded from metabolic studies with 25 g erythritol, that it is readily absorbed. Almost complete absorption was assumed to take place as no increase in hydrogen exhalation after erythritol intake had been observed in humans.

However polyol absorption from disaccharide alcohols like isomalt, on account of the slow liberation of sorbitol and mannitol from isomalt, is retarded compared with the absorption of sorbitol itself, with the consequence that a larger proportion of the hexitols from disaccharide alcohols is transferred into the colon (Langkilde *et al.*, 1994).

Beaugerie *et al.* (1990) compared digestion and absorption of sorbitol, maltitol and a hydrogenated starch hydrolysate in adapted volunteers with the ileal intubation technique by aspiration of the ileal contents after a single meal. Under these experimental conditions sorbitol disappearance from the small intestine was higher when it was given as free sorbitol but lower from maltitol and HSH (Lycasin 80/55). The disappearance of sorbitol, maltitol and Lycasin from the small intestine was generally higher in this study than in other reports.

4.4 Postprandial effects of polyols on blood glucose and polyol concentrations

The metabolism and metabolic capacity for sorbitol, fructose and xylitol have been extensively studied as they are used in parenteral nutrition.

Xylitol, sorbitol and fructose were of great interest in clinical nutrition, especially in post-traumatic situations, as their uptake into cells is insulin-independent. The infusion rates for the respective sugars and polyols were defined. These infusion rates do not represent maximal metabolism and oxidation rates in humans, but reflect the metabolic capacity and turnover rates of the human liver, as the liver is virtually the only metabolic organ for these substances.

4.4.1 Glycemic profiles of polyols

Postprandial increases in blood glucose concentrations depend exclusively on the glucogenic potential of the monosaccharide alcohols such as mannitol, sorbitol and xylitol, which can undergo intrahepatic inter-conversion to glucose subsequently released into the blood stream. For disaccharide alcohols, blood glucose profiles depend on glucose availability from pre-formed glucose as well as from the conversion of the constituents of the polyols into glucose. The rate of glucose release in the small intestine determines the blood glucose kinetics, and is in general slower from disaccharide alcohols than from actively absorbed glucose from starches, maltose or sucrose.

Although healthy persons or diabetics respond in different ways, the data available allow some generalisations on blood glucose increases and insulin requirements. The relatively low metabolic conversion of polyols and fructose to glucose is the basis for their use as sugar replacers in diets for diabetics, as these substances induce only moderate, if any, blood glucose increments, and are metabolised under substantially reduced insulin requirements (Mehnert, 1976).

Oral loading tests such as glucose tolerance tests or test meals and diet studies of different designs with healthy volunteers, diabetic adults and children gave ample evidence for the suitability of these substances in the dietary management of diabetes (Bär, 1991). It is not within the scope of this review to describe all the studies performed on the glycemic profiles of the polyols. Only strictly relevant studies permitting comparisons of the polyols are taken into account here, as publications focusing on dietary management of diabetes are numerous.

Oral tolerance tests with 30 and 50 g xylitol in adapted healthy adults induced only slight increases of blood glucose concentrations (10 mg glucose/100 ml) above baseline levels. Serum xylitol concentrations reached peak concentrations of about 5 mg xylitol/100 ml after 30 g xylitol by mouth and about 20 mg xylitol/100 ml after an oral load of 50 g xylitol. Serum insulin increased only about 10 μU insulin/ml after 50 g xylitol by mouth whereas an increase of 40 μU insulin/ml after 50 g oral glucose had been observed (Müller-Hess *et al.*, 1975).

Nguyen *et al.* (1993) compared the glycemic reactions to 20 g oral loads of glucose, sorbitol, xylitol, maltitol and a hydrogenated starch hydrolysate in healthy volunteers over a period of 3 h. This demonstrated differences between the polyols. No deviation from baseline values in blood glucose and serum insulin concentrations with 20 g of sorbitol or xylitol were observed. However maltitol and hydrogenated starch hydrolysate gave rise to about 50% of the peak blood glucose concentrations of glucose. These substantial changes were reflected by increases in insulin and changes in C-peptide levels (Nguyen *et al.*, 1993).

Kearsley *et al.* (1982) also observed graded glycemic responses to maltitol and hydrogenated starch hydrolysates in healthy volunteers under both nonadapted and adapted conditions. They compared these responses with those to mixtures of glucose and sorbitol which mimicked the composition of maltitol and HSH. Within a ten-day consumption period some adaptation to maltitol and hydrogenated starch hydrolysate was observed with higher blood glucose increments. The authors concluded that high maltitol syrups and HSH are metabolised approximately to the same extent as their basic components given as mixture of glucose and sorbitol (Kearsley *et al.*, 1982).

Wheeler *et al.* (1990) reported from oral loading tests with 50 g of different types of hydrogenated starch hydrolysates in both diabetics and healthy subjects that there was a reduced plasma glucose response compared with glucose, depending on the type of the hydrolysate. Elevations in plasma glucose by these products were in the range 70–80% of that of a 50 g oral glucose load. The HSH or HGS (hydrogenated glucose syrups) tested consisted primarily of maltitol, sorbitol and variable proportions of hydrogenated oligo- and polysaccharides.

Several reports on the glycemic response of maltitol in healthy subjects and diabetics had shown a reduced blood sugar increase compared with an oral challenge of either glucose or sucrose (Atsuji *et al.*, 1982; Nishikawa, 1982). According to Kamoi *et al.* (1975) elevation of blood glucose levels of maltitol reached 25–50% of that of glucose.

Similar observations were made by Felber *et al.* (1987), comparing plasma glucose and insulin peaks from maltitol and sucrose in healthy volunteers. There was a substantial blood glucose increase above baseline after 30 g maltitol ingestion, but it was still lower than that from sucrose. However in the second phase of the observation period, from 120–360 minutes, both blood glucose and serum insulin increments remained slightly higher after maltitol than after sucrose (Felber *et al.*, 1987).

The majority of the data published demonstrate that there is a limited glycemic response after maltitol, somewhat lower than that from glucose, and always statistically significantly increased responses above baseline levels in healthy (Felber *et al.*, 1987; Kearsley *et al.*, 1982) and diabetic subjects (Wheeler *et al.*, 1990; Secchi *et al.*, 1986).

The glycemic responses to isomalt differed from those observed with maltitol and maltitol-based syrups, in that oral loading and diet studies with healthy and diabetic humans showed practically no deviation of blood glucose and serum insulin concentrations from baseline levels (Hütter *et al.*, 1993; Gee *et al.*, 1991; Pometta *et al.*, 1985; Kaspar and Spengler, 1984; Thiébaud *et al.*, 1984; Petzoldt *et al.*, 1982; Drost *et al.*, 1980; Grupp and Siebert, 1978). Similar observations of unchanging blood glucose concentrations after the ingestion of isomalt were reported in diabetics (NIDDM) when the glycemic response of isomalt was compared with that to sucrose and fructose (Bachmann *et al.*, 1984).

Blood glucose profiles were not influenced by an oral dose of 24 g lactitol per day, according to an unpublished report by Doorenbos (1977) as quoted by den Uyl (1987). The lack of impact on blood glucose profiles by lactitol is explained by the negligible hydrolysis of lactitol on its intestinal passage, as well as by the low glucogenic potential of its constituents, galactose and sorbitol. Lactitol has also been termed a 'non-absorbed' sugar, which refers to its unavailability in the small intestine, although it is fermented in the colon (Grimble *et al.*, 1988).

Erythritol at a dosage of 1 g/kg body weight by mouth influenced neither blood plasma glucose nor insulin levels in humans (Bornet *et al.*, 1992). Mannitol also has no influence on blood glucose responses (Ellis and Krantz, 1941).

In summary, the polyols erythritol, sorbitol, xylitol, mannitol, isomalt and lactitol have practically no impact on blood glucose concentrations and if at all, only a moderate influence on postprandial serum insulin profiles. Maltitol and polyol syrups based on maltitol and hydrogenated starch hydrolysates (HSH) have substantially higher glycemic responses, however, which correlates with their glucose contents. A higher proportion of hydrogenated oligosaccharides and polysaccharides in the product induces a higher blood glucose increment.

4.4.2 Effects on serum polyol concentrations

Postprandial effects on plasma polyol concentrations in humans have been investigated primarily with the monosaccharide alcohols xylitol and sorbitol. Serum sorbitol (Adcock and Gray, 1957; Malaisse-Lagae *et al.*, 1984; Macdonald *et al.*, 1978) and xylitol concentrations in blood (Müller-Hess *et al.*, 1975) after oral loading tests or after dietary consumption of a polyol-rich diet were consistently reported to be in the micromolar range, and thus negligibly small. These low blood concentrations of polyols are explained by their slow and incomplete absorption and by the hepatic extraction and metabolism of sorbitol and xylitol, whereas absorbed mannitol is eliminated almost unchanged via the kidneys.

Serum levels of erythritol in humans after consumption or glucose tolerance tests has not yet been reported. According to Hiele *et al.* (1993),

52% of a dose of 25 g erythritol to humans by mouth was eliminated via the urine within 6 h and urinary elimination reached 84% within 24 h. The assumption was made that absorbed erythritol is distributed easily between blood and tissue compartments and that an enormous dilution leads to low serum erythritol levels in humans.

From intestinal perfusion studies in pigs with 440 g of a maltitol-rich HSH containing 6% sorbitol, the appearance of sorbitol in the portal blood was reported to be 60 times lower than glycemia values under these specific conditions. Peak portal blood concentrations of sorbitol were in the range of 3 mg/100 ml when maltitol-rich HSH was infused into the duodenum. The portal blood profile from HSH showed a biphasic behaviour which might reflect the slow absorption of free sorbitol as well as of sorbitol liberated from maltitol and maltotriitol (Rérat *et al.*, 1991).

4.5 Microbial utilisation of undigested and unabsorbed polyols in the colon

The human colon can be seen as representing a fermentation chamber analogous to the rumen. The colonic flora is composed of several hundred different species of microorganisms. Substrates for microbial fermentation in the large intestine comprise macromolecules from both endogenous and dietary sources. Undigested polyols are readily fermented almost quantitatively in the human colon, and the excretion of polyols into the faeces is negligibly low. Studies on the effects of dietary polyols like sorbitol and xylitol in rats for example have shown that the intestinal flora can adapt to utilise these polyols (Salminen *et al.*, 1985, 1986). Oral administration of high doses of sorbitol or xylitol caused changes in the composition of the faecal bacteria. This adaptive process was explained either by a selective proliferation of microorganisms capable of utilising sorbitol or xylitol, or possibly by induction of sorbitol and xylitol dehydrogenases in the bacteria already present (Salminen *et al.*, 1985, 1986). It is well known that fermentation of carbohydrates in the colon results in the formation of short chain fatty acids (SCFA) and gaseous compounds such as carbon dioxide, hydrogen and methane plus a variety of other volatile compounds.

The energetic contribution of short chain fatty acids from the large intestine to the human basal requirement was estimated by Cummings (1981) to be in the range of 6–9% of the digestible energy intake. The relative proportions of acetate, propionate and butyrate vary with the diet and usually remain within the range of 60–75% of acetate, 15–25% of propionate and 10–15% of butyrate, with acetate predominating in man. Short chain fatty acids are absorbed into the portal blood and partly metabolised in the digestive epithelium. Their complete oxidation to carbon dioxide is often the main catabolic pathway.

Turning to energy balance in man, colonic fermentation of polyols is the underlying reason for their lower energetic availability compared with easily metabolisable saccharides. The losses of energy due to intracolonic fermentation are difficult to determine experimentally, as the carbohydrates undergoing fermentation cause the multiplication of faecal microorganisms, leading to gas and heat production. The Nutrition Council of The Netherlands (1987) suggested a conversion efficiency of about 0.2 kJ faecal bacteria per kJ carbohydrate fermented. A value between 0.2 and 0.4 kJ/kJ was reported for non-starch polysaccharides and 0.15 to 0.25 kJ for polyols (Livesey, 1992). The efficiency of conversion of fermentable carbohydrates to molecular hydrogen and methane is low in humans. The Nutritional Council of The Netherlands (1987) assumed that a 50% loss of energy has to be ascribed to intracolonic fermentation of polyols. Livesey (1992) summarised the data on energy losses due to microbial mass, gas and heat production for polyols, unavailable carbohydrates and nonstarch polysaccharides.

The non-invasive technique of breath analysis for exhaled hydrogen and methane is suitable for determining the participation of the lower gut flora in the utilisation of dietary components. Experience with this method demonstrated that if dietary regimes are not strictly controlled with respect to the fibre content of the diet etc., only qualitative data can be obtained. However, under strictly controlled conditions and monitoring exhalation profiles over a whole day, quantitative data also are available (Fritz *et al.*, 1985).

The hydrogen breath tests performed with the polyols revealed differences in hydrogen formation, with higher gas production from sorbitol than from equal quantities of isomalt (Lee *et al.*, 1994). The exhalation of hydrogen after ingestion of lactitol was significantly higher than after isomalt and maltitol (Würsch *et al.*, 1989; Beaugerie *et al.*, 1991).

4.6 Gastrointestinal tolerance

Polyols, although they differ in their chemical nature, have a common problem of tolerance in the gastrointestinal tract. The physiological factors determining tolerance are their limited absorption and digestibility in the small intestine, thus eventually giving rise to symptoms of malabsorption, a phenomenon well known with poorly digestible carbohydrates such as raffinose.

Symptoms of carbohydrate malabsorption may include flatulence, accelerated intestinal transit times, bloating, increased stool frequency, diarrhea or constipation. These symptoms of gastrointestinal intolerance are a general feature of osmotically active substances which are not absorbed by active transport mechanisms in the upper intestinal tract and which are not completely digested on transit through the small intestine.

Delayed digestion and incomplete absorption may lead to shortened intestinal transit times due to the osmotic effects of the unabsorbed polyols. Carbohydrates which have not been digested and absorbed in the upper intestine reach the colon, where they feed the symbiotic colonic flora, thereby increasing the bacterial mass and formation of bacterial fermentation products like short chain fatty acids and gases like hydrogen, and eventually also methane.

This enhanced metabolism of the intestinal flora may contribute to the osmotic load of the colon and may influence the normal water absorptive capacity of the colon, eventually resulting in a so-called overflow-diarrhea. Such an osmotically induced diarrhea has to be differentiated from intolerance reactions of the intestine caused by pathogenic microbial infections. Gastrointestinal intolerance symptoms induced by polyols are transient and readily reversible when the consumption of the individual polyol is stopped.

The laxative thresholds for the osmotic loads may vary, depending on individuals' gastrointestinal tract susceptibility to osmotic loads. The tolerance level may increase after continuous exposure, i.e. adaptation. Adaptation to a polyol means that increased quantities of the product consumed can be tolerated after a short adaptation period.

In summary, the gastrointestinal tolerance of polyols is determined by factors such as their chemical nature and the extent to which they are digested and absorbed, which are influenced by the total quantities and portion sizes consumed, the mode of ingestion (whether as a drink or solid food), as well as by the frequency and time of ingestion. Furthermore other dietary components like dietary fibres and other poorly digested components of the diet might interfere with or contribute to the osmotic load and thus interfere with the tolerance of the ingested polyols.

The contribution of different polyols to the osmotic pressure is related to their molecular weights. In other words, the lower the molecular weight, the higher its contribution to the osmotic pressure.

Table 4.2 (modified, from Menzies *et al.*, 1990) illustrates that quantities of osmotically active carbohydrates required for equal osmotic effects increase with molecular weight. Thus 'second generation' polyols, e.g. of the disaccharide alcohol type, are generally better tolerated than polyols of lower molecular weight.

The evaluation of the gastrointestinal tolerance of polyols by certain scientific bodies resulted in a common classification, avoiding ranking the polyols according to their gastrointestinal tolerance, even though differences in their absorption and metabolism exist.

The Scientific Committee on Food of the European Commission summarised its position on the gastrointestinal tolerance of polyols in 1984, that "laxation may be observed at high intakes; consumption in the order of 20 g/person/day of polyols is unlikely to cause undesirable laxative

Table 4.2 Relative osmotic efficacy of different polyols and undigestible carbohydrates compared to tetritols or hexitols

Polyols		Molecular weight	Equi-osmolar quantities	Relative efficacy compared to tetritols	Relative efficacy compared to hexitols
Tetritol	Erythritol	122	6.7 g	100%	150%
Pentitol	Xylitol	152	8.3 g	80%	120%
Hexitols	Sorbitol Mannitol	182	10 g	67%	100%
Glucosyl-hexitols	Isomalt Lactitol Maltitol	344	20 g	35%	50%
Mixture of sorbitol, maltitol and higher poly-glucosyl-sorbitols	Lycasin	mixture	20 g	35%	50%
Disaccharide	Lactulose	342	20 g	35%	50%
Trisaccharide	Raffinose	504	32.6 g	20%	30%

symptoms" (Report of the Scientific Committee for Food and Sweeteners, 1984).

The gastrointestinal tolerance of some polyols has been very well documented by specific tolerance studies in healthy and diabetic adults and children. However comparative assessments or direct comparisons are limited owing to variations in study designs and criteria.

4.6.1 Erythritol

Information on the gastrointestinal tolerance of erythritol in humans is limited. The only publication on erythritol metabolism in man (Oku and Noda, 1990a) reported about 92% of urinary excretion in 24 h after a 10 g intake of erythritol, and of 91% excretion within 48 h after a 20 g intake of erythritol under fasting conditions as single doses, indicating that approximately 10% of unabsorbed erythritol is transferred to the large bowel. The half-life of erythritol clearance from blood was reported to be 3–4 h after consumption of 20 g (corresponding to 0.3 g erythritol per kg body weight). Any comments on gastrointestinal symptoms are missing from this report (Oku and Noda, 1990a). From rat feeding studies it is known that the intestinal absorption capacity for erythritol is limited and that large amounts may induce a transient diarrhea (Oku and Noda, 1990b; Noda and Oku, 1990, 1992). When the dosage was increased the amount of erythritol transferred into the large intestine increased as well.

Hiele *et al.* (1993) recovered 84% of 25 g erythritol (after 24 h) in non-adapted adults. No increase in pulmonary hydrogen excretion was observed within 6 h after the consumption of erythritol by non-adapted adults. Eventual methane production was not determined. This study report did not comment on the gastrointestinal tolerance of erythritol or on diuretic effects in man. In perfusion studies with rats, the absorption of erythritol was found to be slower than that of galactose (Winne *et al.*, 1985). Erythritol in doses of 1.8 g/100 g body weight rat caused diuresis due to osmotic changes (Beck *et al.*, 1938). Oku and Noda (1990b) reported a significant increase in the animals' water consumption and an increase in the urinary volume, which was 6-fold larger in the 10% erythritol group than in the control group. There was no increase in the rats' relative kidney weight. The diuretic activity of erythritol observed in rats has to be taken into account when assessing the overall tolerance in humans. Whether or not the diuretic activity of erythritol represents a risk of osmotic nephrosis-like lesions for humans remains to be answered. A review by Röper and Goossens (1993) did not discuss these aspects of an eventual erythritol-specific diuresis. The statement that fermentation in the colon does not arise because erythritol is absorbed quantitatively in the small intestine and therefore does not give any intestinal discomfort (Röper and Goossens, 1993) needs to be verified with a clinical tolerance study in man.

Erythritol was considered a difficult fermentation substrate for the colonic microflora, based on the absence of microbial hydrogen production in man. But there are also reports that erythritol can be utilised by *Brucella abortus* (Sperry and Robertson, 1975) and by *Propionibacterium pentosaceum*, which can grow anaerobically with erythritol as the sole carbon source (Waszkiewicz and Barker, 1968).

4.6.2 Xylitol

Tolerance studies with xylitol date back to the 1960s and 1970s (Mäkinen, 1976; Mäkinen *et al.*, 1982; Förster *et al.*, 1982). Dubach *et al.* (1969) studied the maximal quantities tolerated by increasing the initial dose of 5 g xylitol day by day up to 75 g per day and compared it with sorbitol in 26 adults. This was consumed in most cases in 4 or 5 portions per day. It was reported that xylitol was tolerated better than sorbitol. However, symptoms such as diarrhea, flatulence and bloating were also reported. The usefulness of this study is restricted, as the symptoms were not related to the doses ingested, so that later tolerance studies with other polyols could not be compared directly with it.

Åkerblom *et al.* (1982) studied the tolerance of xylitol in 13 Finnish children aged 7 to 16 years. Xylitol was given in sweets in divided portions over the day. The dosage was gradually increased. Occasional abdominal pain was observed with 4 out of 13 children at 25 g xylitol per day.

Increased flatulence was observed with 6 out of 13 children at 45 g xylitol per day, rising to 11 out of 13 children. 65 g Xylitol per day was accompanied by diarrhea (4 cases out of 13 children). The authors concluded that 10 g and 25 g of xylitol were tolerated by children. The criteria for frequency of flatulence after xylitol of the study were not comparable with other studies, as flatulence was evaluated only when the frequency was reported on at least 2 successive days. This assessment procedure deviated significantly from other conventional reporting schemes and criteria for gastrointestinal symptoms and frequencies in clinical tolerance studies, and any comparison with other studies must be limited.

Culbert *et al.* (1986) studied the gastrointestinal tolerance to increasing daily doses of xylitol from 30 g up to 100 g in divided portions in 12 American adults. In general gastrointestinal intolerance was accompanied by varying degrees of diarrhea. The report concluded that gastrointestinal side effects may limit the maximum tolerated xylitol dosage to 20 g, and 60 g per day after adaptation.

4.6.3 Sorbitol

The literature on the gastrointestinal tolerance of sorbitol is conflicting (Koizumi *et al.*, 1983a–c; Jain *et al.*, 1985; Badiga *et al.*, 1990; Remessen and Gudmand-Høyer, 1987). Doses of 5 g to 10 g sorbitol led to digestive symptoms, whereas up to 30 g and 40 g per day, ingested in divided portions throughout the day, were tolerated but with some flatulence after adaptation. Children were more susceptible to the effects of sorbitol than adults. The absorption capacity of ingested sorbitol was reported to be around 10–20 g per hour (Förster, 1974). This led to the assumption that a single dose of 20 g sorbitol might provoke a laxative effect by osmotic diarrhea.

4.6.4 Mannitol

Data on the gastrointestinal tolerance of mannitol were reported by Ellis and Krantz (1941) and Nasrallah and Iber (1969). Single dosages of 28–32 g mannitol were tolerated, but 6 out of 10 patients developed diarrhea at dosages of 40 g mannitol or more. Dosages greater than 40 g generally led to diarrhea. Saunders and Wiggins (1981) reported that healthy volunteers tolerated 20–40 g of mannitol before 'faecal output' of water (diarrhoea?) exceeded 400 ml in 48 h or before the test carbohydrate appeared in the stool. However no information was given on symptoms of discomfort which must have accompanied the faecal output of water. This description by Saunders and Wiggins (1981) is in accordance with the report by Ellis and Krantz (1941), that doses of mannitol of 10–20 g have laxative effects.

4.6.5 *Isomalt*

Specific clinical tolerance studies for assessing the gastrointestinal tolerance of isomalt in adults and healthy and diabetic children of different age groups have been performed, taking different dosages and consumption patterns of food products sweetened with isomalt into account (Spengler *et al.*, 1987; Zumbé and Brinkworth, 1992; Bachmann *et al.*, 1984; Paige *et al.*, 1992; Pometta *et al.*, 1985; Drost *et al.*, 1980; Fritz *et al.*, 1985). Isomalt as a disaccharide alcohol is better tolerated than monosaccharide alcohols in general. These studies in healthy and diabetic adults and children demonstrated that isomalt is well tolerated in amounts up to 50 g after a short adaptation period. Paige *et al.* (1992) reported that isomalt is as well tolerated by adults and children as sucrose when given in the form of sweets in divided portions over the day. A comparative tolerance test (Lee *et al.*, 1994) measuring the exhalation of hydrogen from colonic fermentation, indicated that the gastrointestinal tolerance of isomalt-sweetened chocolate was superior to that of a sorbitol-sweetened milk chocolate, and that the volume of breath hydrogen produced after eating chocolate containing isomalt was significantly lower than that for the sorbitol chocolate.

4.6.6 *Lactitol*

Van Velthuijsen and Blankers (1991) referred to the study by Grimble *et al.* (1988), who had determined the laxative threshold of lactitol after adaptation in 21 normal subjects. It was suggested that a quantity of up to 40 g lactitol per day could be quite acceptable, and that "the amount of lactitol tolerated is equal or somewhat higher than that for sorbitol and xylitol". Lactitol is also used as a cathartic agent in the treatment of constipation (Delas *et al.*, 1991) and chronic hepatic encephalopathy (Blanc *et al.*, 1992). Breath hydrogen excretion after consumption of lactitol is an indicator for gastrointestinal tolerance problems (Griessen *et al.*, 1986).

4.6.7 *Maltitol and hydrogenated starch hydrolysates*

The gastrointestinal tolerance of maltitol and HSH, consisting of maltitol, sorbitol and hydrogenated polysaccharides, has been studied in adults and children (Abraham *et al.*, 1981; Leroy, 1982; Tacquet and Devulder, 1978; Kearsley *et al.*, 1982; Koizumi *et al.*, 1983a–c), demonstrating that in healthy and diabetic subjects they are tolerated up to about 30–50 g/day after adaptation. Higher dosages were reported to cause diarrhea (Zunft *et al.*, 1983).

4.7 Caloric utilisation of polyols

Various methods have been used to determine the physiologic caloric value of polyols. These comprise energy balance studies in man using indirect calorimetry, recovery measurements of polyols and their constituents at the lower end of the small intestine of humans and experimental animals, breath hydrogen exhalation analysis in humans, experiments comparing growth rates and the composition of the carcass of experimental animals.

Several expert committees have addressed the scientific challenge of determining the net energy value of a polyol, analysing the comprehensive literature available. The most recent assessment of the caloric value of polyols was that of the Expert Scientific Panel of the Life Sciences Research Office (LSRO) of the Federation of American Societies for Experimental Biology (FASEB) in 1994. This evaluated the literature on metabolic, digestive and energetic aspects of polyols in man and experimental animals. The results are given in Table 4.3, together with the decisions of the Experts of the Nutrition Council of The Netherlands (1987). The Dutch Nutrition Council was the first to elaborate the principle of the factorial approach for determining the net energy values of polyols, taking their partial absorption in the small intestine, their utilisation and the energy losses due to colonic microbial fermentation into account.

Net energy values were suggested for the polyols already on the market (Table 4.3). In principle, the assessments resulted not in a unitary value for all polyols but in individual net energy values for each one. Compared to the net energy value of easily metabolisable saccharides, like glucose,

Table 4.3 Net energy values (kJ/g) of polyols, starch, unavailable complex carbohydrates (UCC) and non-starch polysaccharides (NSP)

	Dutch Nutrition Council (1987)	LSRO Expert Panel (1994)	Livesey (1992)
Erythritol			
Xylitol	15	~10	
Mannitol	8.5	6.7	
Sorbitol	12.5	7.5–13.8	
Isomalt	10	8.4	
Lactitol	8.5	6.7–9.2	
Maltitol	12	11.7–13.4	
HSH		11.7–13.4	
Starch			17.5
Glucose			15.7
Sucrose	16.5		
UCC in mixed diet			6
Wheat bran NSP			2.9
Apple NSP			5.8
Beta-fibre® NSP			6.5

sucrose or starch, which are within the order of 15.7 kJ (glucose) to 17.5 kJ (starch), the net energy values of the polyols range from 13.8 kJ (3.3 kcal) to 6.6 kJ/g (1.6 kcal/g) and approach values which were reported for unavailable complex carbohydrates and for non-starch polysaccharides (Livesey, 1992). Erythritol was not evaluated, but its net energy value was reported to reach only about 10% of the caloric value of sucrose (0.4 kcal/g or 1.7 kJ/g).

In summary, the polyols can be ranked according to their net energy value in decreasing order compared to sucrose (16.5 kJ/g) > maltitol and hydrogenated starch hydrolysates (11.7–13.4 kJ/g = 2.8–3.2 kcal/g) > xylitol (~10 kJ/g = ~2.4 kcal/g) and sorbitol (7.5–13.8 kJ/g = 1.8–3.3 kcal/g) > isomalt (~8.4 kJ/g = ~2.0 kcal/g) and lactitol (6.7–9.2 kJ/g = 1.6–2.2 kcal/g) > mannitol (6.7 kJ/g = 1.6 kcal/g).

Scientific questions which arose with the evaluation of the net energy values were comprehensively reviewed (LSRO, 1994; Livesey, 1990, 1992). The hypothesis that the net energy value of a polyol might depend on the quantity consumed is discussed in the literature. Livesey (1990), having reviewed the literature on that topic, e.g. on isomalt, concluded that there was no evidence of a dose dependency, and that the evidence cited (Grossklaus, 1987; Krüger *et al.*, 1992) for a higher energy value at lower rather than higher intakes appeared to be due to a failure to recognise that the osmotic effects of the polyol are dependent on its rate of entry into the small intestine or inhibition of this process by food.

The difficulty of reaching an objective assessment on this issue stems from the methodology used, and in the selection of the appropriate test model and animal species to allow extrapolation to man. The claim by Grossklaus (Grossklaus, 1987; Krüger *et al.*, 1992) for higher net energy values for polyols was based on marker distribution studies in rats. These data are corroborated neither by the existing comprehensive literature and the conclusions of the authoritative expert committees (see also LSRO/FASEB, 1994), nor by human energy balance data based on indirect calorimetry, which at a dosage of 0.4 g polyol/kg body weight, reported decreased net energy values in humans for isomalt (Thiébaud *et al.*, 1984), lactitol (van Es *et al.*, 1986) and maltitol (Felber *et al.*, 1987) compared to sucrose, and for xylitol compared to glucose (Müller-Hess *et al.*, 1975).

4.8 Dental health related aspects of polyols

The non-fermentability of polyols by oral microorganisms and dental plaque bacteria is considered as an important criterion for noncariogenicity. The fermentation products of carbohydrates are risk factors for damage to the teeth, as the acids derived from bacterial fermentation may demineralise the apatite structure of the teeth and lead to dental caries.

The non-cariogenicity of sorbitol, mannitol, xylitol, isomalt, lactitol and maltitol is well established in the dental literature and documented in reviews. The important criteria for non-cariogenicity are summarised in Table 4.4. The most important criterion is a very low fermentability or the absence of fermentability by oral microorganisms. The absence of glucan formation *in vitro* and the results of rat caries studies are further criteria. Human clinical studies with polyols have been important, historically, but at present they are no longer mandatory nor feasible for ethical and economic reasons.

The established non-cariogenic properties of xylitol (Birkhed, 1994; Aguirre-Zéro *et al.*, 1993; Bär, 1988; Scheinin and Mäkinen, 1975), sorbitol (Hamilton and Svensater, 1991; Hogg and Rugg-Gunn, 1991; Kalfas and Edwardson, 1990), mannitol, isomalt (Featherstone, 1994), lactitol (Grenby, 1989; Grenby *et al.*, 1989), maltitol (Wolfslehner, 1988; Imfeld and Lutz, 1985) and HSH (Rugg-Gunn, 1989) are well documented in the literature (Imfeld, 1994). These polyols are the basic ingredients for the production of toothfriendly sweets. From experience *in vitro* and *in vivo* it is well known that the proportion of hydrogenated oligosaccharides in HSH should not be too high in order to maintain the nonacidogenic properties of HSH.

Some criteria for erythritol related to dental health were assessed by Kawanabe *et al.* (1992) *in vitro* and in rats. Erythritol is neither utilised as a substrate for lactic acid production nor for plaque polysaccharide formation by oral streptococci. A rat caries feeding experiment resulted in significantly fewer caries lesions compared to sucrose when fed at 26% in the diet or in the form of chocolate sweetened with erythritol (Kawanabe *et al.*, 1992).

The claim for 'anticariogenicity', which had been suggested for instance

Table 4.4 Assessment criteria for the dental health related properties of polyols and results

Polyol	Oral acid fermentation (pH-telemetry)	Insoluble glucan formation	Human clinical caries studies available	Rat caries studies	Safe for teeth/ tooth-friendly
Erythritol	?	?	?	low	?
Xylitol	no	no	yes	low	yes
Sorbitol	low/no	no	yes	intermediate/ low	yes
Mannitol	no	no		low	yes
Isomalt	no	no	no	low	yes
Lactitol	no	no	no	low	yes
Maltitol	no	no	no	low	yes
HSH[a]	low/yes[a]	no	yes	low	yes[a]

[a] Depends on product composition.

for xylitol has been criticised by the dental community, since the unequivocal clinical evidence for active cariostatic properties has not been forthcoming (Imfeld, 1994).

From fermentation studies *in vitro* attempts were made to rank the polyols according to their relative acidogenic potential. Sorbitol and hydrogenated starch hydrolysate were classified as low acidogenic, whereas xylitol was classified as practically non-acidogenic, with the other polyols in between. Although xylitol can be classified definitely as being nonacidogenic and as having no cariogenic potential, xylitol-fermenting properties of oral propioniibacteria have been described (Linke *et al.*, 1992).

In summary, all the polyols in Table 4.4 are regarded as safe for teeth if they fulfil the criteria of toothfriendliness, which means that the pH in the dental plaque must not fall below the threshold of pH 5.7 which is regarded as the critical threshold, as a more acidic milieu on the tooth surface favours the dissolution of enamel. Confectionery products for snacks based on polyols as a bulk sweeteners have gained a role in the prevention of dental caries as they help to maintain the salivary milieu at a neutral pH, avoiding acidic attack and favouring the remineralising properties naturally present in saliva.

4.9 Conclusions

Dietary recommendations on sucrose intake for diabetics have changed during the last few years from a strict to a more liberal attitude. However, these changes should not obscure the scientifically proven impact of polyols and fructose on the improvement of glycemic profiles. Although the fact that sucrose has a smaller glycemic index than glucose or starches is now well recognised, the beneficial role of mono- and disaccharide alcohols on glycemic control in diabetics should not be underestimated, as their metabolic effects have generally been established in comparative studies with sucrose. Polyols behave in this respect like slowly digestible carbohydrates, and they have a substantially reduced caloric value compared with easily metabolisable saccharides. Polyols have no hyperglycemic impact, and compared with other saccharides, no or very little insulinogenic effect. The caries preventive benefits of the polyols make them important ingredients for toothfriendly products.

References

Abraham, R.R., Davis, M., Yudkin, J. and Williams, R. (1981) Controlled clinical trial of a new non-calorigenic sweetening agent. *J. Human Nutrition* **35**, 165–172.

Adcock, L.H. and Gray, C.H. (1957) The metabolism of sorbitol in the human subject. *Biochem. J.* **65**, 554–560.

Adelman, R.C., Ballard, F.J. and Weinhouse, S. (1967) Purification and properties of rat liver fructokinase. *J. Biol. Chem.* **242**, 3360–3365.

Aguirre-Zero, O., Zero, D.T. and Proskin, H.M. (1993) Effect of chewing xylitol chewing gum on salivary flow rate and the acidogenic potential of dental plaque. *Caries Res.* **27**, 55–59.

Åkerblom, H.K., Koivukangas, T., Puukka, R. and Mononen, M. (1982) The tolerance of increasing amounts of dietary xylitol in children. *Int. J. Vitam. Nutr. Res., Suppl.* **22**, 53–66.

Asano, T., Levitt, M.D. and Goeth, F.C. (1973) Xylitol absorption in healthy men. *Diabetes* **22**, 279–281.

Atsuji, H., Asano, S. and Hayashi, S. (1982) A study on the metabolism of maltitol, a disaccharide alcohol. Unpublished report from Keio University, Japan quoted in WHO Food Additives Series 20 (1987).

Bachmann W., Haslbeck, M., Spengler, M., Schmitz, H. and Mehnert, H. (1984) Untersuchungen zur Stoffwechselbeeinflussung durch akute Palatinitgaben. Vergleich zu Fructose und Saccharose bei Type-II-Diabetes. (Studies on Influencing Metabolism Via Acute Doses of Palatinit. Comparison with Fructose and Saccharose in Type II Diabetes). *Aktuelle Ernährungsmedizin* **9**, 65–70.

Badiga, M.S., Jain, N.K., Casanova, C. and Pitchumoni, C.S. (1990) Diarrhea in diabetics: the role of sorbitol. *J. Am. Coll. Nutr.* **9** 578–582.

Bär, A. (1991) Sugar alcohols in the diabetic diet. In *Sugars and Sweeteners* (ed. N. Kretchmer), Cl. Hollenbeck, CRC Press, pp. 131–150.

Bär, A. (1988) Caries prevention with xylitol. *Wld. Rev. Nutr. Diet.* **55**, 183–209.

Bässler, K.H., Stein G. and Belzer, W. (1966) Xylitstoffwechsel und Xylitresorption. Stoffwechseladaptation als Ursache für Resorptionsbeschleunigung. *Biochem. Z.* **346**, 171–185.

Bässler, K.H., Prellwitz, W., Unbehaum, V. and Lang, L. (1962) Xylitstoffwechsel beim Menschen. Zur Frage der Eignung von Xylit als Zuckerersatz beim Diabetiker. *Klin. Wschr.* **40**, 791.

Beaugerie, L., Flourié, B. Marteau, P., Pellier, P., Franchisseur, C. and Rambaud, R.-C. (1990) Digestion and absorption in the human intestine of three sugar alcohols. *Gastroenterology* **99**, 717–723.

Beaugerie, L., Flourié, B., Pellier, P., Achour, L., Franchisseur, C. and Rambaud, R.-C. (1991) Tolérance clinique, absorption intestinale et valeur énergétique de quatre polyols pris à jeun. (Clinical tolerance, intestinal absorption, and energy value of four sugar alcohols.) *Gastroenterol. Clin. Biol.* **15**, 929–932.

Beck, F.F., Carr, J.C. and Krantz, J.C. Jr. (1938) Sugar alcohols XVI. The fate of erythritol and erythritan in the animal body. *Quarterly Journal of Pharmacy and Pharmacology* **11**, 234–239.

Birkhed, D. (1994) Cariologic aspects of xylitol and its use in chewing gum: A review. *Acta Odontol. Scand.* **52**, 116–127.

Blanc, P., Daures, J.P., Rouillon, J.M., Peray, P., Pierrugues, R., Larrey, D., Gremy, F. and Michel, H. (1992) Lactitol or lactulose in the treatment of chronic hepatic encephalopathy: results of a meta-analysis. *Hepatology* **15**, 222–8.

Bornet, F., Dauchy, F., Chevalier, A. and Slama, G. (1992) Etude de devenir metabolique, après ingestion chez l'homme sain, d'un nouvel edulcorant de charge basse calorie: l'erythritol. *Gastroenterol. Clin. Biol.* **16**, A 169.

Bugaut, M. and Bentéjac, M. (1993) Biological effects of short-chain fatty acids in nonruminant mammals *Annu. Rev. Nutr.* **13**, 217–241.

Burns, J.J., Mosbach, E.H., Schulenberg, S. and Reichenthal, J. (1955) Studies on the incorporation of 14C administered as L-sorbose into L-ascorbic acid and D-glucose in rats. *J. Biol. Chem.* **214**, 507–415.

Cori, G.T., Ochoa, S., Slein, M.W. and Cori, C.F. (1951) The metabolism of fructose in the liver. Isolation of fructose 1-phosphate and inorganic pyrophosphate. *Biochim. Biophys. Acta* **7**, 304–317.

Culbert, S.J., Wang, Y.-M., Fritsche, H.A. Jr., Carr, D., Lantin, E. and van Eys, J. (1986) Oral xylitol in American adults. *Nutrition Res.* **6**, 913–922.

Cummings, J.H. (1981) Dietary fibre. *British Medical Bulletin* **37**, 65–70.

Delas, N., Gislon, J., Glikmanas, M., Henri-Biabaud, E., Lemerez, M., Licht, H., Sama, J.L. and Gillaume, P.N. (1991) Le lactitol dans le traitement de la constipation de l'adulte. Etude ouverte, non comparative de son efficacité et de sa tolérance clinique et biologique. (Lactitol in the treatment of constipation in the adult. Open, non-comparative study of its efficacy and its clinical and biological tolerance.) *Ann. Gastroenterol. Hepatol. (Paris)* **27**, 231–3.

Doorenbos, H. (1977) Unpublished report. University of Groningen, The Netherlands, quoted by Den Uyl, C.H. (1987) Technical and commercial aspects of the use of lactitol in foods as a reduced-calorie bulk sweetener. In *Developments in Sweeteners – 3* (ed. T.H. Grenby), Elsevier Applied Science, pp. 65–82.

Drost, H., Gierlich, P., Spengler, M. and Jahnke, K. (1980) Blutglucose und Seruminsulin nach oraler Applikation von Palatinit im Vergleich zu Glucose bei Diabetikern vom Erwachsenentyp. *Verh. Dt. Ges. Inn. Med.* **80**, 978–981.

Dubach, U.C., Feiner, E. and Forgo, I. (1969) Orale Verträglichkeit von Xylit bei stoffwechselgesunden Versuchspersonen. *Schweiz. Med. Wschr.* **99**, 190.

Elia, M., Behrens, R., Northop, C., Wraight, P. and Neale, G. (1987) Evaluation of mannitol, lactulose and ^{51}CR-labelled ethylenediaminetetra-acetate as markers of intestinal permeability in man. *Clinical Science* **73**, 197–204.

Ellis, F.W. and Krantz, H.A. Jr. (1941) Sugar alcohols XXII. Metabolism and toxicity studies with mannitol and sorbitol in man and animals. *J. Biol. Chem.* **141**, 147–154.

Featherstone, J.D.B. (1994) Effects of isomalt sweetener on the caries process: a review. *J. Clin. Dent.* **5**, 82–85.

Felber, J.P., Tappy, L., Vouillamoz, D., Randin, J.P., and Jéquier, E. (1987) Comparative study of maltitol and sucrose by means of continuous indirect calorimetry. *J. Parenteral and Enteral Nutrition* **11**, 250–254.

Forbes, A.L. and Bowman, B.A. (eds.) (1993) Health effects of dietary frutose. *Am. J. Clin. Nutr.* **58** (5 Supplement) 721S–823S.

Förster, H. (1974) Comparative metabolism of xylitol, sorbitol and fructose. In *Sugars in Nutrition* (eds. H.L. Sipple and K.W. McNutt), Academic Press, pp. 591–606.

Förster, H., Quadbeck, R. and Gottstein, U. (1982) Metabolic tolerance to high doses or oral xylitol in human volunteers not previously adapted to xylitol. *Int. J. Vitam. Nutr. Res. Suppl.* **22**, 67–88.

Fosdick, L.S., Englander, H.R., Hoerman, K.C. and Kesel, R.G. (1957) A comparison of pH values of *in vivo* dental plaque after sucrose and sorbitol mouth rinses. *J. Amer. Dent. Assoc.* **55**, 191–195.

Fritz, M., Siebert, G. and Kasper, H. (1985) Dose dependence of breath hydrogen and methane in healthy volunteers after ingestion of a commercial disaccharide mixture, Palatinit®. *Brit. J. Nutr.* **54**, 389–400.

Froesch, E.R. and Jakob, A. (1974) The metabolism of xylitol. In *Sugars in Nutrition* (eds. H.L. Sipple and K.W. McNutt), Academic Press, pp. 241–258.

Furuse, M., Kimura, Ch., Takahashi, H. and Okumura, J.-I. (1991) Prevention of the incidence of diabetes by dietary sorbose in nonobese diabetic mice. *J. Nutr.* **121**, 1135–1138.

Furuse, M., Kimura, Ch., Mabayo, R.T., Takakashi, H. and Okumura, J.-I. (1993) Dietary sorbose prevents and improves hyperglycemia in genetically diabetic mice. *J. Nutr.* **123**, 59–65.

Gee, J.M., Cooke, D., Gorick, S., Wortley, G.M., Greenwood, R.H., Zumbé, A. and Johnson, I.T. (1991) Effects of conventional sucrose-based, fructose-based and isomalt-based chocolates on postprandial metabolism in non-insulin-dependent diabetics. *Eur. J. Clin. Nutr.* **45**, 561–566.

Goto, I., Ihaba, M., Shimizu, D. and Maede, Y. (1994) Mechanism of hemolysis of canine erythrocytes induced by L-sorbose. *Am. J. Vet. Res.* **55**, 291–294.

Grenby, T.H. (1989) Latest state of research on lactitol and dental caries. *Int. Dent. J.* **39**, 25–32.

Grenby, T.H., Philips, A. and Mistry, M. (1989) Studies of the dental properties of lactitol compared with five other bulk sweeteners *in vitro*. *Caries Res.* **23**, 315–319.

Grenby, T.H. and Philips, A. (1989) Dental and metabolic effects of lactitol in the diet of laboratory rats. *Brit. J. Nutr.* **61**, 17–24.

Grieshaber, H. (1936a) Beitrag zum Verhalten der Sorbose im Stoffwechsel. I. Mitteilung: Die Resorption und Ausscheidung der Sorbose und ihre Wirkung auf Blut- und Harnzucker. *Z. Klin. Med.* **129**, 412.

Grieshaber, H. (1936b) Beitrag zum Verhalten der Sorbose im Stoffwechsel. II. Mitteilung: Die antiketogene Wirkung der Sorbose beim Diabetes mellitus. *Z. Klin. Med.* **129**, 423.

Griessen, M., Bergoz, R., Balant, L. and Loizeau, E. (1986) Effet du lactitol sur la production d'hydrogène expire chez l'homme normal. (The effect of lactitol on the production of expired hydrogen in normal subjects.) *Schweiz. Med. Wochenschr.* **116**, 469–472.

Grimble, G.K., Patil, D.H. and Silk, D.B.A. (1988) Assimilation of lactitol, an 'unabsorbed' disaccharide in the normal human colon. *Gut* **29**, 1666–1671.

Grossklaus, R. (1987) Dosisabhängigkeit der energetischen Nutzung von Zuckeraustauschstoffen. *Dtsch. Zahnärztl. Z.* **42**, S154–S158.

Grupp, U. and Siebert, G. (1978) Metabolism of hydrogenated palatinose, an equimolar mixture of α-D-Glucopyranosido-1,6-sorbitol and α-D-Glucopyranosido-1,6-mannitol. *Res. Exp. Med. (Berl.)* **173**, 261–278.

Hamilton, I.R. and Svensater, G. (1991) Sorbitol inhibition of glucose metabolism by *Streptococcus sanguis* 160. *Oral Microbiol. Immunol.* **6**, 151–159.

Havenaar, R. (1987) Dental advantages of some bulk sweeteners in laboratory animal trials. In *Development in Sweeteners – 3* (ed. T.H. Grenby), Elsevier Applied Science, pp. 189–211.

Heinz, F. (1973) The enzymes of carbohydrate degradation. *Progr. biochem. Pharmacol.* **8**, 1–56.

Henry, R.R., Crapo, Ph. A. and Thorburn, A.W. (1991) Current issues in fructose metabolism. *Annu. Rev.* **11**, 21–39.

Hers, H.G. (1952) La fructokinase du foie. *Biochim. Biophys. Acta* **8**, 416–423.

Hiele, M., Ghoos, J., Rutgeerts, P. and Vantrappen, G. (1993) Metabolism of erythritol in humans: comparison with glucose and lactitol. *Br. J. Nutr.* **69**, 169–176.

Hogg, S.D. and Rugg-Gunn, A.J. (1991) Can the oral flora adapt to sorbitol? *J. Dent.* **19**, 263–271.

Huchette, M. and Leroy, P. (1973) Zuckerersatzstoff. German patent application 2 308 163 (filed February 19, 1973).

Hütter, R., Böswart, F. and Irsigler, K. (1993) Insulinverbrauch von Typ-I-Diabetikern nach oraler Gabe von Isomalt. (Insulin requirements of Type I Diabetics after oral administration of isomalt.) *Akt. Ernähr.-Med.* **18**, 149–154.

Imfeld, T.N. (1994) Clinical caries studies with polyalcohols. A literature review. *Schweiz. Monatsschr. Zahnmed.* **104**, 941–945.

Imfeld, T. and Lutz, F. (1985) Malbit®, ein zahnfreundlicher Zuckeraustauschstoff. *Swiss. Dent.* **6**, 44–53.

Jain, N.K., Rosenberg, D.B., Klahann, M.J., Glasser, M.J. and Pitchumoni, C.S. (1985) Sorbitol intolerance in adults. *Am. J. Gastroenterol.* **80**, 678–681.

Kalfas, S. and Edwardson, S. (1990) Sorbitol-fermenting predominant cultivable flora of human dental plaque in relation to sorbitol adaptation and salivary secretion rate. *Oral Microbiol. Immunol.* **5**, 33–38.

Kamoi, M., Shimizu, Y., Kawauchi, M., Fujii, Y., Kikuchi, Y., Mizukawa, S., Yoshioka, H. and Kibata, M. (1975) A study on the metabolism of maltitol (Clinical study). *J. Jap. Diab. Soc.* **18**, 451–460.

Kaspar, L. and Spengler, M. (1984) Wirkung oraler Gaben von Palatinit® auf den Insulinverbrauch bei Typ-I-Diabetikern. (Effect or oral doses of Palatinit® in insulin requirements in type I diabetics.) *Aktuelle Ernährungsmedizin* **9**, 60–64.

Kawanabe, J., Hirasawa, H., Takeuchi, T., Oda, T. and Ikeda, T. (1992) Noncariogenicity of erythritol as a substrate. *Caries Res.* **26**, 358–362.

Kearsley, M.W., Birch, G.G. and Lian-Loh, R.H.P. (1982) The metabolic fate of hydrogenated glucose syrups. *Starch/Stärke* **34**, 279–283.

Keller, P. and Kistler, A. (1978) Acute hemolytic anemia in dogs after administration of sorbose. *Schweiz. Arch. Tierheilk.* **120**, 29–40.

Klein, M., Hofmann, L., Block, H.-J. and Herrmann, U. (1993) Untersuchung der energetischen Verwertung von L-Fructose bei Ratten. (Investigation of energetic utilization of L-fructose in rats.) *Die Nahrung* **27(4)**, 3405–320.

Koizumi, N., Fujii, M., Nonomiya, R., Inoue, Y., Kagawa, T. and Tsukamoto, T. (1983a) Studies on transitory laxative effects of sorbitol and maltitol. 1. Estimation of 50% effective dose and maximum non-effective dose. *Chemosphere* **12**, 45–53.

Koizumi, N., Fujii, M., Nonomiya, R., Inoue, Y., Kagawa, T. and Tsukamoto, T. (1983b) Studies on transitory laxative effects of sorbitol and maltitol. 2. Alterations in serum contents due to laxative effects. *Chemosphere* **12**, 117–124.

Koizumi, N., Fujii, M., Nonomiya, R., Inoue, Y., Kagawa, T. and Tsukamoto, T. (1983c) Studies on transitory laxative effects of sorbitol and maltitol. 3. Differences in laxative effects among various foods containing the sweetening agents. *Chemosphere* **12**, 105–116.

Kroneberg, H.-G., Spengler, M. and Strohmeyer, G. (1979) Zur Resorption von Palatinit®, dem äquimolaren Gemisch von α-D-Glucopyranosido-1,6-Sorbit (GPS) und α-D-Glucopyranosido-1,6-Mannit (GPM) im Dünndarm von colectomierten Patienten. Report from 06.08.1979, updated on 10.11.1983.

Krüger, D., Grossklaus, R., Herold, M., Lorenz, S. and Klingebiel, L. (1992) Gastrointestinal transit and digestibility of maltitol, sucrose and sorbitol in rats: a multicompartment model and recovery study. *Experientia* **48**, 733–740.

Külz, E. (1874) *Beiträge zur Pathologie und Therapie des Diabetes mellitus*. (ed. E. Külz), N.G. Elwert's Verlag, Marburg, pp. 130–156.

Lang, K. (1968) Xylit, Stoffwechsel und klinische Verwendung. *Med. Ernähr.* **9**, 265.

Langkilde, A.M., Andersson, H., Schweizer, T.F. and Würsch, P. (1995) Digestion and absorption of sorbitol, maltitol and isomalt from the small bowel. A study in ileostomy subjects. *European Journal of Clinical Nutrition* **48**, 768–775.

Lauwers, A.-M., Daumeric, C. and Henquin, J.C. (1985) Intestinal absorption of sorbitol and effects of its acute administration on glucose homeostasis in normal rats. *Brit. J. Nutr.* **53**, 53–62.

Lee, A., Zumbé, A. and Storey, D. (1994) Breath hydrogen after ingestion of the bulk sweeteners sorbitol, isomalt and sucrose in chocolate. *British Journal of Nutrition* **71**, 731–737.

Leroy, P. (1982) Report on the tolerance test of Lycasin 80/55 administered in the form of sweets to adult human volunteers. Unpublished report from Roquette Frères, Lestrem, France. Submitted to WHO by Roquette Frères. Quoted according to JECFA: Hydrogenated glucose syrups. Toxicological evaluation of certain food additives and contaminants. *29th Meeting of the Joint FAO/WHO Expert Committee on Food Additives*. Cambridge University Press, 1987.

Leuthardt, F. and Stuhlfauth, K. (1960) Biochemische, physiologische und klinische Probleme des Fruktosestoffwechsels. *Medizinische Grundlagenforschung, Band III*, Thieme, Stuttgart.

Leuthardt, F. and Testa, E. (1951) Die Phosphorylierung der Fructose in der Leber. *Helv. Chim. Acta* **34**, 931.

Linke, H.A.B., Masihesraelian, S. and Visintini, F.D. (1992) Acid profile during fermentation of xylitol by the oral Propionibacterium avidum. *Caries Res.* **26**, 233.

Livesey, G. (1990) The impact of the concentration and dose of Palatinit® in foods and diets on energy value. *Food Sciences and Nutrition* **42F**, 223–243.

Livesey, G. (1992) The energy values of dietary fibre and sugar alcohols for man. *Nutrition Research Reviews* **5**, 61–84.

Lohmann, D., Gehring, F. and Karle, E.J. (1981) Fermentation of L-sorbose by microorganisms of the human dental plaque. *Caries Res.* **15**, 263–271.

LSRO/FASEB (1994) *The Evaluation of the Energy of Certain Sugar Alcohols Used as Food Ingredients*. Life Sciences Research Office/Federation of American Societies for Experimental Biology.

Macdonald, I. (1984) Differences in dietary-induced thermogenesis following the ingestion of various carbohydrates. *Annu. Nutr. Metab.* **28**, 226–230.

Macdonald, I., Keyser, A. and Pacy, D. (1978) Some effects, in man, of varying the load of glucose, sucrose, fructose, or sorbitol on various metabolites in blood. *Am. J. Clin. Nutr.* **31**, 1305–1311.

Mäkinen, K.K. (1976) Long-term tolerance of healthy human subjects to high amounts of

xylitol and fructose: general and biochemical findings. *Int. Z. Vitam. Ernaehrungsforsch. Beih.* **15**, 92–104.
Mäkinen, K.K., Ylikan, R., Mäkinen, P.-L., Söderling, E. and Hämäläinen, M. (1982) Turku sugar studies XXIII. Comparison of metabolic tolerance in human volunteers to high oral doses of xylitol and sucrose after long-term regular consumption of xylitol. *Int. J. Vitam. Nutr. Res. Suppl.* **22**, 29–49.
Malaisse-Lagae, F., Couturier, E., Dufrane, S.P., Sener, A., Spengler, M., and Malaisse, W.J. (1984) Effect of a sorbitol-rich diet on plasma and erythrocyte sorbitol concentrations in normal and diabetic subjects. *Diabetes Research* **1**, 209–211.
Mehnert, H. (1976) Zuckeraustauschstoffe in der Diabetesdiät. (Sugar substitutes in the diabetic diet.) *Int. Z. Vitam. Ernaehrungsforsch. Beih.* **17**, 295–324.
Mehnert, H., Dietze, G. and Haslbeck, M. (1975) Zucker und Zuckeraustauschstoffe in der Diätetik von Störungen des Kohlenhydratstoffwechsels. *Nutr. Metabol.* **18** (Suppl. 1), 171–190.
Mehnert, H., Stuhlfauth, K., Mehnert, B., Lausch, R. and Seitz, W. (1959) Vergleichende Untersuchungen zur Resorption von Glucose, Fructose und Sorbit beim Menschen. *Klin. Wschr.* **21**, 1138–1142.
Mehnert, H. and Förster, H. (1979) Die orale Anwendung von Fructose als Zuckeraustauschstoff in der Diät des Diabetes mellitus. (Oral administration of fructose as sugar substitute in the diet of Diabetes mellitus patients.) *Akt. Ernährungsmedizin* **4**, 178–193.
Mellinghoff, C. (1961) Über die Verwendbarkeit des Xylit als Ersatzzucker bei Diabetikern. *Klin. Wschr.* **39**, 447.
Menzies, I.S., Jenkins, A.P., Heduan, E., Catt, S.D., Segal, M.B. and Creamer, B. (1990) The effect of poorly absorbed solute on intestinal absorption. *Scand. J. Gastroenterol* **25**, 1257–1264.
Mühlemann, H.R. (1976) The effects of a sorbose-sucrose diet on experimental caries and body weight gains. *Schweiz. Mschr. Zahnheilk.* **86**, 1339–1345.
Mühlemann, H.R. and Schneider, P. (1976) The effect of sorbose on pH of mixed saliva and interproximal plaque. *Helv. odont. Acta* **19**, 76–80.
Müller-Hess, R., Geser, C.A., Bonjour, J.P., Jequier, E. and Felber, J.P. (1975) Effects of oral xylitol administration on carbohydrate and lipid metabolism in normal subjects. *Infusionsther. Klin. Ernaehr.* **2**, 247–252.
Nasrallah, S.M. and Iber, F.L. (1969) Mannitol absorption and metabolism in man. *Am. J. Med. Sciences* **258**, 80–88.
Nguyen N.U., Dumoulin, G., Henriet, M.-T., Berthelay, S. and Regnard, J. (1993) Carbohydrate metabolism and urinary excretion of calcium and oxalate after ingestion of polyol sweeteners. *J. Clin. Endocrinol. Metabolism* **77**, 388–392.
Nilsson, U. and Jägerstad, M. (1987) Hydrolysis of lactitol, maltitol and Palatinit® by human intestinal biopsies. *Brit. J. Nutr.* **58**, 199–206.
Nishikawa, K. (1982) Variations of blood and urine sugar values with oral administration of maltitol. Unpublished report from Nishikawa Hospital, Tadera, Japan. Submitted to the WHO by Anic S.p.A. Quoted according to JECFA (1987) Hydrogenated glucose syrups. Toxicological evaluation of certain food additives and contaminants. *29th meeting of the Joint FAO/WHO Expert Committee on Food Additives*. Cambridge University Press.
Noda, K. and Oku, T. (1992) Metabolism and disposition of erythritol after oral administration to rats. *J. Nutr.* **122**, 1266–1272.
Noda, K. and Oku, T. (1990) The fate and availability of erythritol in rats. In *Proceedings of the International Symposium on Caloric Evaluation of Carbohydrates* (ed. N. Hosoya), Research Foundation for Sugar Metabolism; published by The Japanese Association of Dietetic and Enriched Foods, Tokyo, Japan, pp. 51–63.
Nutrition Council of The Netherlands (1987) The energy value of sugar alcohols. *Recommendations of the Committee on Polyalcohols*. The Hague.
Oku, T. and Noda, K. (1990a) Erythritol balance study and estimation of metabolizable energy of erythritol. In *Proceedings of the International Symposium on Caloric Evaluation of Carbohydrates* (ed. N. Hosoya), Research Foundation for Sugar Metabolism; published by The Japanese Association of Dietetic and Enriched Foods, Tokyo, Japan, pp. 65–75.
Oku, T. and Noda, K. (1990b) Influence of chronic ingestion of newly developed sweetener.

erythritol on growth and gastrointestinal function of the rats. *Nutrition Research* **10**, 987–996.

Ooshima, T., Izumitani, A., Minami, Tt., Yoshida, T., Sobue, S., Fujiwara, T. and Hamada (1992) Noncariogenicity of maltitol in specific pathogen-free rats infected with mutans Streptococci. *Caries Res.* **26**, 33–37.

Paige, D.M., Bayless, Th. M. and Davis, L.R. (1992) Palatinit® digestibility in children. *Nutrition Res.* **12**, 27–37.

Patil, D.H., Grimble, G.K. and Silk, D.B. (1987) Lactitol, a new hydrogenated lactose derivative: intestinal absorption and laxative threshold in normal human subjects. *Br. J. Nutr.* **57**, 195–199.

Petzoldt, R., Lauer, R., Spengler, M. and Schöffling, K. (1982) Palatinit bei Typ-II-Diabetikern. Wirkung auf Blutglucose, Seruminsulin, C-Peptid und freie Fettsäuren im Vergleich mit Glucose. (Palatinit in type II diabetics: effect on blood-glucose, serum-insulin, C-peptide and free fatty acids.) *Dtsch. Med. Wschr.* **107**, 1910–1913.

Pometta, D., Trabichet, C. and Spengler, M. (1985) Effects of a 12-week administration of Isomalt (Palatinit®) on metabolic control in type-II-diabetics. *Akt. Ernähr.-Med.* **10**, 174–177.

Rérat, A., Vaissade, P. and Vaugelade, P. (1991) Comparative digestion of maltitol and maltose in unanesthetized pigs. *J. Nutr.* **121**, 737–744.

Röper, H. and Goossens, J. (1993) Erythritol, a new raw material for food and non-food applications. *Starch/Stärke* **45**, 404/405.

Rugg-Gunn, A.J. (1989) Lycasin® and the prevention of dental caries. In *Progress in Sweeteners* (ed. T.H. Grenby), Elsevier Applied Science, pp. 311–330.

Rumessen, J.J. and Gudmand-Høyer, E. (1987) Malabsorption of fructose-sorbitol mixtures. Interactions causing abdominal distress. *Scand. J. Gastroenterol.* **22**, 431–436.

Salminen, S., Salminen, E., Koivistoinen, P., Bridges, J. and Marks, V. (1985) Gut microflora interactions with xylitol in the mouse, rat and man. *Fd Chem. Toxic.* **23**, 985–990.

Salminen, S., Salminen, E., Bridges, J. and Marks, V. (1986) The effects of sorbitol on the gastrointestinal microflora in rats. *Z. Ernährungswiss.* **25**. 91–95.

Saunders, D.R. and Wiggins, H.S. (1981) Conservation of mannitol, lactulose, and raffinose by the human colon. *Am. J. Physiol.* **241** (*Gastrointest. Liver Physiol.* **4**), G397–G402.

Scheinin, A. and Mäkinen, K.K. (1975) Turku sugar studies I–XXI. *Acta Odont. Scand.* **33** suppl. 70, 1–349.

Schwarz, J.-M., Schutz, J., Froidevaux, F., Acheson, K.J., Jeanprêtre, N., Schneider, H., Felber, J.-P. and Jécquier, E. (1989) Thermogenesis in men and women by fructose vs glucose added to a meal. *Am. J. Clin. Nutr.* **49**, 667–674.

Schwarz, J.-M., Acheson, K.J., Tappy, L., Piolino, V., Müller, M.J., Felber, J.P. and Jéquier, E. (1992) Thermogenesis and fructose metabolism in humans. *Am. J. Physiol,* **262** (*Endocrinol. Metab.* **25**), E 591–E 598.

Secchi, A., Pontiroli, A.E., Cammeli, L., Bizzi, A., Cini, M. and Pozza, G. (1986) Effects of oral administration of maltitol on plasma glucose, plasma sorbitol, and serum insulin levels in man. *Klin. Wochenschr.* **64**, 265.

Siebert, G., Romen, W., Schnell-Dompert, E. and Hannover, R. (1980) Wirkungen von L-Sorbose auf die intakte Ratte. Sorbose-Wirkungen. *Infusionstherapie und klinische Ernährung — Forschung und Praxis* **7**, 271–275.

Siljeström, M., Eliasson, A.C. and Björk, I. (1989) Characterization of resistant starch from autoclaved wheat starch. *Starch/Stärke* **41**, 147–151.

Spengler, M., Somogyi, J.C., Pletcher, E. and Boehme, K. (1987) Tolerability, acceptance and energetic conversion of Isomalt (Palatinit®) in comparison with sucrose. *Akt. Ernähr.-Med.* **12**, 210–214.

Sperry, J.F. and Robertson, D.C. (1975) Erythritol catabolism by *Brucella abortus*. *J. Bacteriol.* **121**, 619–630.

Tacquet, A. and Devulder, A.B. (1978) Study of clinical and biological tolerance to Lycasin 80/55 in man. Calmette Hospital, d'Lille, France. Quoted in WHO (1978) *Food Additives Series* **20**, 179–206.

Tamura, J., Furuse, M., Matsuda, S., Shimizu, T. and Okumura, T. (1991) Energy utilization of dietary sorbose in growing rats. *J. Agric. Food Chem.* **39**, 1623–1626.

Tappy, L., Randin, J.P., Felber, J.P. *et al.* (1986) Comparison of thermogenic effect of fructose and glucose in normal humans. *Am. J. Physiol.,* **250**, E 718–24.
Tappy, L. and Jéquier, E. (1993) Fructose and dietary thermogenesis. *Am. J. Clin. Nutr.* **58** (suppl) 766 S–770 S.
Thannhauser, S.J. and Meyer, K.H. (1929) Sorbit als Kohlenhydratersatz für den Diabeteskranken. *Münch. Med. Wschr.* **76**, 356.
Thiébaud, D., Jacot, E., Schmitz, H., Spengler, M. and Felber, J.P. (1984) Comparative study of isomalt and sucrose by means of continous indirect calorimetry. *Metabolism* **33**, 808–813.
Touster, O. (1974) The metabolism of polyols. In *Sugars in Nutrition* (eds. H.L. Sipple and K.W. McNutt) Academic Press, pp. 222–240.
van Es, A.J.H., de Groot, L. and Vogt, J.E. (1986) Energy balances of eight volunteers fed on diets supplemented with either lactitol or saccharose. *Br. J. Nutr.* **56**, 545–554.
van Velthuijsen, J.A. and Blankers, I.H. (1991) Lactitol: a new reduced-calorie sweetener. In *Alternative Sweeteners* (eds. Lyn O'Brien Nabors and R.C. Gelardi), pp. 283–298.
Vinuela, E., Salas, M. and Sols, A. (1963) Glucokinase and hexokinase in liver in relation to glycogen synthesis. *J. Biol. Chem.* **238**, PC 1175–PC 1177.
Wawszkiewicz, E.J. and Barker, H.A. (1968) Erythritol metabolism by Propionibacterium pentosaceum. The over-all reaction sequence. *J. Biol. Chem.* **243**, 1948–1956.
Weerden van, E.J. and Huisman, J. (1993a) The digestion process of the sugar alcohol isomalt in the intestinal tract of the pig. 1. Studies with administration of isomalt in the feed. *Brit. J. Nutr.* **69**, 455–466.
Weerden van, E.J. and Huisman, J. (1993b) The digestion process of the sugar alcohol isomalt in the intestinal tract of the pig. 2. Studies with administration of isomalt as a sweet. *Brit. J. Nutr.* **69**, 467–479.
Wheeler, M.L., Fineberg, S.E. Gibson, R. and Fineberg, N. (1990) Metabolic response to oral challenge of hydrogenated starch hydrolysate versus glucose in diabetes. *Diabetes Care* **11**, 733–740.
Winne, D., Görig, H. and Müller, U. (1985) *In vivo* studies of mucosal-serosal transfer in rat jejunum. *Naunyn-Schmiedeberg's Arch. Pharmacol.* **329**. 70–76.
Wolfslehner, L. (1988) Malbit®, ein zahnschonender, energiereduzierter diabetikerverträglicher Zuckeraustauschstoff. (Malbit, tooth-protective, energy-reduced, low-insulin-dependent sweetener.) *Ernährung/Nutrition* **12**, 788–792.
Würsch, P., Welsch, Ch. and Arnaud, M.J. (1979) Metabolism of L-sorbose in the rat and the effect of the intestinal microflora on its utilization both in the rat and in the human. *Nutr. Metab.* **23**, 145–155.
Würsch, P., Koellreutter, B. and Schweizer, T.F. (1989) Hydrogen excretion after ingestion of five different sugar alcohols and lactulose. *Eur. J. Clin. Nutr.* **43**. 819–825.
Yuasa, H., Jga, T., Hanano, M. and Watanabe, J. (1989) Relationship between *in vivo* first-order intestinal absorption rate constant and the membrane permeability clearance. *J. Pharmaceutical Sciences* **78**, 922–924.
Ziesenitz, S.C. (1986) Zur Verwertung des Zuckeraustauschstoffes Palatinit® im Stoffwechsel. *Beitr. Infusionstherapie klin. Ernähr.* Vol. 16, Karger, Basel, pp. 120–132.
Ziesenitz, S.C. and Siebert, G. (1987) The metabolism and utilization of polyols and other bulk sweeteners compared with sugar. In *Developments in Sweeteners*, Vol. 3, (ed. T.H. Grenby), Elsevier Appl. Sci. Publ., London, pp. 109–149.
Zumbé, A. and Brinkworth, R.A. (1992) Comparative studies of gastrointestinal tolerance and acceptability of milk chocolate containing either sucrose, isomalt or sorbitol in healthy consumers and Type II diabetics. *Z. Ernährungswiss.* **31**, 40–48.
Zunft, H.-J., Schulze, J., Gärtner, H. and Grütte, F.-K. (1983) Digestion of maltitol in man, rat and rabbit. *Ann. Nutr. Metab.* **27**, 470–476.
Zürrer, M., Firestone, A. and Mühlemann, H.R. (1978) Hemolysis *in vitro* by sorbose, sorbitol and xylitol. *SSO Schweiz, Monatsschr. Zahnheilkd.* **88**, 929–933.

5 Versatility of maltitol in different forms as a sugar substitute

M. HEUME and A. RAPAILLE

5.1 Introduction

Bulk sugar substitutes of the 'second generation' are characterised by their non-reducing properties and their disaccharide-derived structure. Among these, maltitol combines outstanding performance with reasonable cost, especially when it is in syrup form.

As the sensorial and technological properties of maltitol are very close to those of sucrose, this disaccharide polyol is one of the most attractive substitutes for traditional bulk sweeteners in non-cariogenic and/or calorie-reduced food products. The availability of an extended range of maltitol syrups has encouraged the confectionery industry to produce a large variety of different types of sugar-free confectionery products. The commercial availability of pure crystalline maltitol powder was necessary for the development of very good tasting sugar-free chocolate, for which processing conditions need only minor adaptation. Further, maltitol is a very interesting material for combination with other low-molecular weight polyols such as xylitol and erythritol.

5.2 Properties

5.2.1 Physical and chemical properties

Pure maltitol is a white crystalline powder. It shows an excellent heat stability up to 200°C and does not take part in Maillard-type browning reactions. Chemically, it is a hydrogenated maltose with the disaccharide structure illustrated in Figure 5.1.

Table 5.1 gives a comparison of the physical properties of maltitol and sorbitol. This shows that the melting point of maltitol is significantly higher than that of sorbitol and is relatively close to that of sugar (186°C). The solubility of maltitol is highly satisfactory, and it does not give crystallisation problems in applications when crystallisation is not desired. The solubility of maltitol exceeds even that of sucrose at a temperature above 62°C (Figure 5.2).

On account of its good solubility, maltitol exists also in the form of stable

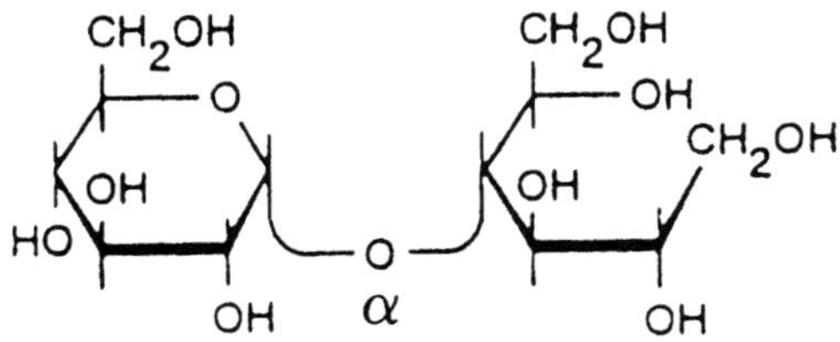

Figure 5.1 Chemical structure of maltitol.

Table 5.1 Physical properties of maltitol in comparison with sorbitol

Property	Sorbitol	Maltitol
Melting point (°C)	93–97	148–152
Solubility (g in 100 g water at 20°C)	235	160
Heat of solution (kJ/kg)	−111	−69
Viscosity of a 70% solution (mPa.s at 20°C)	180	2500

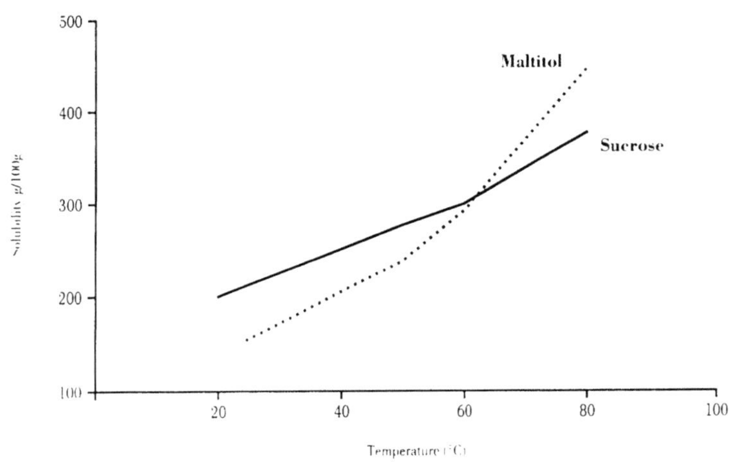

Figure 5.2 Solubility of maltitol (· · ·) and sucrose (——).

non-crystallising syrups. It has only a slight negative heat of solution and gives consequently no cooling effect, unlike sorbitol and xylitol. As maltitol has the highest solubility of all the disaccharide polyols, it has the highest viscosity in solution, significantly higher than sorbitol at the same concentration. Maltitol syrups provide appropriate viscosity profiles for many applications, e.g. delivering the same functions as glucose syrups in traditional confectionery products.

The hygroscopicity of crystalline maltitol is low and comparable to that of sucrose, as the two have similar water sorption isotherms. Maltitol is a non-volatile humectant with a lowering effect on water activity (a_w). It is less hygroscopic than sorbitol, and its equilibrium relative humidity (EHR) is attained more slowly.

When subjected to change in ambient humidity, products stabilised with maltitol syrups are less susceptible to a change in water content than those stabilised with glycerol or sorbitol (Figure 5.3).

5.2.2 Sensorial properties

Maltitol has a clean and pleasant natural sweetness. The relative sweetness of the pure form varies from 80 to 90 (sucrose = 100). Only xylitol among the polyols is sweeter (Table 5.2).

Depending on their maltitol content, the sweetness of maltitol syrups may be lower than this, but in general it is not lower than 65% of the sweetness of sucrose. This sweetness level is almost equivalent to that of

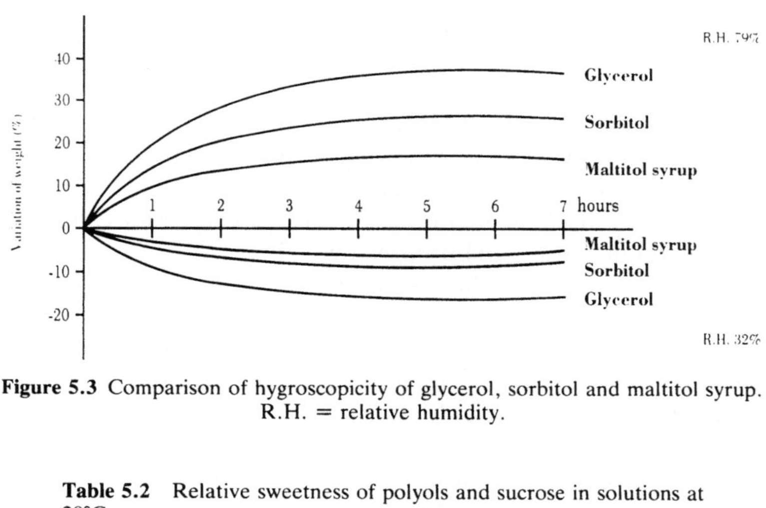

Figure 5.3 Comparison of hygroscopicity of glycerol, sorbitol and maltitol syrup. R.H. = relative humidity.

Table 5.2 Relative sweetness of polyols and sucrose in solutions at 20°C

Sugar	Relative sweetness
Polyol	
xylitol	80–100
sorbitol	50–60
mannitol	50–60
maltitol	80–90
lactitol	30–40
isomalt	50–60
Sucrose	100

sucrose/glucose blends used in manufacturing traditional confectionery products. As with most sweeteners, the sweetness is influenced by factors such as concentration, temperature, acidity and the nature of other nutrients.

Due to its high sweetening power, maltitol can be used as sole sweetener in most sugar-free product applications, without the need to add intense sweeteners. As already mentioned, when maltitol is present in foods in crystalline form it does not give any cooling effect in the mouth.

For this reason it is more compatible with fruit flavours than sorbitol or xylitol. In general it has excellent flavouring properties with good flavour release in foods.

5.2.3 *Physiological properties*

Like the other polyols, maltitol is non-cariogenic, calorie-reduced, non-toxic, and could cause laxation if high doses are ingested.

Caries is initiated with fermentable sugars as substrates, from which dental plaque is formed via dextran and levan production on the surface of the teeth. This is the ideal environment for the growth of microflora containing acidogenic organisms like *Streptococcus mutans* which attack carbohydrates to form acids. The resulting pH-drop can lead to demineralisation of the dental enamel and tooth decay.

Several methods have been developed to measure the cariogenicity of food products, for example by following the pH-drop in the dental plaque after ingestion of the test substance, e.g. the so-called Mühlemann test. All the commercially available maltitol syrup and powder grades pass this test, since they do not cause a pH-drop below 5.7.

Another way to judge the cariogenic potential of a product is the determination of acid production rates. Measurements have been done on sorbitol and maltitol in comparison with glucose. The results clearly demonstrate that maltitol is even less acidogenic than sorbitol.

Turning to the metabolic utilisation of maltitol, due to the fact that no active transport mechanism for penetration of the intestine membranes exists, direct absorption from the gut is only possible by diffusion. This takes place very slowly since the size of the maltitol molecules greatly exceeds the pore size of the gut wall. On the other hand, maltitol can be hydrolysed enzymatically to sorbitol and glucose or fermented by the microflora of the large intestine to volatile fatty acids, carbon dioxide, hydrogen and methane.

It is very difficult to determine the relative proportions of maltitol metabolised by these two pathways, although the estimation of the caloric value of maltitol is based on it. A study based on available scientific information calculates a figure of 3.2 kcal/g for maltitol. On the other hand, a recent study administering ^{14}C-labelled maltitol to rats and following the distribution in the body by measuring the radioactivity, supports a figure of 2.0 kcal/g.

Nevertheless, it is uncontested that maltitol provides fewer calories than glucose or sucrose due to its different metabolic utilisation pattern. Giving an exact figure is impossible, as the caloric value depends on the dose ingested and other factors, but the Commission of the European Community has assessed an energy value of 2.4 kcal/g for all sugar alcohols (EU Directive 90/496/EEC). Several toxicity studies confirm that maltitol products show neither acute toxicity nor genetic or subchronic toxicity. This forced the Scientific Committee for Food of the EU to consider "hydrogenation products of high maltose-containing glucose syrups as acceptable provided the limitations due to their laxative action were kept in mind".

A recently completed chronic toxicity/carcinogenicity study concluded that no adverse effects are expected from ingestion of maltitol (Conz, 1992). As a result of these maltitol evaluations, the European Union recently permitted maltitol and maltitol syrups to be widely used in preparation of foodstuffs (EU Directives 94/35/EC and 95/2/EC). This will now be followed by local implementation of the directives in the European countries, and current legislation will be harmonised. Outside Europe, maltitol is already permitted in most countries.

The laxative effect of polyols originates from their slow transport across the intestine membranes and low water retention. It is extremely difficult to determine the exact tolerance level of individual polyols, since the performance of the digestive tract varies with age and type of food ingested, etc. Absolutely no laxative effect was observed for maltitol up to a 30 g dosage and no significant discomfort was experienced below an ingestion level of 50 g/day. These figures do not take into account the adaptation to maltitol which could raise levels of toleration.

5.3 Commercial forms of maltitol

5.3.1 Production

Like sorbitol, maltitol is derived from starch, which may be enzymatically converted to dextrose (the raw material for sorbitol) or to maltose (the raw material for maltitol) (Figure 5.4). If the enzymic conversion of starch is stopped after a certain period, syrups with a high maltose content are obtained. Besides maltose, minor amounts of dextrose and higher oligosaccharides are present. Maltitol-based products are produced by catalytic hydrogenation of these maltose-containing syrups. Hydrogenation is conducted at elevated temperature and hydrogen pressure in the range of 20–150 bar. After refining the reaction mixture by ion exchange and carbon treatment, the product is concentrated to the required dry

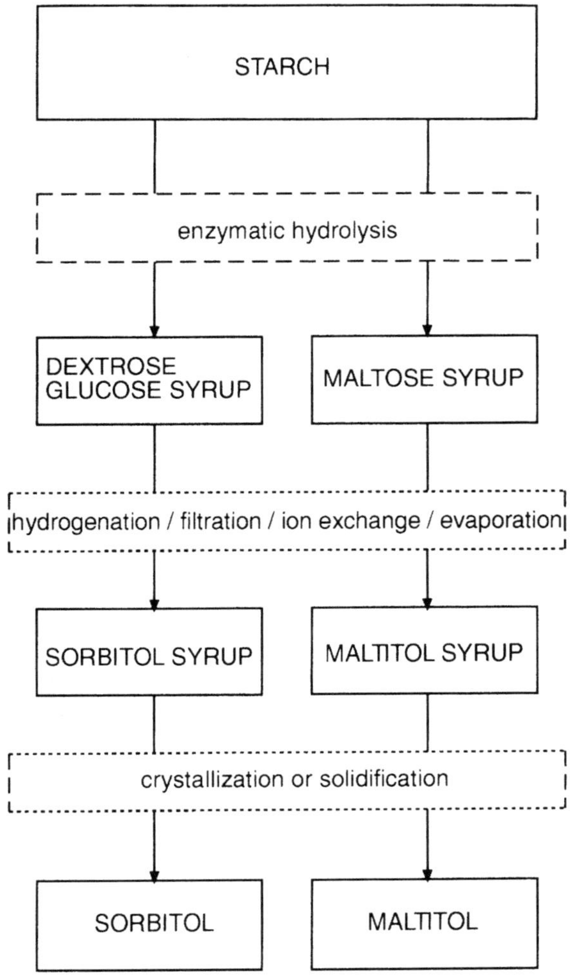

Figure 5.4 Production of sorbitol and maltitol.

substance. Crystallisation or solidification of a very high-maltitol syrup leads to maltitol powder.

5.3.2 Compositions

On the market are several grades of maltitol syrup, with different maltitol contents. Three categories can be distinguished (Table 5.3): those with 50–55% maltitol, the second with 72–77% maltitol and the last with 80–90% maltitol.

An increasing maltitol content intensifies the sweetness of the product.

Table 5.3 Composition of maltitol syrups

Maltitol (%)	Sorbitol (%)	Dry substance (%)
50–55	max. 8	75–85
72–77	max. 5	75–83
80–90	max. 5	70

Other components of the syrups are minor amounts of sorbitol, maltotriitol and hydrogenated higher oligosaccharides. For the grades below 77% maltitol, the dry substance content is normally adjusted to 75%, but syrups of 83 or 85% dry substance are also available on request. In order to avoid crystallisation, syrups with 80–90% maltitol on a dry basis are produced commercially at 70% total dry matter. Maltitol powder is sold in two forms, a solidified grade with 86–90% maltitol (moisture content <1%) and a pure crystalline grade with minimum 99% purity (moisture content <0.5%).

5.4 Applications

5.4.1 Confectionery products

5.4.1.1 Hard-boiled candies. Maltitol is currently used to produce sugar-free hard-boiled candies, its main advantages in comparison with alternative bulk sweeteners being high sweetness level (90% compared to sucrose) and excellent flavour release. These candies are produced from commercially available syrups, with maltitol levels varying between 50 and 88% on dry substance. In some cases the syrups are prepared by dissolving maltitol powder. When high maltitol levels are used there is no need to add intense sweeteners. Maltitol gives a transparent candy with a smooth glossy surface and pleasant sweetness. Due to its excellent heat stability there is no colour development during boiling. Candies containing maltitol have good workability for depositing as well as for plastic moulding. Sugar-free candies with low hygroscopicity – and showing no cold flow – are achieved following these simple rules:

- Cook the candy mass to below 1.0% residual moisture content, preferably 0.5–0.8%. Maltitol-based hard candies can be produced without the need to adapt traditional or high-speed continuous candy lines as long as the equipment allows cooking temperature and vacuum times higher than those generally applied for glucose/sucrose hard candy. When using traditional cooking equipment, the cooking temperature is preferably 168–170°C with 8 min vacuum time.

- Adapt moulding conditions, taking into account the low viscosity and the high cooking temperature of polyols. A longer cooling time than that generally used for sucrose/glucose candies must be allowed in order to reach similar plasticity (moulding temperature *c.* 65°C).
- Acid should only be added to a *cooled* candy mass to avoid syrup decomposition.
- Wrap the sweets as soon as possible after moulding to avoid moisture pick-up. If needed, before wrapping, the sweets should be stored at low RH (below 50% RH) and at low temperature (max. 25°C).
- Ensure good protection by accurate wrapping and/or packaging.

Originally maltitol candies were made from syrups containing 50 or 75% maltitol on a dry basis. In general 5–8% mannitol was added to the syrup before cooking in order to reduce hygroscopicity of the candies. Further studies have been conducted with the aim of reducing the cooking temperature and hygroscopicity level of maltitol hard candies. One of the approaches is to add a mixture of gum arabic and mannitol to the syrup before boiling. Lower boiling temperatures can be applied (e.g. 145°C) but vacuum is still needed.

Recent studies show that improved shelf life properties of candies are obtained with 82–84% maltitol content syrup. This results in a significant decrease in hard candy moisture pick-up in comparison with other compositions, including maltitol/mannitol blends. However, for unwrapped candies exposed to high humidity/temperature conditions during storage, the hygroscopicity levels of candies based on 83% maltitol syrup could still be too high. A further decrease in moisture pick-up is obtained by increasing maltitol content to 87–89%.

This higher maltitol content provokes more rapid stable microcrystallisation, resulting in reduced hygroscopicity but also giving slight cloudiness after 3 weeks wrapped storage. Using this syrup the cooking temperature can be reduced from 168°C to 155°C, without significant hygroscopicity increase in the hard candy. The corresponding residual moisture content becomes 1.8% instead of 0.8%. Using optimum manufacturing conditions, sugar-free hard candies based on syrups with 83–88% maltitol have similar hygroscopicity levels to conventional candies. Shelf life can be further improved by dusting the candies with high purity (min. 99%) crystalline maltitol. In some cases maltitol syrups are combined with other disaccharide polyols.

5.4.1.2 Toffees and chewy candies. Maltitol-based soft-boiled candies are made from maltitol syrups, using conventional processing equipment. High maltitol syrup (e.g. 75% maltitol) is recommended. In this case there is no need to add intense sweeteners to the formulations.

Table 5.4 Toffee

Formulation		(%)
	Maltitol syrup (75% maltitol on d.b.)	73
I	Concentrated whole milk (non-sweetened)	19
	Sodium bicarbonate	0.2
	Hydrogenated cocoa fat	7
II	Glycerol monostearate (GMS)	0.6
	Lecithin	0.2
	Flavour	

Manufacturing process

- Mix GMS and lecithin in cocoa fat (65°C) (II).
- Mix maltitol syrup with the concentrated milk and sodium bicarbonate (I).
- Blend I and II and heat to 70°C.
- Homogenise ~5 min.
- Cook to 125–127°C.
- Cool the mass to ~100°C and add the flavour.
- Forming.

5.4.1.2.1 Toffees. Maltitol syrups are used as sole sweetener. Different types of milk sources can be used. For taste and colour reasons non-sweetened concentrated milk is preferred, but whole milk powder and de-lactosed whey powder are also suitable. As with all polyols, maltitol does not react with proteins and will not give the typical toffee taste and colour. For this reason it is necessary to add toffee flavour and caramel colour to the cooled toffee mass before moulding.

In this way, sweetness, taste and texture characteristics of maltitol-based toffees are made very similar to those of traditional sweetened products, using similar process conditions. In general the cooking is a few degrees Centigrade higher than for standard recipes. This results in a lower final humidity and the right texture. The shelf life properties are very similar to glucose/sucrose-based products. Table 5.4 gives a typical toffee formulation, including the manufacturing process.

5.4.1.2.2 Chewy candies. Maltitol-based chewy sweets are produced like conventionally sweetened products. For the sugarless products slightly higher cooking temperatures have to be applied in order to obtain the right texture. Maltitol-based chewy products need in general a longer pulling time. They are based on maltitol syrups, but in some cases a combination of maltitol syrups and maltitol powder is used. Stickiness can be avoided by adding small amounts of lecithin. Typical formulations for grained and non-grained chewy candy are given in Table 5.5. The grainy type has a shorter and softer texture.

Table 5.5 Chewy candy

Formulations	(%) Nongrained	Grained
Maltitol powder (90% maltitol)		35
Maltitol syrup (75% maltitol on d.b.)	88	53
Gelatine solution		
gelatine 150 bloom	1.5	1.5
water	2.7	2.7
Fat		
hydrogenated cocoa fat (melting point 34–36°C)	7.0	7.0
Glycerol monostearate (GMS)	0.6	0.6
Lecithin	0.2	0.2

Manufacturing process

- Dissolve the gelatine in warm water (60°C) under slow agitation.
- Heat the maltitol syrup to 136–145°C (depending on the desired texture).
- Dissolve the GMS and lecithin in the melted fat at 70°C and add to maltitol syrup at ~100°C.
- Add in the gelatine solution under mixing.
- Cool the mass to 50–60°C and pull the paste.

5.4.1.3 Gums, pastilles and jellies

5.4.1.3.1 Gelatine gums. For sugar-free gelatine gums with taste and sensorial properties equivalent to conventionally sweetened gums, maltitol syrup with 73–75% maltitol can replace all the sugar and glucose syrup.

This maltitol syrup has a high sweetness so the addition of intense sweeteners is no longer necessary. The syrup composition enables the gelatine to achieve its full gelling power; in sugar-free fruit gums containing citric acid it allows a reduction in gelatine level compared with conventional gum, while maintaining an identical texture profile. If standard gelatine levels are maintained the gum will be too hard with possibly insufficient cohesiveness.

On the other hand, for non-acidified gums more gelatine (e.g. +0.5%) is needed to achieve similar texture to the standard. At the same gelatine level it is possible to obtain good texture values with a limed bone gelatine. Maltitol syrups with more than 76% maltitol are not suitable, as crystallisation of the polyol will occur during storage.

The procedure to make sugar-free gelatine gum must be adapted slightly. Maltitol syrup (75% maltitol) must be cooked to a dry substance of ~90% before cooling to 90°C and then adding previously-prepared gelatine solution (at 60°C). After cooling to 80°C, citric acid flavour and colour are added. The mass is then ready to be deposited in moulding starch. In order to reach this very high dry substance more rapidly and thereby to facilitate the manufacturing process, a high dry substance syrup (83%) is preferred. This maltitol syrup presents no viscosity problem even at this high dry substance concentration.

The clarity of the gums is further improved by a short application of vacuum to remove included air. Alternatively, if the mix is allowed to stand for 10 minutes in a warm room, the mixture will begin to clear and foam will rise to the surface. The gums are then moulded in cold, dry starch, stored overnight, and after demoulding, coated with traditional wax or oil.

By varying gelatine level and type, maltitol syrup level and cooking temperature, any required traditional texture profile from very soft to hard pastille-type gums can be obtained (Table 5.6). Due to the low viscosity of maltitol syrup, the mass is easy to handle, and moulding at high dry substance is possible. However in some cases, this high dry substance moulding could result in small hard pieces in the end-product due to non-homogeneous gelling of the gelatine. Moulding at slightly lower d.s. is recommended to avoid this problem.

The slightly higher water activity of maltitol could require a lower final moisture content in the gum. However gums containing maltitol may have additional advantages due to higher hygroscopicity and humectancy, giving better resistance to drying out than traditional gums when exposed to a low humidity atmosphere. In this way stickiness or deformation during storage are avoided. It should be further noted that all fruit gums are slightly more hygroscopic than non-acidified gums.

It should also be mentioned that the difference in hygroscopicity at 75% relative humidity between sucrose- and maltitol-based gums is much smaller than at 65% RH. The shelf life of gums stored under extremely high temperature/humidity conditions can be further improved by addition of 0.5% pectin.

5.4.1.3.2 Gum arabic pastilles. Maltitol is also the ideal sweetener in combination with gum arabic for production of sugar-free pastilles. The

Table 5.6 Gelatine gums

Gum texture		Soft	Medium	Hard	Pastilles
Gelatine	220 bloom	4.8	6.0		
	260			6.0	
	270				9.0
Water		9.6	12.0	12.0	13.0
Maltitol syrup (75% maltitol on d.b.)		85.6	82	82	78
Cooking temperature (°C)		112	112	114	125
End moisture (%)		22	22	21	12

Table 5.7 Gum arabic pastilles

Formulations	Soft	(%) Medium	Hard
Gum arabic solution (50% d.s.)	44.6	55	60
Maltitol syrup (75% maltitol on d.b.)	65.4	45	40
Citric acid	–	as desired	–
Flavour and colour	–	as desired	–

Manufacturing process

- Mix the maltitol syrup with the gum arabic solution.
- Cook the mass to 70/72% d.s. at ~110°C.
- De-areate for minimum 30 minutes at the moulding temperature 80–85°C.
- Add the colour/flavour and if required, the citric acid solution.
- Deposit in dry starch and dry at 60°C during 40–60 h to a minimum d.s. of 90%.
- After demoulding, the pastilles should be coated with wax or oil.

correct sweetness can be obtained using a high-maltitol syrup (75% maltitol) as sole sweetener. Texture, transparency, gloss and storage stability are key properties. Gum arabic pastilles formulated with maltitol syrup have excellent storage stability and better controlled moisture pick-up and retention than sorbitol syrup formulations.

Pastilles having textures from soft to hard are obtained by varying gum arabic level, maltitol syrup level and cooking temperatures. In this way any required traditional pastille texture profile can be produced (Table 5.7)

Another advantage is that with standard equipment, it is possible to add the gum arabic prior to cooking. In traditional gum manufacturing, this is only possible when using very modern equipment. The lower viscosity of the cooked mass, typical for polyols, allows higher dry substance for depositing, leading directly to reduced drying time. Sugar-free pastilles are normally deposited in moulding starch and dried under standard conditions.

5.4.1.3.3 Pectin jellies. High-maltitol syrup is also the preferred product in a range of other confectionery applications such as pectin jellies. Sugar-free pectin jellies are made with this syrup using the same pectin type and levels as with conventionally-sweetened jellies. For optimum results defined procedures should be adhered to during manufacture (Table 5.8).

The traditional texture of pectin jelly is obtained by slightly increasing the final moisture level of the product (5%). The shelf life properties of sugar-free pectin jellies are identical to those of conventionally-sweetened jellies. However for storage under high temperature/humidity conditions a small amount of carrageenan (0.4–0.5%) must be added. As

Table 5.8 Pectin jellies

Formulation	(%)
Maltitol syrup (75% maltitol on d.b.)	75.8
Water	22.4
Pectin (high methoxyl D100°)	1.8
Citric acid 50% solution	6.5
Flavour and colour	as needed

Manufacturing process

- Mix a first part of the maltitol syrup (*c.* 23%) with the water and heat to *c.* 70°C.
- Add the pectin to the syrup/water mix while stirring.
- Heat slowly while stirring until the solution starts to boil. Boil for 2 min.
- Add the remaining maltitol syrup slowly (to avoid too rapid cooking) and cook to 109°C (*c.* 81°Brix).
- Add colour/flavour and citric acid solution and deposit immediately in starch (temperature of solution 90°C).
- After demoulding the jellies are sanded with crystalline maltitol powder.

If needed the shelf life of maltitol syrup-based jellies can be improved by adding carrageenan (0.5%).

for conventional pectin jellies, firmness is slightly increased and the product has improved resistance to more severe storage conditions.

For these products, the high maltitol content helps to prevent the pectin from setting too quickly which would occur if a syrup with 50% maltitol content is used. Rapid pectin setting can also be avoided by depositing in dry starch or plastic moulds at temperatures above 90°C. After demoulding, the jellies can be sanded with high purity maltitol crystals.

5.4.1.4 Chocolate. Both dark and milk chocolate with outstanding flavour can be produced with crystalling high purity (min. 99%) maltitol, by applying normal manufacturing processes. Higher conching temperatures than generally recommended for other sugar alcohols can be used without recrystallisation and problems from any increase in viscosity. Dry conching at high temperature gives additional advantages in terms of production technology and flavour development. Well-tempered moulded maltitol chocolate is non-hygroscopic and has excellent gloss and good breaking characteristics. The smooth melting profile and natural sweetness of chocolate containing crystalline maltitol are very similar to a sucrose-based product and make this product a high-quality sugar-free alternative, allowing *c.* 12–15% calorie reduction.

Chocolate evaluation studies have shown that 50/50 combinations of crystalline maltitol with low-calorie bulking agents such as polydextrose and/or insulin give further calorie reductions up to 23%. In this case the cocoa butter content is also reduced. When combining crystalline maltitol with the above polysaccharides, the conching conditions have to be adapted to maintain high conching temperatures. This is the so-called 'dry/wet' conching method which also gives a good workability and normal chocolate mass rheology, and yields good sensorial properties. Industrial production has confirmed the good workability and sensorial quality of maltitol chocolate. There are now sugar-free chocolates offering 25–30% calorie reduction on the market. To reach such levels it is also necessary to reduce the fat content of the chocolate, reformulating for example by using less cocoa butter and adding defatted cocoa powder. In order to ensure good flow properties of the chocolate mass, addition of a specific emulsifier such as Polyglycerol Polyricinoleate (PGPR) is advised.

A typical maltitol-based chocolate formulation (23% calorie reduced) and manufacturing conditions are given in Table 5.9.

5.4.1.5 Lozenges. High purity maltitol solutions crystallise easily, producing sugar-free lozenges with a hard, crunchy texture. Also there is no need to add intensive sweeteners. Sugar-free lozenges are prepared using conventional manufacturing process, are non-hygroscopic and have excellent storage stability.

Maltitol requires the same drying time as icing sugar to obtain the correct tablet texture with a similar low residual moisture content. Tablet

Table 5.9 Dark chocolate

Formulation	(%)
Cocoa mass	44
Cocoa butter	9.5
Crystalline maltitol	23
Inulin	23
Emulsifier lecithin	0.5
Vanilla	0.02

Manufacturing process

Insert maltitol powder, inulin, cocoa mass and *c.* 25% cocoa butter into the Z-kneader.
Kneading time: 10–15 min
Kneading temperature: 40°C

Refining
The mass is milled on cooled 5-roll refiner.

Conching
Dry conching at 60°C during 3 h.
Wet conching up to 80°C during 15 h.
During wet conching the remaining cocoa butter is added.
Lecithin and vanilla are added after 17 h conching.

Tempering – tableting
Temperature: 28–31°C.

hardness can vary depending on stoving time. Table 5.10 illustrates the production of maltitol-based lozenges.

5.4.1.6 Tablets. Maltitol powder morphology is important in determining its suitability for direct compression. A very good tableting performance is obtained by using a solidified powder with a purity of 85–87%, for example. A moisture content above 1% could lead to capping or breakage of the tablets on the rotary press. The particle size distribution for maltitol powder is less critical than that for sorbitol. Due to the high sweetness there is no need to add intense sweeteners to maltitol tableting compositions. As the tablets give no cooling effect there is a better compatibility with fruit flavours than for sorbitol.

Table 5.11 gives typical conditions for producing maltitol-based fruit flavoured tablets.

5.4.1.7 Chewing gum. The main components of sugar-free chewing gum are gum base, polyol powder(s) as solid phase, and sorbitol or maltitol

Table 5.10 Lozenges

Formulation	(%)
Crystalline maltitol (min. 99% purity)	1.5 kg
Gelatine solution (10% d.s.)	150 ml

Manufacturing process

- Prepare 10% gelatine solution (170 bloom) by slowly adding the gelatine to warm water (~50°C).
- Insert maltitol powder into the Z-blade mixer preheated to 40–45°C.
- Add slowly the warm gelatine solution under mixing.
- Mix for ~10 min to obtain a smooth, homogeneous paste.
- Remove the paste from the kneader, roll-out and cut in shape.
- Stove at 45°C during 24 h.

Table 5.11 Fruit tablets

Formulations	(%)
Maltitol powder	98.2
Mg stearate	0.5
Citric acid	1.3
Fruit flavour	q.s.

Manufacturing process

- Mix maltitol powder with the flavour and citric acid for 2 min.
- Add Mg stearate (lubricant), mixing for 3 min.
- Direct compression at 25 to 30 kN.

Table 5.12 Chewing gum

Formulation	(%)
Gum base	28
Sorbitol powder	51.6
Maltitol syrup (75% maltitol on d.b. – 85% dry substance)	17
Mannitol	3
Lecithin	0.4
Flavour	q.s.
Colour	q.s.

Preparation method

- Preheat the kneader to 40°C.
- Bring the warm gum base (40°C) into the kneader and knead for 1 min.
- Add the maltitol syrup and knead for 1 min.
- Add half of the sorbitol powder and all mannitol and knead for 2 min.
- Add the second half of sorbitol and knead for 2 min.
- Add the lecithin and knead for 1 min.
- Add flavour and colour and knead for 2 min.
- Cool the gum mass to 30°C, extrude and form.
- Dust with mannitol powder.

syrup as liquid phase. The gum base level varies in general between 20 and 30% depending upon the type of base and the required chewiness. Maltitol can be used in the liquid as well as in the solid phase. Although machinability, texture and stability of a chewing gum composition depend on the equilibrium between the three main components, the liquid phase plays an important role in the water exchange with the surrounding air. Maltitol syrups, with maltitol contents from 50–75% of dry substance prevent the chewing gum drying out and hardening during storage so that plasticity is maintained. Maltitol syrups are used at levels of 10–40%. Mannitol is added to prevent stickiness during processing. Table 5.12 gives an example of a typical formulation. Maltitol powder can be used to texturise and to maintain the composition in a substantially anhydrous form. The incorporation level of maltitol powder is preferably between 45 and 55%.

5.4.1.8 Aerated products. Maltitol syrups, preferably with 75% maltitol on a dry basis can be used as the sole sugar-free bulk sweetener in foamed confectionery products such as marshmallows and angel-kisses. The manufacturing processes are similar to those for conventionally-sweetened products but sometimes conditions have to be adapted. Maltitol-based marshmallow compositions have to be whipped at a lower temperature, due to their lower viscosity. In general the texture properties of the products are at least similar to those of the standard formulation, and

Table 5.13 Angel-kiss

Formulation		(%)
(A)	Egg white	2.5
	Gelatine (100 bloom)	0.5
	Maltitol syrup (75% maltitol on d.b.)	22.7
	Water	11.4
(B)	Maltitol syrup (75% maltitol on d.b.)	62.9
Manufacturing process		
(A)	• Mix egg white with gelatine, warm water and maltitol syrup and aerate.	
(B)	• Cook residual maltitol syrup to ~122–124°C (87–88% d.s.).	
	• Mix (A) in (B).	
	• If desired, sweetness can be increased by adding 30–40 mg/kg Acesulfame K.	

sometimes they are improved. A maltitol-based angel-kiss is more tender and creamy than a sucrose/glucose-based product.

During storage, maltitol-based foams have generally a better storage stability than standard compositions. The humectant character of maltitol prevents the foams from drying out, while there is no danger of hardening due to crystallisation. In some formulations intense sweeteners are added. Table 5.13 describes the preparation of a maltitol-based angel-kiss composition.

5.4.1.9 Bars. Maltitol powders and syrups can be used in several types of bars or countlines. Maltitol can be the sole sweetener in the centre of the bar or it can be incorporated in the chocolate covering the centre. Interesting sugar-free components of bar-centres could be maltitol-based caramel or toffee, nougat, fudge, marzipan, marshmallow, etc. (see above and section 5.4.1.4 for chocolate). Maltitol syrup can also be the main component of the binding syrup of a cereal bar or granola bar, which can also be enrobed in maltitol chocolate. Table 5.14 gives a maltitol-based cereal bar composition. Besides sugar-free claims, it is also possible to add nutritional claims for countlines, particularly when there are other specific nutrients present in the composition.

5.4.1.10. Dragee coating. Pure crystalling maltitol allows the production of sugar-free hard coatings with sensorial properties very similar to sugar-based coatings. Using conventional coating equipment, cores (e.g. sugar-free chewing gum) are covered with layers of crystallising maltitol solution. As for all crystallising sweeteners, the coating conditions for maltitol depend upon its solubility and viscosity at saturation. Different concentrations of maltitol can be applied, e.g. 60 and 75% dry substance. In general the core temperature is kept preferably between 25 to 30°C.

Table 5.14 Cereal bar

Formulation	(%)
Binding syrup	
Maltitol syrup (75% maltitol on d.b.)	68
Maltodextrin 10 DE from waxy starch	8
Water	12
Hydrogenated cocoa fat	11.6
Lecithin	0.4

Manufacturing process

- Dissolve maltodextrin in water at 70°C.
- Add maltitol syrup and cook to 117°C (*c.* 88°Brix) for a soft type or to 125°C (*c.* 91°Brix) for a hard type.
- Mix the molten fat and lecithin in the syrup.
- Hold the syrup at 95°C.

Cereal blend
Blend of different cereals such as muesli (10 fruits), almonds, hazelnuts, extruded wheat.

Bar preparation
Mix the cereal blend with the binding syrup in the following proportions:

	Soft type (%)	Hard type (%)
Cereal blend	54	51
Binding syrup	46	49

After forming and cooling the bar can be coated with sugar-free chocolate, based on maltitol.

Compared with other polyols, maltitol coatings have distinct advantages such as high whiteness and high sweetness with a weak cooling effect. The latter makes maltitol more suitable for fruit-flavoured products, in which a cooling effect is not desired. Maltitol coatings have a good, stable hardness and crunchiness.

5.4.2 Bakery products

5.4.2.1 Cakes and sponge cakes. Maltitol powder, preferably a grade containing 80–90% maltitol, is suitable as the sole sweetener in cake and sponge cake. In comparison with standard recipes, identical preparation methods can be applied. For pound cake using maltitol powder together with a 50% fat replacement by a fat-mimetic, an excellent calorie-reduced product is obtained. A formulation and manufacturing process is given in Table 5.15. The incorporation of native wheat starch is responsible for the very regular and fine cell structure. The use of a preservative or modified atmosphere packaging gives the cake a very good shelf life. Table 5.16 gives a formulation of a sponge cake completely sweetened with maltitol powder. Applying this composition and manufacturing process, a well-aerated and good-tasting sugar-free product with a long shelf life can be obtained.

Table 5.15 Pound cake

Formulation	(%)
Wheat flour	19.76
Native wheat starch	9.88
Maltitol powder	23.71
Eggs	24.70
Margarine	8.90
Fat mimetic	8.90
Emulsifier	2.96
Dicalcium phosphate	0.59
Sodium bicarbonate	0.50
Potassium sorbate	0.10

Manufacturing process

A. Fat mimetic

Composition

Spray dried glucose syrup	10
Low DE maltodextrins from potato starch	20
Water	70

Manufacture
Put all ingredients together.
Mix at high speed for 45 sec.
Store the solution at 4°C during 2 h.

B. Cake batter

- Scale all the ingredients in a bowl.
- Mix for 20 sec at low speed and 5 min at high speed.
- Bring 400 g batter in a baking pan.
- Baking: 180°C during 45 min.
- Wrap the cake (eventually MA-packaging).

5.4.2.2 Biscuits. Different types of sugar-free biscuits can be made, replacing all the sucrose by maltitol powder, preferably with 80–90% maltitol. For hard biscuit a combined fat reduction and flour replacement by native maize starch results in a dough with excellent workability (Table 5.17). This combined use is based on the fact that fat reduction increases the hardness of hard biscuits while introduction of starch gives a reduction of hardness. A 100% maltitol-sweetened shortbread dough has normal figures for physical properties such as viscosity and specific volume (Table 5.18). A good open structure together with a sandy and very crispy texture are supplementary advantages.

5.4.2.3 Marzipan. For the production of sugar-free marzipan the best results are obtained by combining high-purity crystalline maltitol with maltitol syrup. Table 5.19 shows a sugar-free marzipan with the highest

Table 5.16 Sponge cake

Formulation	(%)
Wheat flour	11.4
Native wheat starch	14.6
Eggs	34.3
Maltitol powder	24.7
Emulsifier	5.1
GDL	0.6
Trisodium phosphate	0.3
Colouring	0.07
Vanillin	0.03
Sodium bicarbonate	0.3
Water	8.5
Potassium sorbate	0.1

Manufacturing process

- Scale the liquid phase in a bowl.
- Blend the dry ingredients.
- Mix during 5 min at medium speed.
- Scale 300 g batter in a biscuit pan.
- Baking: 20 min at 190–200°C.

Table 5.17 Hard biscuit

Formulation	(%)
Group I	
Shortening	7.4
Maltitol powder	16.4
Native maize starch	9.5
Whole milk solids	1.2
Salt	0.4
Sodium acid pyrophosphate	0.15
Vanillin	0.03
Water	2.7
Group II	
Biscuit flour	50.1
Sodium bicarbonate (SBC)	0.3
Ammonium bicarbonate (ABC)	0.15
Sulphite	0.07
Water	11.6

Manufacturing conditions

- Cream group I during 3 min.
- Dissolve the SBC and ABC, sieve the flour.
- Add the cream to the flour/sulphite.
- Add the dissolved baking salts and the rest of the water.
- Mixing time: 30 min.
- Relaxing time: 30 min.
- Baking: 270°C for 3.5 min (depending on the equipment).

Table 5.18 Shortbread

Formulation	(%)
Wheat flour	44.3
Shortening	28
Maltitol powder	19.2
Egg powder	0.7
Whole milk solids	0.5
Salt	0.5
Sodium bicarbonate (SBC)	0.2
Water	6.6

Manufacturing process

- Cream all the ingredients, except the flour and SBC at high speed.
- Add the flour and SBC, mix until a homogeneous mass is obtained.
- Deposit the shortbreads.
- Baking: 200°C for 13 min.

Table 5.19 Marzipan

Formulation	(%)
Almonds	29.4
Crystalline maltitol (< 300 μm)	58.8
Maltitol syrups (50% maltitol on d.b.)	11.3
Aspartame	0.1
Potassium sorbate	0.5
Water	0

Manufacturing process

- Wash the almonds carefully.
- Fracture the almonds.
- Blend the almonds with the other ingredients.
- Break twice to pieces.
- Knead the whole mass.

possible content of maltitol products. Of course other ratios of almond/polyol also result in excellent products. Addition of a small quantity of intense sweetener is necessary to adjust sweetness, and the use of a maltitol powder granulometry below 300 μm is advised. The manufacturing process is nearly the same as for a sucrose-based standard recipe. The addition of a preservative is necessary to ensure a good shelf-life.

5.4.3 Jams

Maltitol syrup, preferably with a maltitol content of 75% on a dry substance basis, is suitable for preparing different types of jams and fruit spreads. Maltitol intensifies the fruity taste of jams, and its sweetness is sufficient, so the addition of intense sweeteners is not necessary. The preparation method is similar to that for standard formulations based on sucrose and glucose syrup. Good, stable products are obtained providing an appropriate dry substance, e.g. 60% in the end-product, is observed. At lower dry substance levels the addition of preservatives is necessary to ensure good shelf life.

Table 5.20 Jam with 40% calorie reduction

Components	Parts
Fruit	48
Sucrose	14
Glucose syrup 40 DE	10
Maltitol syrup (75% maltitol on d.b.)	33
°Brix	min. 60 min. 45 (with preservatives)

Table 5.21 Ice cream

Formulations	(%) Sugarless recipe	Sugarless fat reduced recipe
Butterfat	8–10	3
Spray dried low DE maltodextrin from potato starch	–	1
Skimmed milkpowder	10	10
Sucrose	–	–
Glucose syrup 40 DE	–	–
Maltitol syrup (75% maltitol on d.b.)	20	22
Stabiliser/emulsifier	0.6	0.7
Natural vanilla flavour	q.s.	q.s.
β-carotene	q.s.	q.s.
Aspartame of Acesulfame K	q.s.	q.s.
Water	up to 100	
Manufacturing process[a]		
Process conditions		
Pasteurisation	84°C	
Homogenisation	150 bar	
Water cooler	18°C	
Ageing temperature	~5°C	
Ageing time	3 h	
Freezing overrun	100%	
Ice cream temperature	−6°C	
Packaging		
Freezing	30°C	
Storage	~−20°C	

[a] The ice cream recipes have been run on a Mix Mark Pilot plant unit.

It is possible to replace all the conventional sweeteners by maltitol syrup, e.g. in recipes for diabetics. However partial replacement can permit a claim of sufficient calorie reduction. A formulation is illustrated in Table 5.20.

5.4.4 Ice cream

Maltitol is a very suitable sweetener for sugar-free ice cream. In general syrups with a maltitol content of *c*. 75% dry substance are optimal. The taste and sweetness are also optimal. In some formulations there is a need to add intense sweeteners. Physical properties such as hardness and melting resistance are excellent. Additional to the replacement of sugar and glucose syrup, further calorie reduction can be achieved by replacing part of the fat by a fat-mimetic, e.g. a low-DE maltodextrin from potato starch. The manufacturing process for all these formulations is similar to that for standard recipes. Table 5.21 gives a typical sugarless and fat-reduced recipe.

5.5 Conclusion

The intensive development that has been carried out on maltitol, its product range and its applications in particular has provided the opportunity to produce confectionery and bakery products, jams and ice cream in sugar-free and toothfriendly versions, often also in calorie-reduced forms, and retaining the taste the consumer desires.

Maltitol is commercially available for individual applications, making use of all the benefits of characteristics such as viscosity, solubility and crystallisation tendency.

The legal approval of maltitol has followed and today all grades of maltitol are permitted in the European Union and most other countries.

The use of this versatile polyol is expected to increase further, possibly in combination with other polyols to benefit from synergistic effects.

Bibliography

Boursier, B. (1986) Production of pharmaceutical or confectionery products with maltitol coating by applying maltitol syrup to moving bed of cores. European Patent 0201412.

Cherukuri, S.R., Wong, L.L., Orlandi, D.A. and Faust, S.M. (1988) Use of maltitol in chewing gum compositions for improved physical stability.

Conz, (1992) Combined chronic toxicity/carcinogenicity study. RBM Istituto di Ricerche Biomediche, Ivrea, Italy.

Gonze, M. and Van der Schueren, F. (1993) Sugar-free Confectionery: Recent Developments Lecture Food Ingredient Asia, Singapore.

Gonze, M., Van der Schueren, F. and Rapaille A. (1992) Process for the production of hard candy. European Patent Application 0533334.

Gonze, M., Van der Schueren, F. and Rapaille, A. (1992) Lozenges. European Patent Application 0530995.

Hirao *et al.* (1990) US Patent 4.408.041.

Pierson, F., Lycasin Lecture, French course at ZDS – Solingen (1988).

Rapaille, A. (1988) Application of Hydrogenated Products. *Stärke*, **40** No. 9, S.356–359.

Rapaille, A., Goossens, J. and Heume, M. (1993) *Sugar Alcohols, Encyclopaedia of Food Science, Food Technology and Nutrition.* Academic Press Limited.

Rapaille, A., Gonze, M. and Van der Schueren, F. (1993) Malbit CH, the ideal sweetener for sugar-free chocolate. FIE Seminar at Paris.

Serpelloni, M. and Bussiere, G. (1985) Sugarless ice cream containing polyol as sweetener and texture agent preferably as hydrogenated starch hydrolysate containing sorbitol and maltitol. European Patent 0152351.

Van der Schueren, F., Rapaille, A. and Gonze, M. (1991) Process for the production of chocolate. European Patent Application 0.489515.

6 Basic structure and metabolism of isomalt

S.C. ZIESENITZ

Summary

Isomalt is a sweetener with versatile properties due to its noncariogenicity, its practically insulin-independent metabolism and its reduced physiological caloric value compared to sugar. Isomalt is produced by hydrogenation of its precursor disaccharide isomaltulose, which is obtained from sucrose as raw material. The hydrogenation of the ketosugar isomaltulose gives a mixture of the two glucosylhexitols glucosyl-α-(1→1)-mannitol and glucosyl-α-(1→6)-sorbitol. Isomalt is now used in the formulation of toothfriendly sweets, in the production of calorie-reduced foods and in special products for diabetics. Isomalt sweetened products are sensorically not distinguishable from sucrose-based products.

Isomalt is noncariogenic, as was shown with rat caries model experiments. This was corroborated by plaque pH-telemetry in humans demonstrating the absence of any relevant intraplaque acid production from isomalt. Isomalt is not a substrate for plaquepolysaccharide synthesizing glucosyltransferases from Streptococci and was found to inhibit glucan formation of glucosyltransferase from *Streptococcus mutans*.

The limited digestion and absorption of isomalt in the small intestine and its symbiotic bacterial utilization in the colon are the basis for the reduced energetic utilization of isomalt compared to sugar. Most experiments for the assessment of the energetic values of polyols gave evidence, that isomalt has less than 50% of the calories of sucrose. Dietetic studies and oral glucose tolerance studies demonstrated that isomalt is a suitable sweetener for diabetics, without any adverse metabolic effects. Isomalt is due to its specific nutritional qualities a promising alternative to sucrose, with versatile properties for many food and pharmaceutical applications.

6.1 Introduction

Since isomalt has been approved as sweetener in many countries its use and application in foods, confectionery and pharmaceuticals has grown. The

Isomalt is the generic name of Palatinit®, the registered trademark of SÜDZUCKER AG.

increasing demand for isomalt is a consequence of its versatility, favourable properties and specific nutritional qualities, so that it is becoming the preferred choice in the production of calorie-reduced and noncariogenic sweets and confectionery, which are suitable for diabetics. Isomalt now competes with other polyols available on the market.

This review describes the production, structural and metabolic characteristics of isomalt and summarizes its health and nutritional advantages compared to sucrose which are of major importance in the choice of an alternative sweetener. The sensoric and chemical characteristics as well as the aspects of application of isomalt in food technology are reviewed by Willibald-Ettle and Schiweck in chapter 7 of this volume.

6.2 Production of isomalt

Isomalt (Palatinit®) is manufactured from sucrose in a two step procedure (Schiweck, 1986, 1993, 1994). The first step (Figure 6.1) comprises the enzymatic transglycosylation of the non-reducing sucrose to the reducing keto-disaccharide isomaltulose (Palatinose®). In a second step (Figure 6.1) isomaltulose is hydrogenated to isomalt, an approximately equimolar mixture of D-glucopyranosyl-α-(1→1)-D-mannitol (abbreviated to GPM) and D-glucopyranosyl-α-(1→6)-D-sorbitol (GPS or GPG). The enzymatic glucosyl transfer from the nonreducing α1→β2 glycosidic bond in sucrose to the reducing α(1→6) bond of isomaltulose is catalyzed by immobilized *Protaminobacter rubrum* CBS 574.77, a microorganism with the enzymatic specificity to convert sucrose preferably and in a very high yield (Schiweck and Munir, 1992) to isomaltulose. The safety of the strain of *P. rubrum* CBS 574.77 used has been established in comprehensive studies (Porter *et al.*, 1991).

Analogous to the hydrogenation of the ketosugar fructose, resulting in a mixture of the two polyols mannitol and sorbitol (glucitol), the hydrogenation of isomaltulose leads to a mixture of two disaccharide alchols, D-glucosyl-α-(1→1)-D-mannitol and D-glucosyl-α-(1→6)-D-sorbitol. Both disaccharide alcohols can be crystallized as colorless crystals, appearing white. Whereas glucosyl-α-(1→6)-sorbitol crystals are anhydrous, glucosyl-α-(1→1)-mannitol crystallizes as a dihydrate.

The safety of isomalt had been assessed by JECFA which allocated an 'ADI not specified' in 1985 [FAO/WHO, 1986/1987; Report TRS (WHO Technical Report Series) 733-JECFA 29/34; Tox monograph FAS (WHO Food Additive Series) 20-JECFA 29/207], taking account of reports from different safety studies (de Groot, 1987; Smits-van Prooije *et al.*, 1990; Waalkens-Berendsen *et al.*, 1989, 1990). The Scientific Committee on Food of the European Union arrived at the corresponding conclusion and classified isomalt as 'acceptable' in 1984 (SCF, 1984, 1987). The sweetener

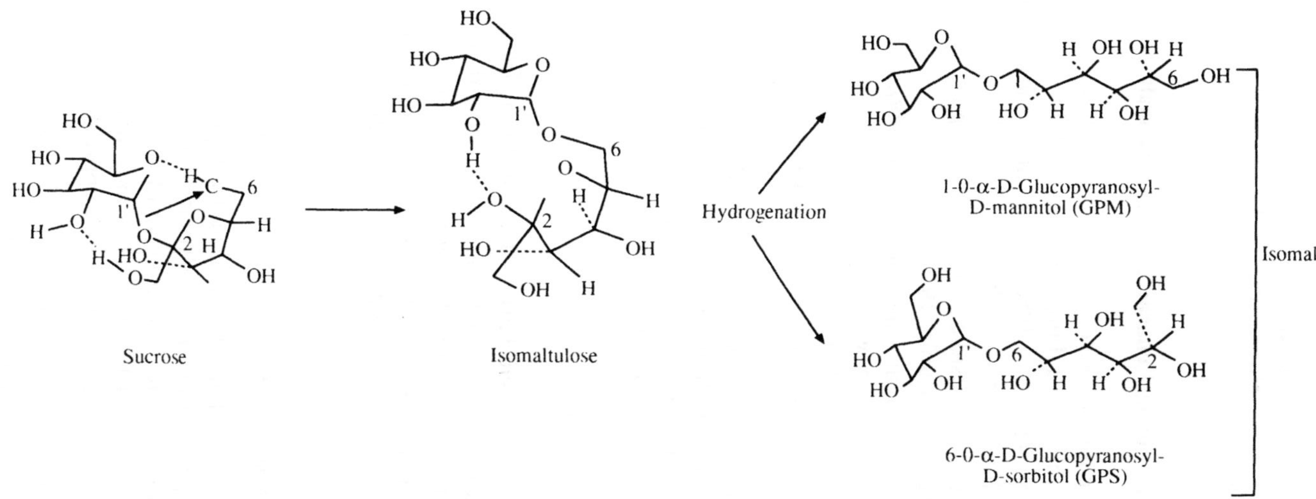

Figure 6.1 Production of isomalt by enzymatic transglycosidation of sucrose to isomaltulose and hydrogenation of isomaltulose to isomalt (adapted from Schiweck and Munir, 1992).

directive of the European Union (94/35/EC), which became effective in 1994 regulates the use of intense and nutritive sweeteners including isomalt in Europe. In the USA, a petition for GRAS affirmation of isomalt had been filed (1990, 55 FR 42484).

6.3 General physico-chemical characteristics of isomalt

Isomalt is nonhygroscopic, even less so than sucrose, and is commercially available as a dry, white crystalline powder, in different grain sizes for different requirements (Sträter, 1986). Both disaccharide alcohol components are chemically extremely stable, as 12.7 kJ/mole are required for the hydrolysis of the glycosidic bond linking glucosyl and hexitol constituents (Schiweck and Munir, 1992).

The sweetness of isomalt in water solution is about 50% of that of sucrose. It is chemically inert with a pure sweet taste and no off- or aftertaste. In food and beverages the sweetness of isomalt depends on its concentration and matrix in which it is used (texture, pH, temperature, etc). Synergistic sweetness effects result from the combination of isomalt with other polyols and/or intense sweeteners (Paulus and Fricker, 1980). Physical properties similar to those of sucrose (Schiweck, 1993, 1994) make this disaccharide alcohol mixture a versatile bulk sweetener which can be processed in similar ways to sucrose.

6.4 Metabolism of isomalt

Before being metabolized disaccharides have to be hydrolyzed into their constituent monosaccharides. These are then absorbed through the intestinal mucosa into the portal blood and become available for metabolism. Thus isomalt has to be hydrolyzed into its components, glucose, sorbitol and mannitol, which are then available for absorption, metabolism and bacterial utilization.

6.4.1 Dental health aspects: caries prevention with isomalt

Replacing cariogenic carbohydrates with noncariogenic sweeteners is one strategy for reducing the cariogenicity of foods. Experimental evidence for the noncariogenicity of isomalt was important for its use as a sweetener in toothfriendly and/or safe-for-teeth sweets, and was demonstrated in a comprehensive set of different experimental approaches comprising:

(i) fermentation and growth experiments *in vitro* with relevant cariogenic microorganisms, either pure strains or mixtures of dental plaque or salivary microorganisms;
(ii) effects of isomalt on streptococcal glucosyltransferase activity *in vitro*;

(iii) rat caries studies;
(iv) studies in humans to assess intra-oral and intraplaque reactions and acid production after exposure to isomalt (pH-telemetry).

6.4.1.1 In vitro *fermentation studies with isomalt.* The first report on the potential use of isomalt as a nonacidogenic sweetener dates back to the 1970s (Gehring, 1973), when sorbitol and xylitol were already of interest for dental reasons. Tests *in vitro* demonstrated that isomalt is neither fermented nor used as carbon source for growth by cariogenic streptococci, nor was there formation of extracellular polysaccharides, which is a typical sign of the utilization of sucrose by streptococci.

The apparent nonfermentability of isomalt by oral streptococci was also demonstrated with other cariogenic microorganisms. Gehring and Karle (1981) screened different strains of *Streptococcus mutans* and lactobacilli and mixed human dental plaque *in vitro* for fermentation and growth with 1% isomalt. In contrast to sucrose, glucose or fructose, isomalt remained practically unfermented by *S. mutans*, *S. salivarius*, *S. sanguis*, mixed dental plaque or salivary flora, and combinations of cariogenic streptococci, comprised of different strains of *S. mutans* serotypes a, b, c, d, as well as of *S. sanguis*, *S. salivarius*, *S. viridans*, *S. milleri* and *S. mitis* (Gehring and Karle, 1981). Only a very limited number of incubations of bacteria *in vitro* showed a small pH-drop, but such experiments registered only final pH, providing no information on fermentation rates.

Van der Hoeven (1979) made similar observations while screening about 50 oral bacterial strains. The majority of the cultures tested did not ferment isomalt. Only very few strains of Streptococci and Actinomyces showed a small pH-drop.

However, more recent studies, with the focus on measuring fermentation rates from human dental plaque suspensions, based on a very sensitive method for determining acid production rates, again demonstrated the absence of microbial utilization of isomalt. Fermentation rates by cariogenic dental plaque microorganisms were less than about 2% of those of easily fermentable sugars, and were thus negligible for isomalt, sorbitol and maltitol, whereas hydrogenated starch hydrolysate (HSH) reached about 23% of the fermentation rate of glucose (Ziesenitz and Siebert, 1987). A slight *in vitro* pH-drop indicating some acid production need not automatically be of intra-oral relevance. The results from intra-oral plaque pH tests are used to decide on the final classification of a sweetener. Tests of this type demonstrated that isomalt does not cause plaque acid production below the dentally critical pH-threshold (see section 6.4.1.4).

6.4.1.2 Inhibition of glucan synthesis by isomalt. Extracellular glucosyltransferases of streptococci catalyze the synthesis of plaque polysaccharides from sucrose, providing the matrix for the adhesion of microorganisms to the tooth surface. Due to its structure, isomalt is not a substrate for plaque polysaccharide synthesis. Using purified glucosyltransferase from *S. mutans* AHT, Siebert (1987) demonstrated that isomalt is not used as a substrate, but inhibits the synthesis of soluble glucan from sucrose in a noncompetitive manner. These observations might foster speculation that this might be of intra-oral relevance in limiting plaque synthesis from sucrose after the ingestion of isomalt.

6.4.1.3 Noncariogenicity of isomalt in caries models. Rat models may be used for assessing the cariogenic potential of foods and food ingredients. Dental caries can develop in a relatively short time if the appropriate cariogenic challenges are provided, e.g. superinfection with cariogenic microorganisms and a frequency-controlled intake pattern of experimental diets. Various tests with isomalt in rats were carried out, and studies using either superinfected rats, superinfected desalivated rats or gnotobiotic rats proved the noncariogenicity of isomalt.

Karle and Gehring (1978) reported on the noncariogenic properties of isomalt, assessed in SPF-Sprague Dawley rats (3–4 weeks of age) which had been superinfected with two strains of *Streptococcus mutans*, representing isolates from the rat's mouth. The rats were fed for 6 weeks with 18 meals per day. During the first three weeks they were slowly adapted to the test diets. The concentrations of sucrose and isomalt were gradually raised from 10% up to 30% by the third week. After 6 weeks the teeth of the lower mandibles were scored for fissure caries according to the method of König (1966). The rats (12 per group) fed sucrose (30%) had the highest fissure caries score, but fissure caries in isomalt fed rats were not statistically different from the negative control fed the basic diet ($p < 0.30$), although highly significantly reduced when compared with the sucrose fed rats ($p < 0.001$).

Similar trends of lower fissure caries scores were also demonstrated in isomalt versus sucrose fed desalivated rats (Karle and Gehring, 1979). In general xerostomized rats developed more caries than did non-desalivated animals, and fissure caries in isomalt fed rats was highly significantly reduced compared with sucrose in respect to both test models. Fissure caries scores for xylitol tended to be lower than those for isomalt, but the two sets of figures were not statistically different.

Diets with 30% isomalt and 30% sucrose were compared in a gnotobiotic rat caries model. Germ-free rats which had been inoculated with *S. mutans* were gradually adapted to the 30% levels of isomalt or sucrose in the diets. Feeding sucrose for 8 weeks induced severe fissure caries, whereas isomalt caused only a few lesions classified as slight decalcifications, not caries lesions (Karle and Gehring, 1981).

The conclusions of the practical absence of a cariogenic potential of isomalt were corroborated by further experiments assessing the cariogenic

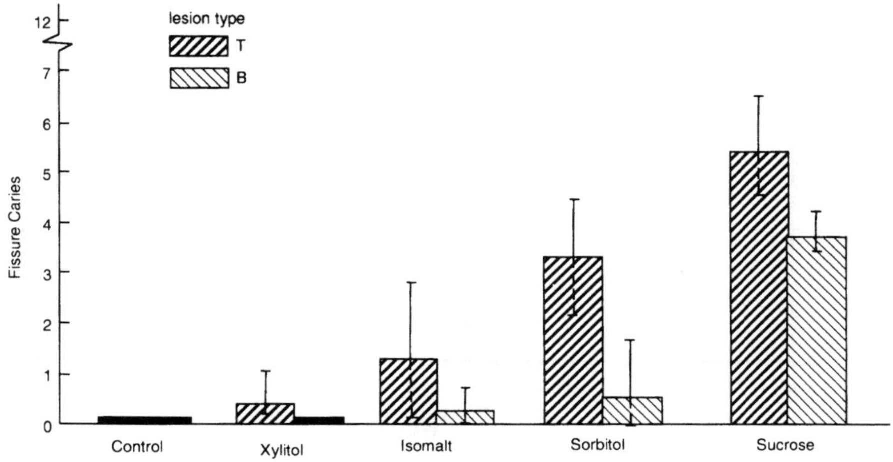

Figure 6.2 Fissure caries in rats fed 6 weeks with 18 meals per day with sucrose or isomalt or sorbitol or xylitol chocolate containing diets, 12 rats per group (Gehring and Karle, 1981). Reproduced by permission.

potential of chocolates sweetened with sucrose, isomalt, sorbitol or xylitol in *S. mutans* infected rats (Figure 6.2), which demonstrated the non-cariogenicity of the chocolate sweetened with isomalt as well as of chocolate sweetened with xylitol (Gehring and Karle, 1981).

The number of carious fissures in rats fed the isomalt sweetened chocolate was lower than that from a sorbitol-sweetened chocolate, although sorbitol is usually classified as non-cariogenic. Rats fed xylitol-containing chocolates had the lowest caries score (Gehring and Karle, 1981).

Using conventional SPF-Osborne–Mendel rats, van der Hoeven (1980) studied the cariogenicity of isomalt under *ad libitum* feeding conditions with diets containing either 16% isomalt or 16% sucrose. In one experiment, the development of fissure caries over a period of 14 weeks was monitored, while every two weeks 3 rats in each group were scored for dentinal fissure caries lesions by the method of König (1966).

Figure 6.3 demonstrates that caries had progressed steadily in the sucrose group but not in the isomalt group, showing striking differences between the dental properties of isomalt and sucrose.

The question whether the oral flora might eventually adapt to isomalt was studied in the rat (van der Hoeven, 1979). Osborne–Mendel rats were fed *ad libitum* on diets containing either 16% sucrose or 16% isomalt for 14 weeks. Plaque samples were removed at fortnightly intervals from the fissures and approximal, buccal and lingual surfaces, and pooled. Suspensions of the plaque were cultured on tryptone yeast agar supplemented with isomalt or glucose. Different morphological types of microorganisms

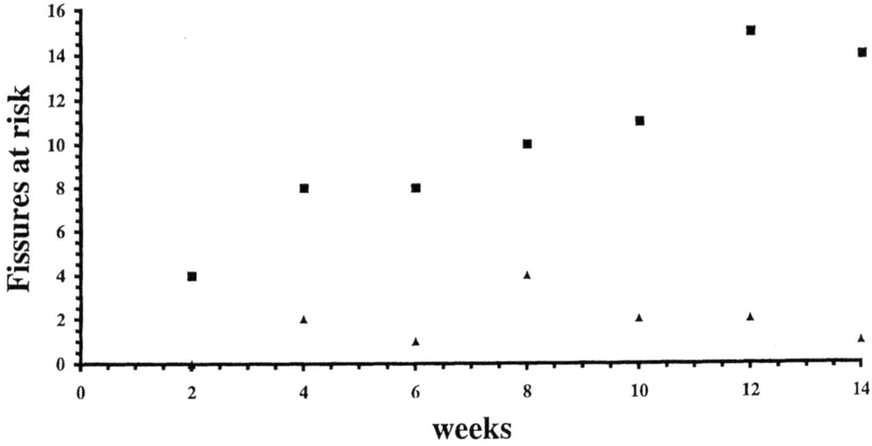

Figure 6.3 Fissure caries (sum of caries scores for 3 rats) in SPF–Osborne–Mendel rats fed diets containing either sucrose or 16% isomalt (van der Hoeven, 1980). ▲, isomalt; ■, sucrose. Reproduced by permission of S. Karger AG, Basel.

were counted and subcultured on blood agar. The predominant biotypes of *Streptococcus*, *Actinomyces viscosus* and *Veillonella* were expressed as a percentage of the total count on blood agar. The isolates were also tested for fermentation of isomalt in a broth culture. No striking differences were observed in the bacterial composition of the plaque between sucrose and isomalt fed rats. These results demonstrated that no adaptation to isomalt when fed to rats for over 3 months had occurred, as indicated by the absence of any accumulation of isomalt fermenting microorganisms (van der Hoeven, 1979).

This was confirmed by van der Hoeven (1980) in a second experiment, in which fissure caries and dental plaque development were compared in Osborne–Mendel rats fed for 4 weeks on diets containing either 16% sucrose or 16% isomalt or uncooked wheat flour (native starch) diets. The microorganisms for superinfection at the start of the experiment were *Streptococcus mutans* C67-1 and *Actinomyces viscosus* Ny 294, which had demonstrated some limited fermentation *in vitro* of isomalt to a final pH of 4.9 and 5.1 respectively, although no isomalt splitting activity had been detectable in these strains (van der Hoeven, 1979).

Isomalt and wheat flour (native starch) showed similar low cariogenic potentials under conventional conditions or after superinfection, although the number of carious lesions was higher in conventional rats. The inoculated *S. mutans* established well in the sucrose group, but not in the isomalt and starch groups.

The results of these studies testing whether adaptation to isomalt might be detectable, demonstrated that isomalt was highly unlikely to play an important part in caries development in rats, and it did not stimulate the growth or proliferation of inoculated *Streptococcus mutans* strains in the rats (van der Hoeven, 1979, 1980).

6.4.1.4 Isomalt in human studies. According to the report of the Scientific Consensus Conference on *Methods for the Cariogenic Potential of Foods*, the true cariogenicity of a food can only be established by determining the extent of tooth decay associated with a given food in man (Stamm *et al.*, 1986). Since such clinical caries trials are no longer ethically feasible, carefully standardized indirect methods for assessing a food's propensity to foster human caries must be relied on (Stamm *et al.* 1986).

One approach for estimating the cariogenic potential of foods is to measure intra-orally the pH of dental plaque after food consumption. These measurements represent an important test procedure for assessing food cariogenicity. The acidogenic response of plaque to ingestion of a test food can be extrapolated to the cariogenicity of the food in human subjects.

The effect of isomalt on human dental plaque acidogenicity had been studied by a variety of different techniques:

(i) by the plaque sampling method, in which plaque is scraped from accessible tooth surfaces before and after ingestion of the test substrate, then suspended in a small volume of water to measure the pH *in vitro*;
(ii) by the touch electrode method to measure the plaque pH *in situ*, in the aerobic, anaerobic or microaerophilic areas of the dental plaque;
(iii) by the in-dwelling electrode method, registering continuously the plaque pH in an interproximal space.

This method of intra-oral plaque pH telemetry has been accepted since 1969 by the Swiss Government as the basis of a 'safe-for-teeth' or toothfriendly claim (Imfeld, 1983, 1987), and is regarded by many as the benchmark for plaque pH testing (Harper *et al.*, 1986).

Results based on human dental plaque scraping techniques were summarized in section 6.4.1.1, demonstrating that human dental plaque microorganisms do not utilize isomalt.

Using the touch electrode technique, the acidogenic response of isomalt in human dental plaque was studied in five male and female volunteers (Hufnagel, 1981; Gehring and Hufnagel, 1983). All refrained from oral hygiene for at least 24 h. After a five-minute period of paraffin chewing, the volunteers rinsed for 2 minutes with a solution of isomalt in water (20% w/v). The plaque-pH was measured every 2 minutes for up to 32 minutes after this, and the results for sucrose, glucose, fructose and isomalt were compared. Whereas the intra-plaque pH dropped rapidly from pH

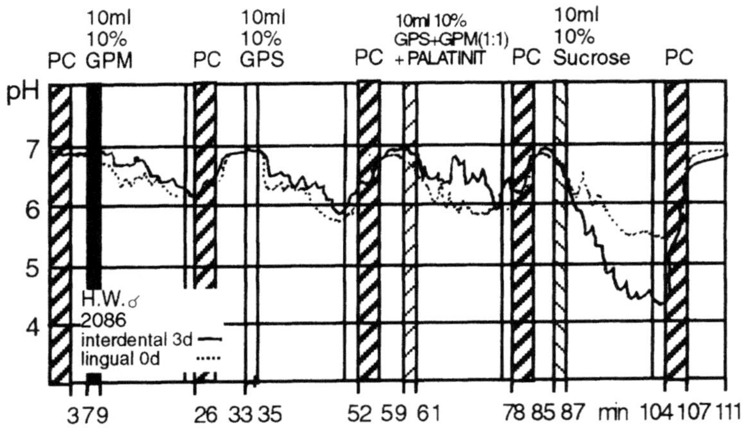

Figure 6.4 Intra-oral pH-telemetry with isomalt (Imfeld, 1983). Reproduced by permission of S. Karger AG, Basel.

6.7 to around pH 4.6 after the sugars, the intraplaque pH after isomalt remained in the dentally-safe range.

Isomalt, its components and isomalt-based products have all been tested by intraoral plaque pH telemetry. One of the advantages of this technique is that potential acidogenic effects of a food or of food components in the dental plaque can be monitored continuously while chewing, thus measuring acid development under realistic conditions taking the complex responses in the mouth, the effects of the food and the oral clearance time into account.

Intraoral plaque pH changes were studied in 3- to 4-day-old plaque from human volunteers after rinsing for 2 minutes with isomalt or its constituent disaccharide alcohols glucosyl-α-(1→1)-mannitol and glucosyl-α-(1→6)-sorbitol, or after ingestion of isomalt-based confectionery products.

Figure 6.4 (Imfeld, 1983) shows the telemetrically recorded pH-profiles of a 3-day interdental plaque and pH of the oral fluid-saliva of a male volunteer, who had rinsed with 10 ml each of 10% solutions of isomalt or with solutions of glucosyl-α-(1ξ1)-mannitol and glucosyl-α-(1→6)-sorbitol. There was no pH fall to below the critical pH of 5.7 in interdental plaque nor in salivary pH. The pH drop after the control rinse with sucrose demonstrates the existence of actively fermenting bacteria in the plaque, thus confirming the non-acidogenicity of isomalt in an actively fermenting dental plaque.

More important for the concept of caries prevention by nonfermentable sweeteners are plaque pH-telemetry test results with food products manufactured from isomalt recorded under realistic conditions of consumption. Plaque pH and salivary pH profiles after sucking isomalt candies were always above the critical pH of 5.7, the threshold limit for the classification of toothfriendly versus acidogenic products.

All these intra-plaque pH data demonstrate that isomalt is a non-acidogenic and thus noncariogenic sweetener, suitable for toothfriendly sweets (Featherstone, 1994).

6.5 Digestion and absorption of isomalt

The general principles of the digestion and absorption of polyols have been reviewed (Ziesenitz and Siebert, 1987), and are therefore only briefly described here. Concentration-dependent uptake of disaccharides in the human gut is negligible. This applies to isomalt, since only negligible traces are absorbed as glucosylhexitols, which are excreted unchanged via the kidneys (Grupp and Siebert, 1978). The digestion patterns of both disaccharides and disaccharide alcohols follow the same principle, as both have to be hydrolyzed into their monomers to make them absorbable into the portal blood.

6.5.1 Digestion of isomalt in the small intestine

Disaccharides as well as disaccharide alcohols are substrates of digestive carbohydrases, the sucrase–isomaltase and the glucoamylase–maltase complexes of the intestinal brush border. However, major differences exist in the affinities of disaccharides and isomalt to these digestive mucosa enzymes, as isomalt is hydrolyzed much more slowly than maltose or sucrose (Ziesenitz and Siebert, 1987). These basic differences of higher affinity and a lower cleavage rate, have a major influence on the intestinal digestibility of isomalt.

Digestibility of isomalt in the small intestine and absorption of the liberated polyols are both limited, as has been shown in rats (Grupp and Siebert, 1978), pigs (van Weerden and Huisman, 1993a,b) and in humans (Kroneberg *et al.*, 1979; Langkilde *et al.*, 1994). Ingested isomalt is incompletely digested and absorbed in the small intestine. Substantial amounts of intact isomalt, as well as unabsorbed sorbitol and mannitol, reach the colon and are fermented to volatile fatty acids, hydrogen and eventually methane by colonic microorganisms.

Studies *in vitro* with intestinal disaccharidases from rat, pig and human mucosa demonstrated that isomalt is cleaved at a much slower rate than sucrose or maltose. Kinetic studies with human mucosa enzyme preparations demonstrated that isomalt and its components have a strong affinity to carbohydrases and are split *in vitro* at about 1–2% of the maximal rates of the usual substrates such as maltose and sucrose (Ziesenitz and Siebert, 1987). Biopsy samples of human intestinal mucosa had a very low capacity to digest isomalt (Nilsson and Jägerstad, 1987). Also an inhibitory effect of isomalt on the hydrolysis of maltose was observed *in vitro* (Grupp and Siebert, 1978; Nilsson and Jägerstad, 1987; Ziesenitz and Siebert, 1987; Heymann, 1991).

Studies on the small intestinal transit of isomalt in rats, pigs and humans demonstrated that the enzymatic hydrolysis of isomalt and absorption of the liberated sorbitol and mannitol during the passage through the small intestine is incomplete.

In a study with colectomized patients ($n = 4$) receiving 30 g isomalt (with 95% dry matter) in 250 ml tea under fasting conditions, the excretion of 57% of the ingested dosage into the ileal bag was observed within 10 hours. Under these circumstances a maximum of 43% of ingested isomalt had disappeared, indicating that in humans at least 57% of the ingested isomalt is not absorbed from the small intestine and it is thus transferred to the colon (Kroneberg *et al.*, 1979).

The reduced intestinal digestibility of isomalt was also reported from studies with fistulated pigs (van Weerden and Huisman, 1993a,b). Based on analytical data for isomalt, sorbitol and mannitol analyzed in the ileal chyme, the disappearance of isomalt was slightly less than 40% (van Weerden and Huisman, 1993a,b), in line with data from colectomized patients (Kroneberg *et al.*, 1979).

The study in fistulated pigs gave the additional information that, based on calculations of the energy contents of the ileal chyme, only approximately 10% of isomalt had been digested in the small intestine. The very limited digestibility (about 10%) of isomalt was ascribed to the effects of isomalt on the digestibility of other dietary components such as dietary starches or disaccharides. As this would be difficult to prove *in vivo*, it might be speculated that the inhibitor function of isomalt recognized *in vitro* on maltase activity of human jejunal mucosal enzyme preparations might help to explain the interaction (Grupp and Siebert, 1978; Nilsson and Jägerstad, 1987; Heymann, 1991).

A recent study into the intestinal digestion and absorption of isomalt in ileostomy subjects demonstrated that the digestibility and absorption of isomalt is reduced to a much lower level than that of maltitol and sorbitol (Langkilde *et al.*, 1994). The design of this study was different from that reported by Kroneberg *et al.* (1979) in that the polyols were given in the form of chocolate bars after a meal and not in a drink under fasting conditions. The first chocolate bar, containing either 15 g isomalt, sorbitol or maltitol, was consumed 30 minutes after breakfast, and the second was consumed 6½ hours after the first. After 8 and 24 hours the ileal polyol effluents were quantified. At both times the disappearance of the ingested polyols from the small intestine was lowest with isomalt, followed by sorbitol and maltitol. After the first chocolate bar, corresponding to 15 g isomalt, about 37% of the dosage had been excreted into the ileal effluent. Thus 63% of the ingested isomalt had disappeared within 8 hours.

Within 24 hours 43% of the isomalt given with the two chocolate bars (corresponding to 30 g isomalt) was excreted into the ileal effluents. Analysis of the ileal effluents of the nine ileostomy patients demonstrated that 53–67% (mean 60%) of the ingested isomalt, when given in association with other nutrients, had disappeared from the small intestine. The products arriving in the ileal effluents, in the order of increasing quantities, were sorbitol, glucosyl-α-(1→6)-sorbitol, mannitol and glucosyl-α-(1→1)-mannitol. These observations in man are in line with earlier findings in experimental animals, that the two glucosylhexitols of isomalt are not digested to the same extent in the small intestine, and the absorption of sorbitol is more efficient than that of mannitol. This comparative study also gave evidence that α-(1→4) glucosidic linkages, e.g. like that of maltitol, are cleaved more efficiently than the α-(1→6) and α-(1→1) glucosidic linkages of isomalt. The data from colectomized humans (Langkilde *et al.*, 1994) corroborated earlier observations from studies with fistulated pigs that isomalt increased substantially the excretion of dry matter into the ileal effluents, compared with the polyol-free test diet. The excess dry matter excreted in the ileal effluents exceeded even the quantities of isomalt recovered. The same phenomenon was also observed with maltitol and sorbitol (Langkilde *et al.*, 1994).

Taking into account the energetic equivalents of the increased dry matter effluxes into the colon, the data gave evidence that polyols may interfere with the small intestinal availability of dietary energy. Comparing the combustion energy values of the increased ileal effluents, the data demonstrated that in relation to the ingested polyol, the most pronounced effect in increasing the energy transfer rate of ileal contents into the large intestine was observed with isomalt. The excess energy transferred into the colon was even higher than the combustion energy equivalent of isomalt and of its components excreted into the ileal effluents, and approached 50% of the gross energy intake of isomalt. Thus isomalt has a reduced small intestine digestibility and also reduces the small intestinal utilization of dietary energy by transfer to the large bowel. This effect of increasing the energy transfer rate of ileal contents into the large intestine has to be considered in the energetic evaluation of isomalt.

In summary, different modes and patterns of consumption were tested under colectomized conditions to quantify the fraction of isomalt absorbed in the small intestine and the transfer of intact and digested isomalt into the colon. The different study designs with ileostomized subjects demonstrated that approximately 40% of the ingested isomalt (30 g) given as a drink under fasting conditions had disappeared in the small intestine, so that about 60% of isomalt, consisting of intact isomalt and of polyols, liberated from isomalt but not quantitatively absorbed on passage through the small intestine, were transferred into the colon (Kroneberg *et al.*, 1979).

When isomalt was consumed postprandially with a solid food item, its intestinal digestibility was reduced to almost the same extent compared with consumption under fasting conditions. The incomplete small intestinal digestibility of isomalt was accompanied by the additional transfer of dietary energy from the small intestine into the colon, representing an energetic equivalent of almost 50% of the gross energy of isomalt (Langkilde *et al.*, 1994).

The question of dose dependency on the small intestinal digestion and absorption of two polyols, sorbitol and isomalt, was experimentally addressed by Langkilde *et al.* (1994) in a pilot study with two ileostomy patients. The recovery of three different dosages of sorbitol and isomalt consumed postprandially as a drink in quantities of 5, 10 or 15 g in 200 ml, 30 min after breakfast was measured. Results presented in absolute and in relative terms demonstrated a linear correlation between the amount of polyol consumed and the amount excreted into the ileal effluents.

Within the range of 5–20 g of polyol, reflecting realistic consumption patterns, the results demonstrated that per g of isomalt consumed, about 0.6 g was excreted into the ileal effluents and 0.4 g was absorbed in the small intestine, whereas per gram sorbitol consumed, about 0.4 g was excreted into the ileal effluents and 0.6 g sorbitol was absorbed (Langkilde *et al.*, 1994). The results demonstrated a linear correlation between the quantity of polyol consumed and that transferred into the colon. The data suggested a linear correlation between the amount of polyol consumed and the quantity disappeared during the small intestinal transit under colectomized conditions.

It is well known that oral loads of sorbitol and mannitol cause no obvious postprandial changes in peripheral serum sorbitol and mannitol concentrations. They are absorbed passively by a concentration-dependent process. Sorbitol is extracted quantitatively by the liver from the portal system and metabolized rapidly to fructose. Oral sorbitol loads in dosages up to 0.75 g sorbitol/kg body weight were without effect on plasma glucose and fructose levels (Förster, 1972; Macdonald *et al.*, 1978). Sorbitol-rich diets supplying healthy and diabetic subjects with 44 g of sorbitol per day did not cause any obvious postprandial changes in peripheral blood glucose and sorbitol concentrations (Malaisse-Lagae *et al.*, 1984; Vaaler *et al.*, 1987). From oral loading tests with 5 g mannitol, plasma mannitol concentrations 1–2 hours later reached about 7 mg/100 ml and were thus negligibly small (Laker *et al.*, 1982).

As peripheral postprandial serum sorbitol and mannitol concentrations after consumption of large dosages of sorbitol and mannitol remained always in the micromolar range (Young *et al.*, 1980), the absence of any substantial influence of isomalt on postprandial serum polyol concentrations has to be expected, as the small intestinal liberation of polyols is retarded and incomplete. The reduced digestibility of isomalt, as derived from analytical data and its digestive products in the intestinal contents at the end of the ileum, is also reflected by small, if any, postprandial glycaemic reactions observed after consumption of isomalt.

Oral loading tests with varying amounts of isomalt in humans did not induce any spectacular increases in serum glucose concentrations (Grupp and Siebert, 1978; Thiébaud *et al.*, 1984; Bachmann *et al.*, 1984). An oral loading test in 8 healthy volunteers with 50 g isomalt, a quantity which is conventionally used for oral glucose tolerance tests or for determining the glycaemic index of a given food item (Jenkins *et al.*, 1981), demonstrated no significant deviations of a 3-hour glucose profile from baseline values (Bachmann *et al.*, 1984). These observations in healthy humans (Figure 6.6) gave evidence, that isomalt, although it consists of 50% glucose, has no glycaemic contribution of relevance (Bachmann *et al.*, 1984).

Data on glucose-bioavailability from isomalt, according to the procedure described by Karimzadegan *et al.* (1979), determined in growth experiments with ketotic rats under carbohydrate limitation, demonstrated that the conversion rate for glucose from isomalt was about 39–42% of that of glucose itself. Under these experimental conditions, the metabolic glucose pool of the rat did not therefore receive the full glucose complement of isomalt. Furthermore even the preformed glucose with the theoretical value of 50% of the isomalt was not all fully bioavailable as glucose (Ziesenitz, 1983). This incomplete and reduced glucose availability indicated that a substantial portion of preformed glucose from isomalt was not available in the small intestine, but must have reached the colon where it must have been fermented.

In summary, isomalt does not induce any substantial glycemia increase, and can be classified as having a glycemic index of practically zero.

6.5.2 Digestion of isomalt in the large intestine

The colon and its symbiotic flora play a part in the diminished physiological caloric value of isomalt. The mechanisms have been described in a comprehensive review (Ziesenitz and Siebert, 1987). Colonic fermentation transforms isomalt, sorbitol and mannitol into bacterial mass and fermentation end-products which include short-chain fatty acids, primarily acetate, propionate and butyrate, and gases, primarily CO_2, H_2 and CH_4. The volatile short-chain fatty acids are absorbed and metabolized by the host, and thus contribute some energy (Bugaut and Bentéjac, 1993; Cummings *et al.*, 1989; Macfarlane and Cummings, 1991). The gases may be partially absorbed and can be exhaled. Exhalation of hydrogen and/or methane in humans is an indicator of the colonic contribution in the utilization of dietary substrates. Isomalt reaching the colon is completely cleaved and its components are utilized by the intestinal microflora.

The limited absorption and digestibility in the small intestine may eventually give rise to some symptoms of malabsorption, as with poorly digestible carbohydrates. Symptoms of carbohydrate malabsorption may comprise flatulence, accelerated intestinal transit times, bloating, increased stool frequency, diarrhea or possibly constipation. These symptoms of gastrointestinal intolerance are a general feature of osmotically active substances which are not absorbed by active transport mechanisms in the upper intestinal tract and not completely digested on transit through the small intestine. Gastrointestinal side effects such as flatulence, laxation, or diarrhea are generally observed after consumption of excessive doses of slowly absorbed carbohydrates, e.g. lactose and sugar alcohols. The severity depends upon the individual, the dose consumed, the mode of ingestion, the type of the substance, and any prior period of adaptation.

The gastrointestinal tolerance of isomalt has been determined carefully in specific clinical tolerance studies, in adults, in diabetics and in children of different age groups, taking different consumption patterns of isomalt into account (Lee *et al.*, 1994; Paige *et al.*, 1992; Zumbé and Brinkworth, 1992; Gee *et al.*, 1991; Spengler *et al.*, 1987; Fritz *et al.*, 1985; Pometra *et al.*, 1985; Bachmann *et al.*, 1984; Petzoldt *et al.*, 1982). In summary, these studies demonstrated that isomalt is generally well tolerated, both in single dosage once a day and in total quantity, when it was given in divided portions spaced over a day, or after adaptation.

The study by Paige *et al.* (1992) demonstrated that isomalt is tolerated as well as sucrose by children, in single doses of up to 35 g isomalt under non-adapted conditions and up to 50 g after adaptation, when given in two or three portions per day. Zumbé and Brinkworth (1992) reported from comparative studies of gastrointestinal tolerance with sorbitol and isomalt-sweetened chocolates, that isomalt is marginally superior to sorbitol, and that among diabetics isomalt chocolate was accepted significantly better than sorbitol chocolate.

6.6 Isomalt: a bulk sweetener for diabetics

Based on the promising results of animal experiments showing no substantial hyperglycaemic reaction of isomalt (Grupp and Siebert, 1978), and the absence of any adverse toxicological effects (Smits-van Prooije *et al.*, 1990; Waalkens-Berendsen *et al.*, 1989, 1990a,b), the suitability of isomalt as a sweetener for diabetics has been assessed.

In healthy subjects no significant increase, or at most a marginal increase, in blood glucose was observed even when isomalt dosages of 75 g and up to 100 g per person were given as in oral glucose tolerance tests (Grupp and Siebert, 1978). Similar observations were made in diabetics. In two randomized cross-over studies, non-insulin dependent diabetics (NIDDM) received either 50 g isomalt or glucose dissolved in 250 ml tea after fasting (Drost *et al.*, 1980; Petzoldt *et al.*, 1982). In both studies blood sugar and insulin levels remained essentially stable and were not influenced by the ingestion of 50 g isomalt. Whereas the frequency of glucosuria occurred in 18 out of 22 patients after glucose, glucosuria after isomalt was sharply reduced in one study (Petzoldt *et al.*, 1982), cut down to 5 out of 23 cases in the other study (Drost *et al.*, 1980).

Similar observations were made in another crossover study with 23 non-insulin dependent diabetics, comparing the effects of 50 g isomalt to fructose or sucrose (Bachmann *et al.*, 1984). Figure 6.5 demonstrates the effect of isomalt on blood glucose and serum insulin profiles in healthy subjects and in NIDDM patients. Another difference was that after ingestion of sucrose or fructose significant increases in the levels of serum pyruvate and lactate were observed, while after ingestion of isomalt they did not change significantly from base-line levels (Bachmann *et al.*, 1984).

Insulin requirements in insulin-dependent diabetics (IDDM) were determined by the Biostator technique in two separate studies. In one of these glucose clamp studies, insulin requirements after oral administration

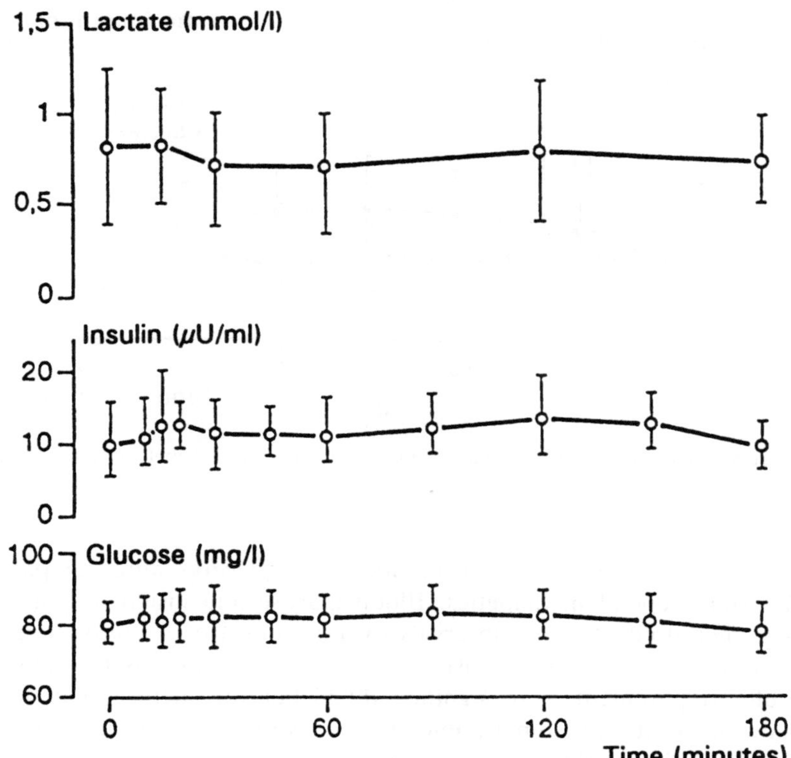

(a)

Figure 6.5 Blood glucose and serum insulin levels after oral isomalt loading tests. (a) 50 g isomalt (50 g/400 ml) in healthy volunteers, $n = 8$, mean $\pm$ s.

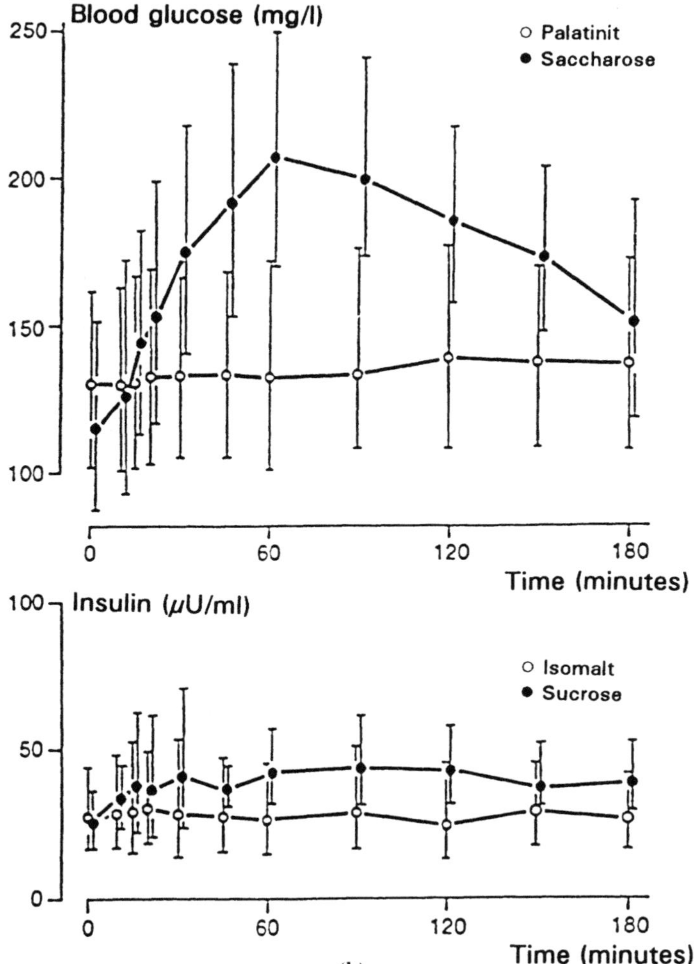

(b)

Figure 6.5 *continued* (b) 50 g isomalt in non-insulin dependent diabetic (NIDDM) patients (n = 12, mean ± s). (Bachmann *et al.*, 1984). Reproduced by permission.

of either 20 g sucrose, sorbitol, isomalt or placebo were compared in 24 insulin-dependent diabetics. Blood glucose and insulin requirements were measured at 1 min intervals over a period of 4 hours after each administration. The mean insulin requirement as well as the maximum insulin requirement were significantly higher for sucrose than for the placebo, sorbitol or isomalt, among which were no significant differences (Kaspar and Spengler, 1984).

In another study, the insulin requirement after ingestion of 24 g isomalt or glucose was quantified in insulin-dependent diabetics (IDDM). The kinetics of insulin requirements were quite different for the two carbohydrates, with a pronounced insulin requirement within the first three hours after glucose, whereas the insulin requirement for isomalt was not statistically significantly different from that for the placebo (Hütter *et al.*, 1993).

Insulin requirement 3 hours after ingestion of isomalt was only 39% of that for the same amount of glucose. Even in the extremely long observation period of up to 8 hours postprandially a difference in insulin requirement up to a factor of two was evident. After ingestion of isomalt the insulin requirement as determined in IDDM with the Biostator technique was only about 51% of that for glucose (Hütter *et al.*, 1993).

The insulin-sparing effect of isomalt was also observed in 60 IDDM diabetic children, who had received either a standardized diet or a diet supplemented with an extra 20 g isomalt per day. This additional quantity of isomalt did not increase the children's insulin requirements and was without any adverse effects on blood glucose and metabolic control (Dorchy *et al.*, 1983).

Pometta *et al.* (1985) reported essentially the same from a 12-week diet-study with NIDDM diabetics, comparing the effects of a daily supplement of 24 g isomalt to the usual diabetic diet or the diabetic diet only. No statistically significant differences were found between these two groups, either in fasting and postprandial blood sugar, haemoglobin A1C, cholesterol, triglycerides and phospholipids, HDL-cholesterol, nor in liver and kidney function tests (Pometta *et al.*, 1985).

Gee *et al.* (1991) compared the effects of either 45 g sucrose, fructose or isomalt in chocolate bars of 75 g weight on the postprandial metabolism of 6 NIDDM diabetics. Postprandial glycaemic responses were significantly the highest with the sucrose chocolate bar. The mean glycaemic response to isomalt was the lowest of all, and was significantly different from that to the sucrose-containing chocolate bar. The same tendency was seen with insulin responses although there was considerable intersubject variation of postprandial insulin levels throughout the study (Gee *et al.*, 1991).

The scientific evidence for the overall beneficial effects of isomalt on blood glucose control and insulinaemia justifies its use in diabetic foods, especially as isomalt is a calorie-reduced sweetener.

6.7 Energetic aspects: isomalt as a sweetener for calorie-reduced foods

Polyols which are not absorbed in the small intestine but are fermented in the colon have a reduced caloric value compared with better-absorbed carbohydrates. The extent of the caloric reduction is determined by the extent of digestion and absorption in the small intestine. Incomplete digestion, colonic fermentation, increased biomass and heat formation are the factors responsible for the energy loss of isomalt, but the assessment of energetic aspects of the polyols is complicated by the circumstance that polyols may interact with other dietary nutrients and lower the energy availability of the diet. This phenomenon was originally observed in experimental animals (van Weerden and Huisman, 1993a,b) and has also been reported recently in humans and seems to be common to the polyols, sorbitol, maltitol and isomalt (Langkilde *et al.*, 1994).

For the assessment of the physiological caloric value of isomalt various techniques and methodologies were applied in different species e.g. rat, pig and man (Ranhotra *et al.*, 1993; Livesey, 1990, 1992; Février and Pascal, 1991; Beaugerie *et al.*, 1991; Krüger *et al.*, 1991; Berschauer and Spengler, 1987; Staudacher and Kirchgessner, 1984; Kirchgessner and Müller, 1983; Kirchgessner *et al.*, 1983; Thiébaud *et al.*, 1984). In rats growth rates and feed efficiencies were reduced, and pigs and rats had leaner body composition (Février and Pascal, 1991; Kirchgessner and Müller, 1983; Zinner and Kirchgessner, 1982; Staudacher and Kirchgessner, 1984).

Indirect calorimetry in man showed reduced caloric utilization compared with sucrose (Grupp and Siebert, 1978). Thiébaud *et al.* (1984) determined net energy by indirect calorimetry, comparing realistic dosages of isomalt and sucrose. After ingestion of 30 g isomalt or sucrose, the carbohydrate and lipid oxidation rates were studied over 6 hours in 10 healthy volunteers. The total carbohydrate oxidation was 79.5 mg/kg after isomalt compared with 185.4 mg/kg after sucrose (Figure 6.6). Also the total lipid oxidation after isomalt was reduced compared with that after sucrose. From these carbohydrate oxidation data, the net energy for humans was calculated as 41–43% that of sucrose, resulting in a physiological caloric value for isomalt in man of 1.64 kcal/g (7 kJ/g).

These calorimetric data of Thiébaud *et al.* (1984) are in line with those of Langkilde *et al.* (1994) determining the physiological caloric value of isomalt from recovery measurements of the ileal effluents and using these analytical data according to the factorial approach of the Nutrition Council of The Netherlands (1987). Langkilde *et al.* (1994) proposed from this a net energy value of 9 kJ/g (2.1 kcal/g) isomalt.

The review of the literature on the energetic evaluation of polyols by Livesey (1990, 1992), taking experimental results as well as factorial-based evaluations of the available energy from polyols into account, concluded that all the data on the rat and the pig suggest a net energy value of about 8.4 kJ/g (or 2 kcal/g) isomalt. However, based on changes in body composition on feeding up to 20% of isomalt in a basal diet, the net energy of isomalt was determined as 8.1 kJ/g. Livesey (1992) concluded that the experimental studies on the energetic assessment of isomalt gave no evidence for a non-linear dose dependency of the caloric value of isomalt.

In the USA, the approach by the LSRO (Life Sciences Research Office)

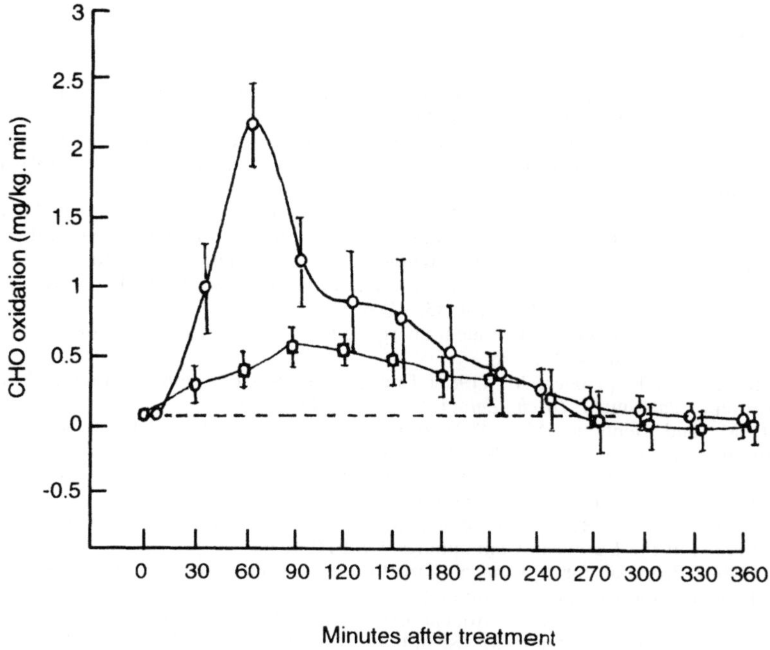

Figure 6.6 Differences in carbohydrate oxidation from basal rates after ingestion of 30 g isomalt or sucrose. □, isomalt; ○, sucrose. (mean ± SD). (Thiébaud *et al.*, 1984). Reproduced by permission.

Expert Panel (1994) for defining the caloric values of polyol sweeteners takes the individual polyol and its metabolization into account. The caloric value of isomalt was determined as 2 kcal/g. This represents the most realistic value, and is supported by experimental evidence from different species.

For the purpose of nutrition labeling, the Scientific Committee of Food (EU) has ascribed to all polyol sweeteners a common caloric value of 2.4 kcal/g or 10 kJ/g. This represents 60% of the energy ascribed to carbohydrates, but it overestimates the real caloric value of isomalt.

6.8 Conclusions

Isomalt is an established polyol sweetener produced from sucrose, with technological properties very similar to those of sucrose, which make it a versatile alternative to sucrose as a bulk sweetener. Isomalt is calorie-reduced and noncariogenic, and has practically no impact on blood glucose concentrations. Isomalt is now widely used in the production of tooth-friendly confectionery and in products for diabetics.

References

Bachmann, W., Haslbeck, M., Spengler, M., Schmitz, H. and Mehnert, H. (1984) Untersuchungen zur Stoffwechselbeeinflussung durch akute Palatinitgaben. Vergleich zu Fructose und Saccharose bei Typ-II-Diabetes. (Studies on influencing metabolism via acute doses of Palatinit. Comparison with frucose and saccharose in Type II diabetes.) *Aktuelle Ernährungsmedizin* **9**, 65–70.

Bässler, K.-H. (1989) Die energetische Nutzung von Zuckeraustauschstoffen, eine Übersicht mit Schwerpunkt auf den Disaccharidalkoholen. *Ernährungs-Umschau* **36**, 395–401.

Beaugerie, L., Flourié, B., Pellier, P., Achour, L., Franchisseur, C. and Rambaud J.-C. (1991) Tolérance clinique, absorption intestinale et valeur énergetique de quatre polyols pris à jeun. *Gastroenterol. Clin. Biol.* **15**, 929–932.

Bernier, J.J. and Pascal, G. (1990) Valeur énergetique des polyols (sucres-alcools). *Medicine et Nutrition XXVI* **No. 4**, 221–238.

Berschauer, F. and Spengler, M. (1987) Energetische Nutzung von Palatinit®. (Energy utilisation of Palatinit®). *Dtsch. Zahnärztl. Z.* **42**, S145–S150.

Bugaut, M. and Bentéjac, M. (1993) Biological effects of short-chain fatty acids in nonruminant mammals. *Annu. Rev. Nutr.* **13**, 217–241.

CEC (1985) *Sweeteners.* Reports of the Scientific Committee for Food (Sixteenth Series), Commission of the European Communities.

CEC (1989) *Sweeteners.* Reports of the Scientific Committee for Food (Twenty-first Series), Commission of the European Communities. CS/EDUL/56.

CEC (1990a) Proposal for a Council Directive on Sweeteners for use in Foodstuffs, Commission of the European Communities, ADD-263/90E.

CEC (1990b) Commission of the European Communities Council Directive (94/35/EU) (10 September 1994) on Sweeteners for use in Foodstuffs of 30 June 1994. ABl. G Nr. L237/3-

Cummings J.H., Gibson, G.R. and Macfarlane, G.T. (1989) Quantitative estimates of fermentation in the hindgut of man. *Acta Veterinaria Scandinavia Suppl.* **86**, 76–82.

Dorchy, H and Ernould, Ch. (1983) Tolerance study with Palatinit® in form of sweets for diabetic children. Unpublished report.

Drost, H., Gierlich, P., Spengler, M. and Jahnke, K. (1980) Blutglucose und Seruminsulin nach oraler Applikation von Palatinit im Vergleich zu Glucose bei Diabetikern vom Erwachsenentyp. *Verh. Dt. Ges. Inn. Med.* **80**, 978–981.

van Es, A.J.H. (1991) Dietary energy density on using sugar alcohols as replacements for sugars. *Proceedings of the Nutrition Society* **50**, 383–390.

FAO/WHO (1986) Evaluation of certain food additives and contaminants. *Report of JECFA*. WHO-Technical Report Series 733, 34–35.

FAO/WHO (1987) Joint FAO/WHO Expert Committee on Food Additives, Meeting (29th: 1985: Geneva) Toxicological evaluation of certain food additives and contaminants. (WHO Food Additives series; 20). *Isomalt* pp. 207–237.

Featherstone, J.D.B. (1994) Effects of isomalt sweetener on the caries process: a review. *J. Clin. Dent.* **5**, 82–85.

Federal Register (1990), Vol. **55**, No. 203, 19 October 1990, Notice of Filing.

Février, Cl., Pascal, G. *et al.* (1991) Comparative energy utilization of isomalt and sucrose in fattening pigs. *Sciences des Aliments* **11**, 367–382.

Fritz, M., Siebert, G. and Kasper, H. (1985) Dose dependence of breath hydrogen and methane in healthy volunteers after ingestion of a commercial disaccharide mixture, Palatinit®. *Brit. J. Nutr.* **54**, 389–400.

Förster, H. (1972) Grundlagen für die Verwendung der drei Zuckeraustauschstoffe Fructose, Sorbit und Xylit. *Medizin und Ernährung* **1**, 7–16.

Gee, J.M., Cooke, D., Gorick, S., Wortley, G.M., Greenwood, R.H., Zumbé, A. and Johnson, I.T. (1991) Effects of conventional sucrose-based, fructose-based and isomalt-based chocolates on postprandial metabolism in non-insulin-dependent diabetics. *Eur. J. Clin. Nutr.* **45**, 561–566.

Gehring, F. (1973) Über die Säurebildung kariesätiologisch wichtiger Streptokokken aus Zuckern und Zuckeralkoholen unter besonderer Berücksichtigung von Isomaltit und Isomaltulose. *Z. Ernährungswiss. Suppl.* **15**, 16–27.

Gehring, F. and Hufnagel, H.-D. (1983) Intra- und extraorale pH-Messungen an Zahnplaques des Menschen nach Spülung mit einigen Zucker- und Saccharoseaustauschstoff-Lösungen. (Intra- and extraoral pH measurements in human dental plaques after rinses with solutions of sugar substitutes.) *Oralprophylaxe* **5**, 13–19.
Gehring, F. and Karle, E.J. (1981) Der Saccharoseaustauschstoff Palatinit® unter besonderer Berücksichtigung mikrobiologischer und kariesprophylaktischer Aspekte. (The sugar substitute Palatinit® with special regard to microbiological and caries prophylactic aspects.) *Z. Ernährungwiss.* **20**, 96–106.
de Groot, A.P. (1987) Isomalt: Chronic toxicity, carcinogenicity, and reproduction studies. In *Low Digestibilitry Carbohydrates.* (eds. Leegwater, D.C., Feron, V.J. and Hermus, R.J.J.) Pudoc Wageningen. pp. 61–70.
Grupp, U. and Siebert, G. (1978) Metabolism of hydrogenated palatinose, an equimolar mixture of α-D-Glucopyranosido-1,6-sorbitol and α-D-Glucopyranosido-1,6-mannitol. *Res. Exp. Med. (Berl.)* **173**, 261–278.
Harper, D.S., Abelson, D.C. and Jensen, M. E. (1986) Human plaque acidity models. *J. Dent. Res.* **65** (*Spec. Iss.*), 1503–1510.
Heymann, H. (1991) Isolierung und Charakterisierung von Di- und Oligosacchariden aus Dünndarmmukosa. Dissertation, Universität Hannover.
van der Hoeven, J.S. (1979) Influence of disaccharide alcohols on the oral microflora. *Caries Res.* **13**, 301–306.
van der Hoeven, J.S. (1980) Cariogenicity of disaccharide alcohols in rats. *Caries Res.* **14**, 61–66.
Hufnagel, H.-D. (1981) pH-Messungen in Situ auf der Zahnoberfläche mit einer Spezialelektrode nach Spülung mit verschiedenen Substraten. Dissertation, Universität Würzburg, Medizinische Fakultät.
Hütter, R., Böswart, F. and Irsigler, K. (1993) Insulinverbrauch von Typ-I-Diabetikern nach oraler Gabe von Isomalt. (Insulin requirements of Type I Diabetics after oral administration of isomalt.) *Akt. Ernähr.-Med.* **18**, 149–154.
Imfeld, Th.N. (1983) Non-nutritive sweeteners, sugar substitutes, and confectionery products. In *Identification of Low Caries Risk Dietary Components.* Karger, Basel. pp. 117–141.
Imfeld, Th.N. (1987) "Zahnschonende" Süßwaren und Schokoladenwaren. *Deutsche Zahnärztliche Zeitschrift* **42**, S107–S112.
Jenkins, D.J.A., Wolever, Th.M.S., Taylor, R.H., Barker, H., Fielden, H., Baldwin, J.M., Bowling, A.C., Newman, H.C., Jenkins, A.L. and Goff, D.V. (1981) Glycaemic index of foods: a physiological basis for carbohydrate exchange. *Am. J. Clin. Nutr.* **34**, 362–366.
Karimzadegan, E., Clifford, A.J. and Hill, F.W. (1979) A rat bioassay for measuring the comparative availability of carbohydrates and its application to legume foods, pure carbohydrates and polyols. *J. Nutr.* **109**, 2247–2259.
Karle, E.J. and Gehring, F. (1978) Palatinit® – ein neuer Zuckeraustauschstoff und seine kariesprophylaktische Beurteilung. *Dtsch. Zahnärtl. Z.* **3**, 189–191.
Karle, E.J. and Gehring, F. (1979) Kariogenitätsuntersuchungen von Zuckeraustauschstoffen an xerostomierten Ratten. *Dtsch. Zahnärztl. Z.* **34**, 551–554.
Karle, E.J. and Gehring, F. (1981) Palatinit und Xylit im gnotobiotischen Rattversuch. *Dtsch. Zahnärztl. Z.* **36**, 673–676.
Kaspar, L. and Spengler M. (1984) Wirkung oraler Gaben von Palatinit® auf den Insulinverbrauch bei Typ-I-Diabetikern. (Effect of oral doses of Palatinit® in insulin requirements in Type I diabetics.) *Aktuelle Ernährungsmedizin* **9**, 60–64.
Kirchgessner, M. and Müller, H.L. (1983) Palatinit®-Verdaulichkeit, Umsetzbarkeit und Verwertung der Energie im Modellversuch an Sauen. *Z. Ernährungswiss.* **22**, 234–240.
Kirchgessner, M., Zinner, P.M. and Roth, H.-P. (1983) Energiestoffwechsel und Insulinaktivität bei Ratten nach Palatinitfütterung. *Internat. J. Nutr. Res.* **53**, 86–93.
König, K.G. (1966) *Möglichkeiten der Kariesprophylaxe und ihre Untersuchung im kurzfristigen Rattenexperiment.* Verlag Hans Huber, Bern.
Kroneberg, H.-G., Spengler, M. and Strohmeyer, G. (1979) Zur Resorption von Palatinit®, dem äquimolaren Gemisch von α-D-Glucopyranosido-1,6-Sorbit (GPS) und α-D-Glucopyranosido-1,6-Mannit (GPM) im Dünndarm von colectomierten Patienten. Report from 06.08.1979, updated on 10.11.1983.

Krüger, D., Grossklaus, R., Klingebiel, L., Ziese, T. and Koch-Gensecke, S. (1991) Caloric availability of Palatinit® in the small intestine of rats: Implications of dose dependency on the energy value. *Nutr. Res.* **11**, 669–678.

Laker, M.F., Bull, H.J. and Menzies, I.S. (1982) Evaluation of mannitol for use as a probe marker of gastrointestinal permeability in man. *European Journal of Clinical Investigation* **12**, 485–491.

Langkilde, A.M., Andersson, H., Schweizer, T.F. and Würsch, P. (1995) Digestion and absorption of sorbitol, maltitol and isomalt from the small bowel. A study in ileostomy subjects. *European Journal of Clinical Nutrition* **48**, 768–775.

Lee, A., Zumbé, A. and Storey, D. (1994) Breath hydrogen after ingestion of the bulk sweeteners sorbitol, isomalt and sucrose in chocolate. *British Journal of Nutrition* **71**, 731–737.

Livesey, G. (1990) The impact of the concentration and dose of Palatinit® in foods and diets on energy value. *Food Sciences and Nutrition* **42F**, 223–243.

Livesey, G. (1992) The energy values of dietary fibre and sugar alcohols for man. *Nutrition Research Reviews* **5**, 61–84.

LSRO/FASEB (1994) *The Evaluation of the Energy of Certain Sugar Alcohols Used as Food Ingredients.* Life Sciences Research Office/Federation of American Societies for Experimental Biology.

Macdonald, I., Keyser, A. and Pacy, D. (1978) Some effects, in man, of varying the load of glucose, sucrose, fructose, or sorbitol on various metabolites in blood. *Am. J. Clin. Nutr.* **31**, 1305–1311.

Macfarlane, G.T. and Cummings, J.H. (1991) The colonic flora, fermentation and large bowel digestive function. In *The Large Intestine*, Ravens Press.

Malaisse-Lagae, F., Couturier, E., Dufrane, S.P., Sener, A., Spengler, M. and Malaisse, W.J. (1984) Effect of a sorbitol-rich diet of plasma and erythrocyte sorbitol concentrations in normal and diabetic subjects. *Diabetes Research* **1**, 209–211.

Nilsson, U. and Jägerstad, M. (1987) Hydrolysis of lactitol, maltitol and Palatinit® by human intestinal biopsies. *British Journal of Nutrition* **58**, 199–206.

Nutrition Council of The Netherlands (1987) *The Energy Value of Sugar Alcohols.* Recommendations of the Committee on Polyalcohols. The Hague.

Paige, D.M., Bayless, Th.M. and Davis L.R. (1992) Palatinit® digestibility in children. *Nutrition Res.* **12**, 27–37.

Paulus, K. and Fricker, A. (1980) Zuckerersatstoffe: Anforderungen und Eigenschaften am Beispiel von Palatinit®. *Zeitschrift für Lebensmittel Technologie und Verfahrenstechnik (ZFL)* **3**, 128–132.

Petzoldt, R., Lauer, R., Spengler, M. and Schöffling, K. (1982) Palatinit® bei Typ-II-Diabetikern. Wirkung auf Blutglucose, Seruminsulin, C-Peptid und freie Fettsäuren im Vergleich mit Glucose. (Palatinit® in type II diabetics: effect on bloodglucose, serum-insulin, C-peptide and free fatty acids.) *Dtsch. Med. Wschr.* **107**, 1910–1913.

Pometta, D., Trabichet, C. and Spengler, M. (1985) Effects of a 12-week administration of Isomalt (Palatinit®) on metabolic control in type-II-diabetics. *Akt. Ernähr.-Med.* **10**, 174–177.

Porter, M.C., Kuijpers, M.H.M., Mercer, G.D., Hartnagel, R.E. Jr. and Koëter, H.B.W.M. (1991) Safety evaluation of Protaminabacter rubrum: intra venous pathogenicity and toxigenicity study in rabbits and mice. *Fd Chem. Toxic.* **29**,(10), 685–688.

Ranhotra, G.S., Gelroth, J.A. and Glaser, B.K. (1993) Usable energy value of selected bulking agents. *Journal of Food Science* **58**, 1176–1178.

SCF/EEC (1984) Report of the Scientific Committee for Food on Sweeteners. December 1984 (EC-DOC III/1316/84–CS/EDUL/27 rev.).

SCF/EEC (1987) Report of the Scientific Committee for Food on Sweeteners. 11 December 1987 (EC-DOC III/B/s – CS/EDUL/56-Final).

Schiweck, H. (1980) Palatinit® – Herstellung, technologische Eigenschaften und Analytik palatinithaltiger Lebensmittel. *Alimenta* **19**, 5–16.

Schiweck, H. (1993) Isomalt. In *Encyclopedia of Food Science Food Technology and Nutrition*, Volume 4, Fumigants–Malnutrition (eds. R. Macrae, R.K. Robinson and M.J. Sadler). Academic Press. pp. 2608–2611.

Schiweck, H. (1994) Isomalt. In *Ullmann's Encyclopedia of Industrial Chemistry*, Volume A25. pp. 426–429.
Schiweck, H. and Munir, M. (1992) Isomalt (Palatinit®, a versatile alternative sweetener – Production, properties and uses). In *Carbohydrates in Industrial Synthesis*, Verlag D. Albert Bartens, Berlin. pp. 5–55.
Siebert, G. (1987) Zuckeraustauschstoffe – Neue Substanzen. *Dtsch. Zahnärztl. Z.* **42**, S128–S134.
Siebert, G., Grupp, U. and Heinkel, K. (1975) Studies on Isomaltitol. *Nutr. Metabol.* **18** Suppl. 1, 191–196.
Smits-van Prooije, A.E., Dreef-van der Meulen, H.C., Sinkeldam, E.J. and de Groot, A.P. (1990) Chronic toxicity and carcinogenicity study of isomalt in rats and mice. *Fd Chem. Toxic.* **28**, 243–251.
Spengler, M., Somogyi, J.C., Pletcher, E. and Boehme, K. (1987) Tolerability, acceptance and energetic conversion of Isomalt (Palatinit®) in comparison with sucrose. *Akt. Ernähr.-Med.* **12**, 210–214.
Stamm, J.W. *et al.* (1986) Integration of methods – Working group consensus report. *Journal of Dental Research* **64** (Spec. Iss.) 1537–1539.
Staudacher, W. and Kirchgessner, M. (1984) Protein- und Fettansatz sowie Energieverwertung bei Verfütterung von Palatinit® an wachsende Ratten. *Z. Tierphysiol., Tierernährung und Futtermittelkunde* **52**, 272–283.
Sträter, P.J. (1986) Palatinit®: Technological and processing characteristics. In: *Alternative Sweeteners.* (eds. L. O'Brien Nabors and R.C. Gelardi), Marcel Dekker, Inc. pp. 217–244.
Thiébaud, D., Jacot, E., Schmitz, H., Spengler, M. and Felber, J.P. (1984) Comparative study of isomalt and sucrose by means of continuous indirect calorimetry. *Metabolism* **33**, 808–813.
Vaaler, S., Bjørneklett, A., Jelling, I., Skrede, G., Hanssen, K.F., Fansa, O. and Aagenaes, Ø. (1987) Sorbitol as a sweetener in the diet of insulin-dependent diabetes. *Acta. Med. Scand.* **221**, 165–170.
Waalkens-Berendsen, D.H., Koëter, H.B.W.M., Schlüter, G. and Renhof, M. (1989) Developmental toxicity of isomalt in rats. *Fd Chem. Toxic.* **27**, 631–637.
Waalkens-Berendsen, D.H., Koeter, H.B.W.M. and van Marwijk, M.W. (1990a) Embryotoxicity/teratogenicity of isomalt in rabbits. *Fd Chem. Toxic.* **28**, 1–9.
Waalkens-Berendsen, D.H., Kiëter, H.B.W.M. and Sinkeldam, E.J. (1990b) Multigeneration reproduction study of isomalt in rats. *Fd Chem. Toxic.* **28**, 11–19.
van Weerden, E.J. and Huisman, J. (1993a) The digestion process of the sugar alcohol isomalt in the intestinal tract of the pig. 1. Studies with administration of isomalt in the feed. *British Journal of Nutrition* **69**, 455–466.
van Weerden, E.J. and Huisman, J. (1993b) The digestion process of the sugar alcohol isomalt in the intestinal tract of the pig. 2. Studies with administration of isomalt as a sweet. *British Journal of Nutrition* **69**, 467–479.
Würsch, P., Koellreutter, B. and Schweizer, T.F. (1989) Hydrogen excretion after ingestion of five different sugar alcohols and lactulose. *Eur. J. Clin. Nutr.* **43**, 819–825.
Young T.K., Lee, S.C. and Tai, L.M. (1980) Mannitol absorption and excretion in uraemic patients regularly treated with gastrointestinal perfusion. *Nephron* **25**, 112–116.
Ziesenitz, S.C. (1983) Bioavailability of glucose from Palatinit®. *Z. Ernährungswiss.* **22**, 185–194.
Ziesenitz, S.C. (1986) Zur Verwertung des Zuckeraustauschstoffes Palatinit® im Stoffwechsel. *Beitr. Infusionstherapie klin. Ernähr.* **16**, pp. 120–132 (Karger, Basel)
Ziesenitz, S.C. and Siebert, G. (1987) The metabolism and utilization of polyols and other bulk sweeteners compared with sugar. Vol. 3 (T.H. Grenby, ed.), Elsevier Applied Science, London. pp. 109–149.
Zinner, P.M. and Kirchgessner, M. (1982) Zur energetischen Verwertung von Palatinit®. *Z. Ernährungswiss.* **21**, 272–278.
Zumbé, A. and Brinkworth, R.A. (1992) Comparative studies of gastrointestinal tolerance and acceptability of milk chocolate containing either sucrose, isomalt or sorbitol in healthy consumers and Type II diabetics. *Z. Ernährungswiss.* **31**, 40–48.

7 Properties and applications of isomalt and other bulk sweeteners

I. WILLIBALD-ETTLE and H. SCHIWECK

7.1 Introduction

Beside traditional (bulk) sweeteners such as sucrose, glucose, fructose and corn or glucose syrups, newly developed sugar-free bulk sweeteners are nowadays widely used in the field of 'tooth-friendly' and calorie-reduced confectionery, baked products and pharmaceutical products. This category of bulk sweeteners comprises polyols such as sorbitol, mannitol, xylitol (all mono-saccharide alcohols), and isomalt, lactitol and maltitol (disaccharide alcohols).

Like sucrose, sugar-free bulk sweeteners give body and texture to the final product, but their sweetening power is less than that of sucrose (with the exception of xylitol), and they are metabolized differently from sucrose by the human body.

Polyols show three distinct physiological properties:

- kind-to-the-teeth or 'tooth-friendly'
- low in calories
- suitable for diabetics

More detailed information about the metabolism of bulk sweeteners can be found in Schiweck (1994a).

Polydextrose and inulin are also classified as bulk sweeteners and are used in certain applications, such as chocolate. In most cases their function is to substitute fat, and their physico-chemical properties differ greatly from those of the polyols. They will therefore not be covered in this chapter.

7.2 Manufacturing processes of mono- and disaccharide alcohols

Mono- and disaccharide alcohols are manufactured by hydrogenation under specific conditions (concentration of approx. 50%, high temperature of 100–150°C, high pressure of approx. 100 bar, Raney nickel catalyst). Depending on the starting material, hydrolysis precedes hydrogenation, e.g. if the starting material is starch. Maltitol syrups are obtained by

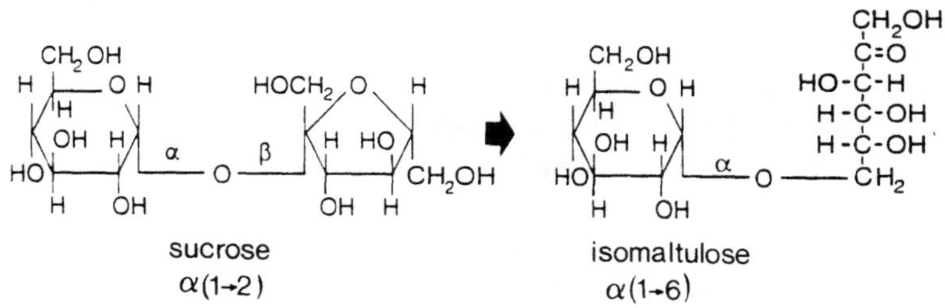

Figure 7.1 Production of isomaltulose by enzymatic transglucosidation.

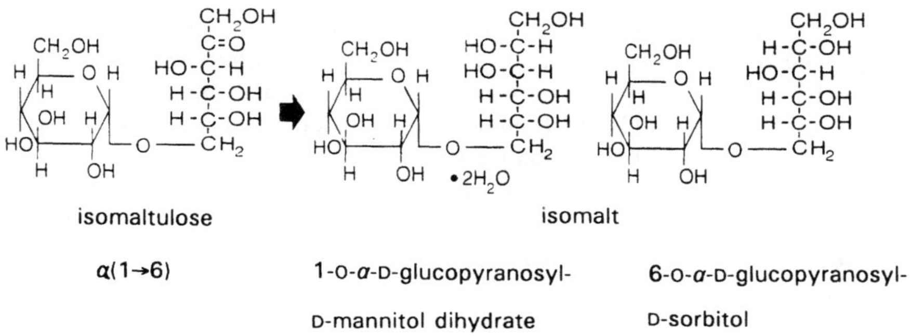

Figure 7.2 Production of isomalt by hydrogenation.

hydrogenation of high-maltose corn syrups. If the starting material is a ketose, then two isomers are obtained in a 1:1 ratio of sorbitol and mannitol. In the case of xylitol a pentose is used as starting material. A good summary of the manufacturing process of polyols can be found in *Ullmann* (Elvers *et al.*, 1994).

The manufacture of isomalt is an exception. Isomalt is produced in two steps from sucrose, the first step comprising enzymatic transglucosidation to the intermediate isomaltulose, which is then hydrogenated to isomalt (Schiweck, 1992; Kunz, 1994). Isomalt is a mixture of two isomers: 1-*O*-α-D-glucopyranosyl-D-mannitol dihydrate (1,1-GPM dihydrate) and 6-*O*-α-D-glucopyranosyl-D-sorbitol (1,6-GPS). The process is illustrated in Figures 7.1 and 7.2

7.3 Safety aspects

7.3.1 Regulatory situation

The use of polyols in the European Union is regulated by two directives: the Council Directive on Sweeteners for use in Foodstuffs (Sweeteners

Directive) (European Union, September 1994), which was adopted in June 1994, and the Directive on Food Additives other than Colours and Sweeteners (Miscellaneous Directive) (European Union, Common Position, March 1994), which was adopted in December 1994.

For sweetening purposes, which is the major sector of application, the Sweeteners Directive regulates the use of polyols in 'no sugar added' and 'energy-reduced' (30% reduction) products. The polyols sorbitol, mannitol, xylitol, isomalt, maltitol and lactitol may be used in a wide range of foodstuffs (all their major applications) as laid down in a positive list in the directive according to the *quantum satis* principle.

The use of polyols for technological purposes (purposes other than sweetening) is regulated in the Miscellaneous Directive. Here polyols are approved for use in foodstuffs other than drinks, in *quantum satis*.

In the USA, xylitol and mannitol are regulated as food additives, and sorbitol has affirmed GRAS (generally recognized as safe) status. Isomalt, lactitol, maltitol and hydrogenated starch hydrolysates have GRAS status on the basis of self-determination. Petitions for them have been accepted for filing by the FDA. A variety of products, mainly sugar-free confectionery, are on the market in the USA.

7.3.2 Caloric value

According to the European Union Nutrition Labelling Directive (European Union, 1990) the energy value of polyols for food labelling purposes is laid down as 10 kJ/g or 2.4 kcal/g. This is the average of the individual scientifically based values of the various polyols, which vary between 8 kJ/g and 15 kJ/g (Nutrition Council, 1987).

Contrary to the European Union system of a unitary value for all polyols, in the USA an individual approach was taken. The scientific evidence was reviewed and each polyol was evaluated by the American Society for Experimental Biology (FASEB, 1994). On the basis of this scientific evaluation, the FDA has issued letters stating that the agency will not object to the use of reduced caloric values for the various polyols as laid down in Table 7.1 (Calorie Control Council, Oct. and Dec. 1994).

7.4 Overview of bulk sweeteners

Table 7.2 gives a summary of the commonly used bulk sweeteners showing their chemical names and different grades. The polyols are available in powder form (crystalline). Sorbitol and maltitol are also produced as concentrated aqueous solutions with different purity levels.

Depending on their properties, polyols are used in different kinds of sugar-free confectionery, baked goods, pharmaceutical applications and

Table 7.1 Review of the current situation of energy values for polyols in the USA

Polyol	FASEB net energy (kcal/g)	FDA (kcal/g)	References
Isomalt	approx. 2	not >2.0	CCC[a] (Oct. 1994)
Lactitol	1.6–2.2	2.0	CCC[a] (Oct. 1994)
Maltitol	2.8–3.2	no request from petitioner to FDA	
Mannitol	1.6	2.0	
Sorbitol	1.8–3.3	2.6	CCC[a] (Dec. 1994)
Xylitol	approx. 2.4	2.4	CCC (Dec. 1994)
HSH[b]	2.8–3.2	3.0	CCC[a] (Dec. 1994)

[a] CCC = Calorie Control Councils.
[b] HSH = hydrogenated starch hydrolysate.

Table 7.2 Overview of bulk sweeteners

Chemical name	Grade
Monosaccharide alcohols	
Sorbitol	crystalline
Hydrogenated glucose syrup (HSH)	syrup, 70% dry matter
Xylitol	crystalline
Mannitol	crystalline
Disaccharide alcohols	
Isomalt	crystalline
Maltitol	crystalline
Hydrogenated corn syrup	syrup, 70–75% dry matter
Lactitol (mono- and dihydrate)	crystalline
Polysaccharides	
Polydextrose	powder/syrup, 70% dry matter
Inulin	powder

even non-food applications. In the following paragraphs the main properties and possible applications are described.

More detailed information is given in *Alternative Sweeteners* (Nabors *et al.*, 1991) and *Ullmann* (Elvers *et al.*, 1994).

7.4.1 Hygroscopicity

The hygroscopicity of a raw material is an important factor in determining storage conditions, shipment and internal handling in a plant. Last but not least, this information helps to estimate the final product's shelf-life.

The hygroscopicity of a raw material is described by a sorption isotherm.

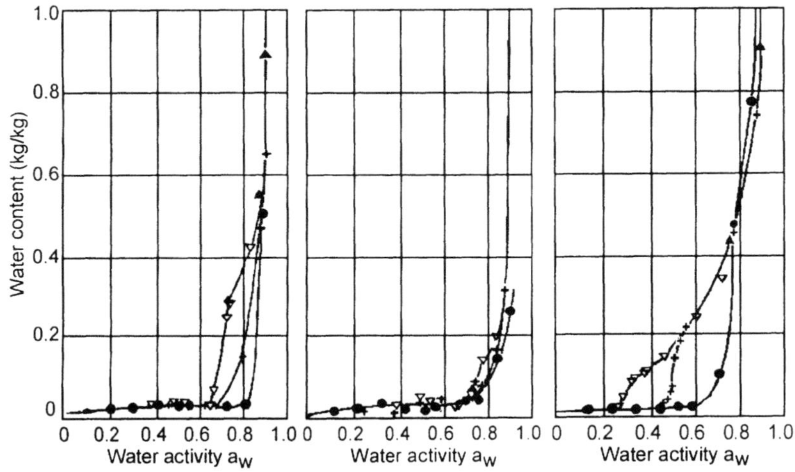

Figure 7.3 Water activity of three different polyols (from left to right): isomalt, mannitol, sorbitol. ●, 25°C hygrostat; +, 60°C SAP = sorption apparatus; ▼, 80°C SAP = sorption apparatus; ▲, 80°C hygrostat. *Source*: Weisser *et al.* (1982).

Mannitol, isomalt, lactitol, maltitol (crystalline) and xylitol show different levels of hygroscopicity, in increasing order. Sorbitol (crystalline) and maltitol syrup, followed by sorbitol syrup, exhibit the highest hygroscopicity levels (Weber, 1990). Figure 7.3 (Weisser *et al.*, 1982) shows sorption isotherm examples of three different bulk sweeteners. Sorbitol, used as a humectant, tends to pick up moisture at 25°C, starting at a relative humidity of 60%. Under the same conditions mannitol and isomalt virtually absorb no moisture. Noticeable moisture pick-up occurs only above a relative humidity of 80%.

Owing to this, isomalt is widely used in the field of hard candy processing. Isomalt-based melts have a very low hygroscopicity and remain stable over a long period. If 100% isomalt is used in hard candies, the amount of packaging material can be reduced significantly. No twist-wrapping is required. A convenient flip-top carton box is sufficient as packaging. Other polyols, such as maltitol syrup or sorbitol syrup, call for more protection. These include multi-layer-wrapping materials such as aluminium foils or single-flow packs etc. Due to the very high hygroscopicity of maltitol syrup, candies based on it exhibit a cold-flow formation during storage. Sorbitol syrup-based candies, however, tend to recrystallize easily. Candies solely based on lactitol or crystalline maltitol tend to recrystallize although their hygroscopicity is low. These polyols therefore need to be blended with another polyol which is non-crystallizing, e.g. maltitol syrup.

The shelf-life of 100% maltitol syrup-based candies can be improved significantly by blending isomalt with maltitol syrup. An isomalt content above 60% permits the twist-wrapping of candies instead of the use of single-flow-packs. Blends of 80% isomalt and 20% maltitol syrup result in a good shelf-life. However, these candies also have to be twist-wrapped.

Due to its very low hygroscopicity, mannitol is mainly used as a dusting agent in the manufacture of uncoated chewing gum to overcome stickiness. In the field of coated products, like coated chewing gum, isomalt provides excellent shelf-life. The coating keeps its crispy texture on account of reduced moisture pick-up. Crystalline sorbitol and sorbitol syrup are widely used in chewing gum base as humectants as well as in baked products or pharmaceutical and cosmetic applications.

7.4.2 Solubility

The solubility of the various polyols in comparison with sucrose as the temperature increases is shown in Figure 7.4 (Schiweck, 1994). It can be seen that the solubility of all the polyols increases with rising temperature. Sorbitol shows the highest solubility, followed by xylitol. The solubilities of maltitol and lactitol are lower. Mannitol exhibits the lowest and isomalt the second lowest solubility.

If the polyols are dissolved and boiled to a glass-like structure as in hard boilings, the solubility does not play a major role in the final product. Here the solubility of the raw material during processing is important – e.g. isomalt has to be predissolved at a high temperature to prepare a solution of approximately 80% dry matter required for the manufacture of hard candies. By contrast, maltitol- or sorbitol-syrup with a dry matter of approx. 75% can be used immediately.

The influence of solubility is important in food systems with a high moisture level, e.g. fruit preserves, gums (confectionery), pectin-based products or baked products. In these applications polyols such as sorbitol, lactitol or maltitol, which have a high solubility, are suitable.

In low-boiling applications the combination of a highly soluble polyol (non-crystallizing phase) and less soluble one (crystalline phase) is required to achieve the desired chewable product texture. A combination of maltitol syrup with e.g. isomalt is therefore recommended. The addition of a less soluble crystalline polyol, e.g. isomalt or mannitol, as seeding material after the boiling process, influences the low-boiling texture by slowing down the hardening process. Furthermore surface stickiness is reduced.

In applications like chewing gum, based on sorbitol for instance, polyols with a low solubility are added to improve texture. Both isomalt and mannitol provide a flexible chewing gum texture because they remain crystalline during processing and storage. The hardening process of an exclusively sorbitol-based chewing gum will therefore be slowed.

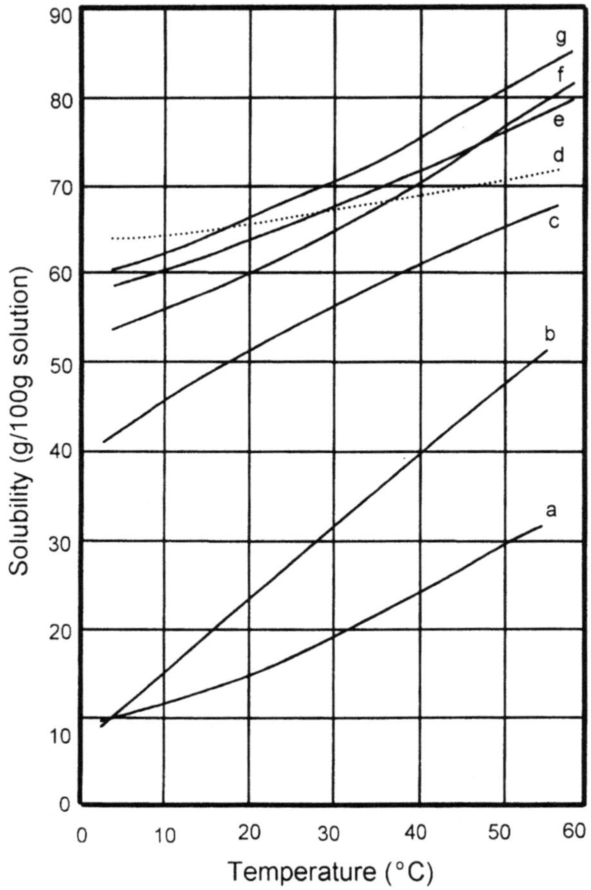

Figure 7.4 Solubility of different polyols with increasing temperature. a = mannitol; b = isomalt; c = lactitol; d = sucrose; e = maltitol; f = xylitol; g = sorbitol. *Source*: Schiweck (1994a).

7.4.3 *Viscosity*

Viscosity plays a decisive role in the manufacture of hard boilings and in coating processes.

Aqueous mono-saccharide alcohol solutions and melts generally show a lower viscosity than aqueous disaccharide alcohol solutions and melts. Compared with aqueous sugar solutions and melts, all polyols, whether in aqueous solutions or as melts, show a lower viscosity. Figure 7.5 shows the viscosities of isomalt compared with sugar melts at a water content $< 2\%$ as the temperature is raised (Mende, 1990).

The low viscosity of the isomalt melt in candy processing calls for slight modifications, such as cooling time extension for solidification of the melt.

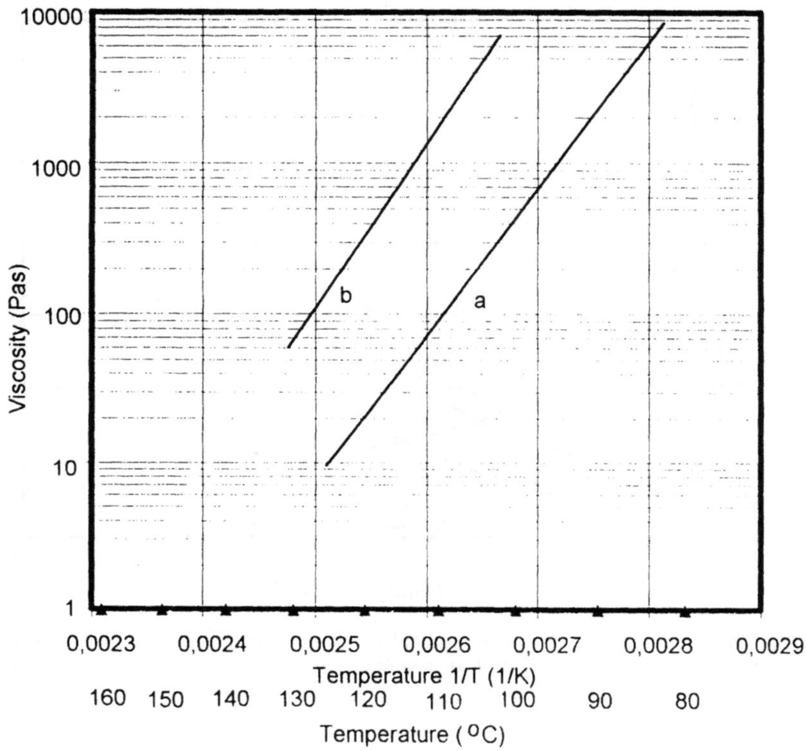

Figure 7.5 Viscosity of an isomalt melt in comparison to a sucrose melt measured with an oscillating rheometer. a = isomalt melt (H_2O = 1.13%); b = sucrose melt (H_2O = 2.1%). *Source*: Mende (1990).

To meet the need for extended cooling devices such as a cooling drum are used in addition to a cooling table or belt. An alternative method to overcome the low viscosity of the melt would be to boil the solution under vacuum as described by Mergelsberg (1989). With the aid of this technology the isomalt solution can be boiled at approx. 140°C as in sugar candy boiling, resulting in a lower temperature of the discharged melt. Within this temperature range the viscosity of an isomalt melt is comparable with that of sugar melt.

In the case of sorbitol hard boilings cannot be made using stamping technology because the melt would not solidify in a suitable time. In fact sorbitol candies can only be manufactured applying moulding technology. Xylitol also strongly influences the viscosity of the melt so its use in hard candies is limited.

The viscosity of coating solutions should not exceed 200 mPas, which is considered to be the critical value. This allows good distribution of the solution on the centres to be coated.

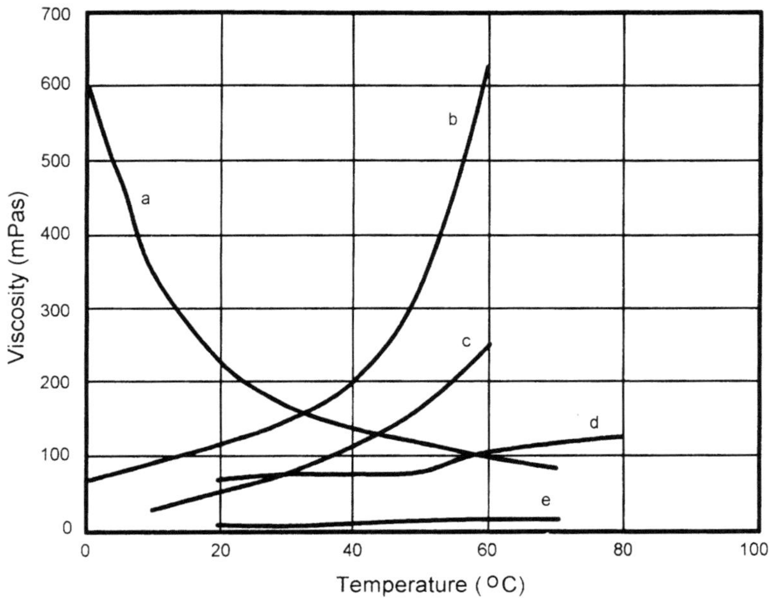

Figure 7.6 Viscosity of different saturated polyol solutions with increasing temperature. a = sucrose; b = sorbitol; c = xylitol; d = maltitol; e = isomalt. *Source*: Weber (1994).

Figure 7.6 (Weber, 1994) shows the changes in viscosity of different, saturated aqueous polyol solutions as a function of temperature compared with sucrose. A sorbitol solution exhibits an increase in viscosity above 200 mPas proportional to the increase of temperature. The viscosity of a maltitol and isomalt solution stays below the critical level, allowing a proper distribution of the solution on the centres to be coated.

7.4.4 Water activity

Water activity influences enzyme action, Maillard reaction, fat oxidation, microbial spoilage and finished product texture, which affect the shelf-life.

Water activity and osmotic pressure of aqueous disaccharide alcohol solutions are similar to those of comparable sucrose solutions. Aqueous monosaccharide alcohol solutions show lower water activity but higher osmotic pressure. Figure 7.7 (Schiweck, unpublished) shows the water activity of aqueous solutions of sorbitol and sucrose in comparison with glycerol, which has the lowest water activity and serves as a standard. The sorbitol solution exhibits a lower water activity than the sucrose solution, so it is widely used in high-moisture products in order to reduce their water activity.

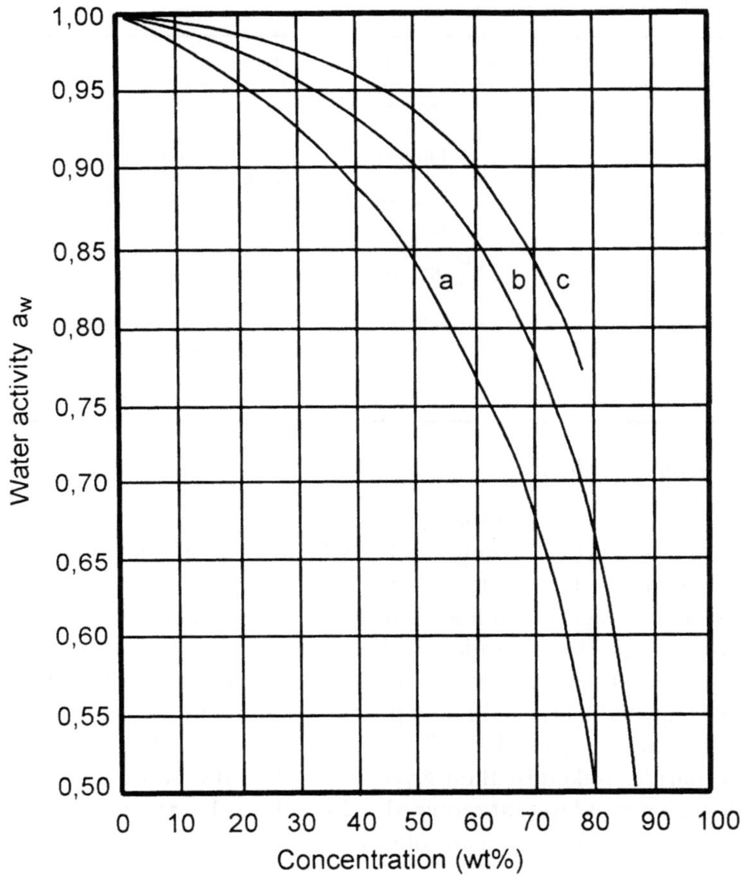

Figure 7.7 Water activity of different substances with increasing concentration. a = glycerol; b = sorbitol; c = sucrose. *Source*: Schiweck, unpublished.

7.4.5 Melting range

The melting ranges of all commonly used polyols are summarized in Table 7.3.

In descending order, mannitol, followed by isomalt and maltitol, show the highest melting ranges and are therefore the most stable to heat. Isomalt for example can therefore be used in extrusion processes to produce hard boilings etc.

Due to its low melting range sorbitol takes on a melt-like texture leading to a 'china-like' hardness and smooth surface of the tablet during the compressing process.

Table 7.3 Melting range of polyols

Chemical name	Melting range (°C)	References
Mannitol	165–168	*Handbuch Süßungsmittel* (Weber, 1990)
Maltitol	148–151	*Sugar-free Confectionery* (Deladrière, 1992)
Isomalt	145–150	
Lactitol (monohydrate)	121–123	*Alternative Sweeteners* (Nabors and Gelardi, 1991)
Sorbitol	97–101	*Sugar-free confectionery (Deladrière, 1992)*
Xylitol	92–96	*Handbuch Süßungsmittel* (Jäger, 1990)
Lactitol (dihydrate)	76–78	*Alternative Sweeteners* (Nabors and Gelardi, 1991)

7.4.6 Chemical stability

7.4.6.1 Acid stability. Mono- and disaccharide alcohols are stable to acids within the pH 3–7 range and stable to heat at the temperatures commonly employed in the food industry.

Under extreme acidic conditions, monosaccharide alcohols tend to form anhydrides. With disaccharide alcohols a splitting of the disaccharide linkage may occur. The energy required to hydrolyse a disaccharide, either by acid or enzymatically, is directly proportional to the energy of the bond linking the two saccharides. The energy required to hydrolyse two monosaccharides linked by their glycosidic hydroxyl groups, as in sucrose, is very low compared to that required to hydrolyse (1→6) linkages, such as are found in isomaltulose, gentiobiose and isomalt (Sträter and Irwin, 1991).

Isomalt and its two isomers, 1,1-GPM dihydrate and 1,6-GPS (see section 7.2 for full names) are very stable to acids. Figure 7.8 (Sträter and Irwin, 1991) shows the acid-stability of isomaltitol (1,6-GPS).

Sucrose dissolved in a 0.01 N solution of hydrochloric acid at 100°C is completely hydrolysed in less than 30 min., whereas to completely hydrolyse 1,6-GPS a 1.0 N solution of hydrochloric acid at 100°C for 2 h is needed. When 1,6-GPS is subjected to 0.1 N solution of hydrochloric acid at 100°C, it is only two-thirds hydrolysed after 10 h. Under the severest conditions that can be encountered during food production (pH 2; heating at 100°C for more than 1 h), less than 10% of 1,6-GPS will be hydrolysed. The disaccharide linkage of 1,1-GPM dihydrate is even more stable than that of 1,6-GPS. 1,1-GPM dihydrate dissolved in a 1.0 N solution of hydrochloric acid at 100°C requires more than 3 h for complete hydrolysis to occur (Sträter and Irwin, 1991).

In the pH 5–7 range maltitol and maltitol solutions are stable up to approx. 150°C. Lower pH values, temperatures above 50°C, and longer

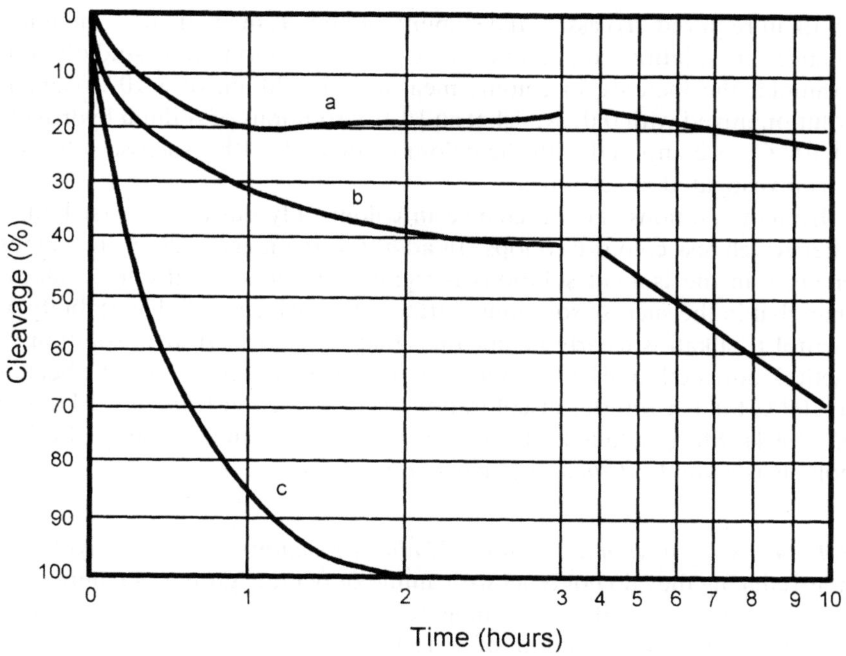

Figure 7.8 Hydrolysis of isomaltitol (1,6-GPS) with different acid concentrations measured over time. a = 0.01 N HCl; b = 0.1 N HCl; c = 1 N HCl. *Source*: Sträter and Irwin (1991).

contact times lead to cleavage of the disaccharide bond (Schiweck, 1994c).

The stability of lactitol to acids is similar to that of lactose. When heating 10% solutions to 100°C for 4 h at pH 1 or 2, 5.6% or 1.4% respectively of the lactitol is hydrolysed (Schiweck, 1994b).

7.4.6.2 Enzymatic stability. Disaccharide alcohols in general are not susceptible to enzymatic hydrolysis, and the rate is very slow.

The disaccharide alcohols must be cleaved to monosaccharides before microbiological activity begins. It is not surprising, therefore, that isomalt, lactitol or maltitol do not serve as a substrate for most organisms (yeast, mould and bacteria) associated with food. For example, the cleavage rate of 1,6-GPS and 1,1-GPM dihydrate, the two isomers of isomalt, by yeast-α-glucosidases is approx. 10^3 times less than that of maltose. Under normal conditions, the enzymatic cleavage of 1,6-GPS and 1,1-GPM dihydrate can be considered negligible (Sträter and Irwin, 1991).

Lactitol also can be cleaved only with great difficulty. The cleavage rate was found to be about 10% of lactose (Schiweck, 1994b).

7.4.6.3 Alkaline stability. In most cases the preparation of food takes place in an acidic or a neutral range. In only a few applications is stability to alkali required. To assess the stability of polyols in an alkaline medium, changes in solution colour and lactic acid formation were studied. For example, the increase in colour, measured at 420 nm of a 10% isomalt solution mixed with 0.01 N, 0.1 N and 1.0 N solutions of sodium hydroxide at 100°C, as compared with the colour change of a 10% sucrose solution, was investigated at set times.

Isomalt solutions show a change in colour only during the first hours, after which less colour develops. In addition, the magnitude of the colour changes in the isomalt solutions is significantly less (in absolute values) than that in the sucrose solutions (Sträter and Irwin, 1991). The stability of lactitol to alkali is clearly higher than that of lactose. A 10% solution of lactitol, adjusted to pH 13 with a sodium hydroxide solution, can be heated to 100°C for 1 h without discoloration, as described in *Ullmann* (Elvers *et al.*, 1994). More detailed information about the alkaline stability of other polyols is given by *Ullmann* (Elvers *et al.*, 1994).

7.4.6.4 Non-enzymatic reaction (Maillard reaction). As all polyols are non-reducing they do not react with amino or peptide groups, so there is no browning reaction. If a browning reaction is required e.g. for baked products, then the addition of a reducing sugar, such as fructose, is recommended.

Generally the Maillard reaction and Strecker's degradation are responsible for the typical caramel-type flavour of the final product in food processing. This typical flavour, which also arises during sugar candy processing, can disguise added flavours such as apple or peach in candy.

Among polyols which do not undergo browning reactions, isomalt in particular allows the added flavour compounds to develop their full strengths, so that new flavours such as apple or peach are now possible in the sugar-free candy market.

7.5 Sensorial properties

In addition to not contributing to the caramel flavour development, the sweetening profile of each individual polyol is of great importance for the product developer. The sweetening profile of a substance depends on its sweetening power and its negative heat of solution value, or cooling effect. This is observed during the dissolving process in which the substances draw energy from their environment, the mouth, to dissolve.

Table 7.4 (Bär, 1991) summarizes the sweetening power of the polyols. Xylitol shows the highest sweetening power of all polyols, comparable in sweetness to sucrose. Lactitol shows the lowest sweetening power. Combinations of polyols have a synergistic effect. This is also apparent

Table 7.4 Relative sweetness of sugar alcohols in water (*c.* 10 w/v) (sucrose = 1.0)

Polyol	Relative sweetness
Xylitol	1.0
Maltitol	0.8
HSH	0.6
Sorbitol	0.6
Isomalt	0.55
Mannitol	0.5
Lactitol	0.35

Table 7.5 Negative enthalpy of solutions of polyols in comparison with sucrose

Polyol	kJ/mol	kJ/kg	References
Sucrose	−6.21	−18.16	Counsell (1978)
Isomalt	−14.60	−39.40	
Lactitol (monohydrate)	−18.86	−52.10	Uyl and Velthuijsen (1990)
Lactitol (dihydrate)	−22.10	−58.10	Uyl and Velthuijsen (1990)
Maltitol	−27.24	−79.12	Deladrière (1992)
Sorbitol	−20.20	−110.99	Benson (1978)
Mannitol	−22.00	−120.88	Benson (1978)
Xylitol	−23.27	−153.07	Counsell (1978)

when polyols are combined with intensive sweeteners such as acesulfame-K or aspartame, etc. The amounts of intensive sweeteners to be used depend on the amount of polyols in each application.

Table 7.5 summarizes the negative heats of solution of different polyols. Isomalt is closest to sucrose, and dissolved in the mouth, does not have any cooling effect, so it is suitable for applications such as chocolate, baked products or any fruit-flavour applications, when no cooling effect is desired. However for peppermint or menthol products, in which a high cooling effect is needed, xylitol is widely used, especially in chewing gum.

7.6 Market acceptance

In recent years there has been a constant demand for products which are sugar-free, kind-to-the-teeth, calorie-reduced and suitable for diabetics. Such products should provide pleasure and enjoyment based on their taste and appearance. The success of a sugar-free confectionery product is therefore strongly dependent on its taste which, in turn, depends on the bulk sweetener used.

The best developed sugar-free confectionary market is that for chewing gum, particularly in Scandinavian countries, where the sugar-free chewing gum share of the market exceeds 90% of the total value (Nielsen, 1993), and is primarily consumed for its property of curbing tooth decay.

A similar trend can be observed in the hard candy market segment. Recent studies of the Swiss confectionery market show that almost 50% by value of the total hard candy market is sugar-free (IHA, 1993). The German sugar-free candy market share rose from 17.9% in the first half of 1993 to 20% in 1994. Sugar-free hard candy sales increased by almost 14% (GfK, 1994). Due to the advantages of an excellent sugar-like taste and its sweetness profile, one of the most widely used sugar replacers in the hard candy field is isomalt, with low hygroscopicity permitting the use of convenient packagings.

References

Bär, A. (1991) Xylitol, in *Alternative Sweeteners*, 2nd edn (eds L.O'B. Nabors and R.C. Gelardi), Marcel Dekker, Inc., New York, p. 351.

Benson, F.R. (1978) Alcohols, Polyhydric (Sugar), in *Kirk-Othmer Encyclopedia of Chemical Technology*, 3rd edn. Vol. 1 (eds M. Grayson and D. Eckroth), John Wiley & Sons, p. 754.

Calorie Control Council (Oct. 1994) FDA OK's self-determined caloric value for lactitol, isomalt, in *Calorie Control Council Focus*, Atlanta.

Calorie Control Council (Dec. 1994) FDA Okays self-determined caloric value for hydrogenated starch hydrolysates, sorbitol and xylitol, in *Calorie Control Council Focus*, Atlanta.

Counsell, J.N. (1978) *Xylitol*, Applied Science Publishers, London, p. 7.

Deladrière, N. (ed.) (1992) *Cerestar Sugar-Free Confectionery*, RCB 496.749, Cerestar International Sales, Brussels.

Elvers, B., Hawkins, S. and Russey W. (eds) (1994) *Ullmann's Encyclopedia of Industrial Chemistry*, 5th edn. Vol. A 25, VCH Verlagsgesellschaft, Weinheim, pp. 413–437.

European Union (1990) Council Directive Nutrition Labelling for Foodstuffs, No. 90/496/EEC, *Official Journal of the European Communities*.

European Union (March 1994) Miscellaneous Directive, Common Position on food additives other than colours and sweeteners, No. EC/17/94, *Official Journal of the European Communities*.

European Union (September 1994) Council Directive on sweeteners for use in foodstuffs, No. 94/35/EC, *Official Journal of the European Communities*.

FASEB (Federation of American Societies for Experimental Biology) (June 1994) *The Evaluation of the Energy of Certain Sugar Alcohols used as Food Ingredients*, Life Science Research Office, Bethesda, Maryland, USA.

GfK (1994) *Handelspanel der GfK über den zuckerfreien Bonbonsmarkt in Deutschland*, GfK Panel Services GmbH, Nürnberg.

IHA (1993) *Konsumentenpanel der IHA über den zuckerfreien Bonbonsmarkt in der Schweiz*, IHA Institut für Marktanalysen AG, Hergiswil, Switzerland.

Jäger, A. (1990) Zuckeralkohole, Kapitel IV, in *Handbuch Süßungsmittel, Eigenschaften und Anwendungen* (eds G.-W. von Raymon Lipinski and H. Schiweck), Behr's Verlag, Hamburg, p. 297.

Kunz, M. (1994) Sugar Alcohols, Chapter 5, in *Ullmann's Encyclopedia of Industrial Chemistry*, 5th edn. Vol. A 25, (eds B. Elvers, S. Hawkins and W. Russey), VCH Verlagsgesellschaft, Weinheim, pp. 426–429, 436.

Mende, K. (1990) *Thermodynamische und rheologische Untersuchungen an Saccharidlösungen und -schmelzen*. Diss. TU Berlin.

Mergelsberg, R. (1989) Zuckerfreie Hartkaramellen mit neuem Verfahren herstellen, *Zucker- und Süßwaren Wirtschaft*, **11**, 362–363.

Nabors, L.O'B. and Gelardi, R.C. (eds) (1991) *Alternative Sweeteners*, Marcel Dekker, Inc., New York.

Nielsen, A.C. (1993) *Int. panel data of sugarfree chewing gum market shares*, Nielsen GmbH, Frankfurt.

Nutrition Council (June 1987) *The Energy Value of Sugar Alcohols*, Recommendation of the Committee on Polyalcohols, The Hague, The Netherlands.

Schiweck, H. (1992) Isomalt (Palatinit®), A Versatile Alternative Sweetener – Production, Properties and Uses, in *Carbohydrates in Industrial Synthesis* (ed. M.A. Clarke), Verlag Dr. Herbert Bartens, Berlin, pp. 37–54.

Schiweck, H. (1994a) Sugar Alcohols, Chapter 1, in *Ullmann's Encyclopedia of Industrial Chemistry*, 5th edn. Vol. A25, (eds B. Elvers, S. Hawkins and W. Russey), VCH Verlagsgesellschaft, Weinheim, pp. 413–416, 432–433.

Schiweck, H. (1994b) Sugar Alcohols, Chapter 6, in *Ullmann's Encyclopedia of Industrial Chemistry*, 5th edn, Vol. A 25, (eds B. Elvers, S. Hawkins and W. Russey), VCH Verlagsgesellschaft, Weinheim, pp. 429–430, 436.

Schiweck, H. (1994c) Sugar Alcohols, Chapter 7, in *Ullmann's Encyclopedia of Industrial Chemistry*, 5th edn. Vol. A 25, (eds B. Elvers, S. Hawkins and W. Russey), VCH Verlagsgesellschaft, Weinheim, pp. 431–432, 436–437.

Sträter, P.J. and Irwin, W.E. (1991) Isomalt, in *Alternative Sweeteners*, 2nd edn, (eds L.O'B. Nabors and R.C. Gelardi), Marcel Dekker, Inc., New York, pp. 309–332.

Uyl, H. den and Velthuijsen, J.A. van (1990) Zuckeralkohole, Kapitel IV, in *Handbuch Süßungsmittel, Eigenschaften und Anwendungen* (eds G.-W. von Raymon Lipinski and H. Schiweck), Behr's Verlag, Hamburg, p. 359.

Weber, W. (1990) Zuckeralkohole, Kapitel IV, in *Handbuch Süßungsmittel, Eigenschaften und Anwendungen* eds G.-W. von Raymon Lipinski and H. Schiweck), Behr's Verlag, Hamburg, p. 325–351.

Weber, W. (1994) Naschen ohne Karies, *Süßwaren 1–2*, 12–13.

Weisser, H., Weber, H. and Loucin, M. (1982) Wasserdampf-Sorptionsisothermen von Zuckeraustauschstoffen im Temperaturbereich von 25 bis 80°C, *Internationale Zeitschrift für Lebensmittel-Technologie und -Verfahrenstechnik* **33**(2), 89–97.

8 Nutritional properties and applications of erythritol: a unique combination?

J. GOOSSENS and M. GONZE

8.1 Introduction

In pursuit of well-being food manufacturers try to fulfil our desire for tasty and convenient foods. Modern lifestyle however has changed eating behaviour and advances in medical sciences have extended life expectancy. These two major changes have meant that we have become increasingly vulnerable to so-called 'civilised world' diseases such as caries, diabetes, cancers and rheumatoid diseases.

It has recently been recognised that the change in our food supply may have played an important role in this development (O'Connor, 1992). There has been increasing awareness of the potential role of foods in the body's delicate equilibrium between health, enjoyment and performance, which help to determine the quality of life. Over the last 20 years, food science, nutrition and medicine have played a part in the search to discover the underlying mechanisms through which nutrition plays a role in preventive health care, but despite some progress our understanding remains fragmentary. One of the earliest findings was that calorie intake tends to be excessive, resulting in an increase in obesity. Nutrition guidelines therefore advise us to reduce calorie intake, especially from fats, and to eat more carbohydrates, particularly complex ones (e.g. fibrous foods instead of mono- or disaccharides). Long before this however, consumers were already attracted by low-calorie and sugar-free foods albeit not primarily for health reasons but rather in search of modern society's ideals of beauty and slimness. The importance of this trend is clear from the number of new 'Light' foods launched. Of all new food products introduced over a period of one year, almost 20% are claimed to have one or more health advantages, ranging from simple calorie reduction to enrichment with ingredients to prevent disease. So far, probably the most successful field where health benefits have created a new market is in the sugar-free product market sector.

Sugar replacement not only achieves a reduction in calories, but the products are generally also tooth-friendly, which means that unlike sugar, they do not play a role in the development of dental caries. The excessive consumption of sugar-based products by children, especially in the confectionery market, and the detrimental effect this has on dental health is of major concern. Sugar-free products offered an alternative. Although a wide range of sugar replacements is now available and some, like xylitol can even claim a caries-reducing effect (Bär, 1988), their use with adapted processing technology still yields finished products of insufficient quality in many cases.

Some of the main shortcomings are:

- limited calorie reduction (EU requires minimum 30%)
- side effects which limit the level of use
- too low a solubility
- difference in taste quality (e.g. in combination with intense sweeteners)
- lack of storage stability
- insufficient texture and bulking properties

In order to resolve these problems, new ingredients are required, one of which is erythritol. This offers new possibilities, both in functional as well as nutritional properties.

8.2 Chemical and physical properties

Erythritol is a linear carbohydrate molecule of four carbon atoms each carrying one hydroxyl group (Figure 8.1).

Chemically, erythritol therefore is classified as a monosaccharide polyol like sorbitol, mannitol, xylitol and glycerol. Erythritol is also a symmetrical molecule and therefore exists only in one form, the *meso*-form. It forms anhydrous crystals with a moderately sweet taste without off-taste or

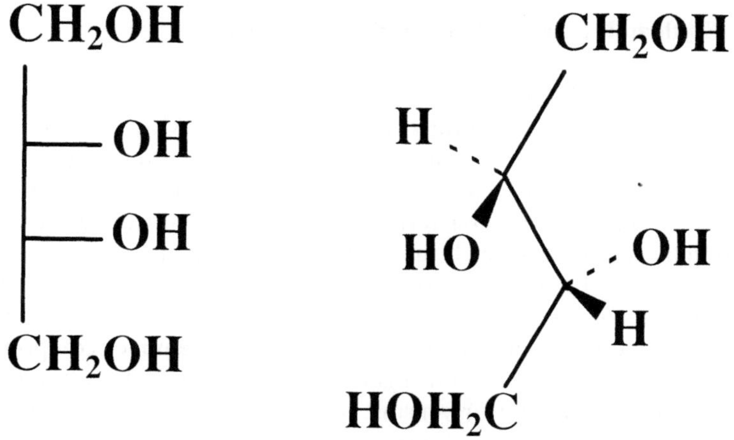

Figure 8.1 Molecular structure and chemical formula of erythritol.

	Erythritol	Xylitol	Mannitol	Sorbitol	Maltitol	Isomalt	Lactitol
Carbon n°	4	5	6	6	12	12	12
Molecular weight	122	152	182	182	344	344	344
Melting point °C	126	94	165	97	150	145–150	122
Heat of solution	−23.3	−36.5	−28.5	−26	−18.9	−9.4	−13.9
Heat stability °C	>160	>160	>160	>160	>160	>160	>160
Acid stability pH	2–12	2–10	2–10	2–10	2–10	2–10	<3
Solubility ww%(25°)	37	64	20	70	60	25	57

Figure 8.2 Main chemical and physical properties of erythritol in comparison to other low calorie bulk sweeteners.

odours. The powder has a transparent white, brilliant appearance and dissolves in water to give a colourless, non-viscous solution. The crystals melt at 122°C to form a colourless and brilliant non-viscous melt.

Erythritol's chemical properties are similar to those of other polyols in that it has no reducing end-groups and thus has excellent heat and acid stability. It differs in having a low solubility, a property which among these simple polyols it shares only with mannitol. Heat of solution is very low, a property shared only with xylitol. However, compared with the group of polyols at present used as sugar-replacers, erythritol has the lowest molecular weight, which of course furnishes it with different properties, such as higher osmotic pressure and water activity in solution.

The most important and special nutritional properties which differentiate erythritol from other polyols are due to its small molecular size. A summary of some of these properties and a comparison with six other polyols are set out in Figure 8.2.

8.3 Erythritol, a naturally occurring substance

Erythritol is present in a wide range of microorganisms, plants and animals as well as occurring naturally in many foods. Most of its reported occurrence seems to be as an end-product of the metabolism of fungi (Yoshida *et al.*, 1984), yeasts (Spencer, 1968; Hattori and Suzuki, 1974; Sponholz *et al.*, 1986) and bacteria (Firme *et al.*, 1994; Cunha *et al.*, 1992). Osmophilic organisms such as yeasts of the genera *Moniliella* and *Leuconostoc* or the fungus *Aureobasidium* have been reported to produce large amounts of erythritol under conditions of oxygen limitation. In many other fungi and yeasts erythritol is often detected as a minor compound in a complex mixture of sugars and sugar-alcohols. Erythritol is often found in minor amounts in higher organisms, for example in the pollen of flowers, fruits like melon, grapes and pears (Shindou *et al.*, 1989) or in the blood and amniotic fluids of cows and other mammals (Roberts *et al.*, 1976). In the human body, traces of erythritol have been detected in the lens tissue (Tomana *et al.*, 1984), in cerebrospinal fluid (Servo *et al.*, 1977), in seminal plasma (Størsel *et al.*, 1978) and in blood serum (Roboz *et al.*, 1984). Like other sugar alcohols, it is also a minor constituent of urine (Leutner, 1981). It is the most abundant polyol in human urine, with reported concentrations of 10–30 mg/l (see Figure 8.3).

At the current state of our knowledge, it is not clear whether erythritol is a normal by-product from plant or mammalian carbohydrate metabolism or the result of fermentation by integrated microorganisms such as the gut microflora.

As a microbial metabolite, it is not surprising to find erythritol in a wide range of foods, especially fermented foods such as wine and sake, soy sauce and miso bean paste (Shindon *et al.*, 1988). More recently, detailed analysis of the sugar alcohol spectrum of wines, sherries and other spirits has been used as a tool to establish authenticity of products (Gomez-Cordoves and Hernandez, 1987; Estrella *et al.*, 1986: Sponholz and Dittrich, 1985; Ribereau-Gayon and Bertrand, 1972). Figure 8.4 lists some concentrations detected in various foods, giving an indication of our erythritol intake from regular foods.

Polyols in human urine	Excretion in μmol/day
Sorbitol	53
Mannitol	153
Ribitol	53
Xylitol	53
Arabitol	293
Erythritol	912
Threitol	10

Figure 8.3 The natural occurrence of polyols: daily output in mg in human urine.

Foods	Erythritol content
Wine	130–300 mg/l
Sherry wine	70 mg/l
Sake	1550 mg/l
Soy sauce	910 mg/l
Miso bean paste	1310 mg/kg
Melons	22–47 mg/kg
Pears	0–40 mg/kg
Grapes	0–42 mg/kg

Figure 8.4 The natural occurrence of polyols in various foods.

8.4 An entirely biotechnological process

Polyols are currently produced industrially by acid or enzymatic conversion of the parent carbohydrate followed by final catalytical hydrogenation under conditions of high pressure/temperature. For some polyols, e.g. xylitol, alternative methods based on fermentation and enzymatic conversion have been described (Meyrial *et al.*, 1991).

Erythritol is the first polyol to be manufactured by an entirely biotechnological process, based on the ability of some yeasts or fungi, especially osmophilic species, to produce it in large amounts. The starting material is a glucose-rich substrate obtained by enzymatic hydrolysis from natural raw materials such as starch or sugar. Glucose is then fermented by an osmophilic yeast (*Moniliella* sp., *Trichosporonoides* sp.) or fungus (*Aureobasidium* sp.) to yield a mixture of polyols containing mainly erythritol, glycerol and ribitol, with other polyols in trace amounts (Aoli *et al.*, 1993; Ishizuka *et al.*, 1989; Sasaki, 1989).

The ratio of these polyols may be influenced by carefully controlling the growth conditions. Because of the osmophilic organisms, the fermentation can be extended by continuous feeding of glucose to reach a very high dry substance of up to 45% in the fermentation broth, which obviously reduces process costs. Erythritol is crystallised at over 99% purity from the filtered and concentrated fermentation broth.

8.5 Nutritional properties

Although chemically a polyol, erythritol is such a small molecule that it behaves very differently from other polyols in the way it passes through the animal and human digestive systems. Its resulting nutritional properties are therefore unique and offer new opportunities for food formulation.

Four key nutritional properties are united in erythritol. These are:

- very low caloric value, less than 0.4 kcal/g (only 10% of the calorific value of sugar)
- higher digestive tolerance
- suitability for diabetics
- non-cariogenicity

Each of these properties is related to its utilisation, absorption, fermentation, metabolism and excretion by the human digestive system, in which erythritol behaves differently from all other bulk sweeteners. Much scientific evidence has been accumulated on its nutritional profile, and this article will give an overview of our present knowledge.

8.5.1 Carbohydrates in the digestive system

The fate of dietary carbohydrates in the digestive system varies depending on their molecular size and chemical nature. The simple monosaccharides are directly absorbed through the mucosal cell layer of the intestine and their rate of absorption may depend upon both active and passive uptake mechanisms. Glucose for example is actively – and therefore rapidly – absorbed compared to fructose which is absorbed through a facilitated diffusion mechanism. Disaccharides first need to be hydrolysed by the intestinal enzymatic systems before they can be absorbed. Polysaccharides such as starch or glycogen, due to their complex, often branched, structures, are absorbed only to the extent that the digestive enzymatic system is capable of liberating their monosaccharide building blocks. Absorbed monosaccharides are transported in the body and are oxidised to CO_2, providing energy, or are metabolised into other substances such as glycogen or fats for energy storage. The remaining undegraded polysaccharide structures or oligosaccharide residues in the gut are then subjected to microbial fermentation in the large intestine, where the resulting production and subsequent absorption of volatile fatty acids (VFA) may contribute additional energy. At this stage gases such as methane and hydrogen are also produced which can be detected in the breath but are mainly excreted as flatus (Flourié, 1989; Southgate, 1989; Levin, 1989). Any undigested carbohydrate is excreted with the microbial biomass in the faeces. Figure 8.5 gives a schematic representation of the different routes carbohydrates can follow.

The metabolism of new food ingredients needs to be considered carefully, as it will determine both calorific value and tolerance compared with traditional ingredients. The wide range of sugar-replacers available today, e.g. polyols or polydextrose, has led to the successful development of a range of sugar-free confectionery products. However, it becomes more and more difficult to expand their range of application because intake levels can reach the limit where side-effects such as flatulence or diarrhoea cause physical or social inconvenience.

The reason is that the mucosal cells in the digestive system cannot actively transport sugar-replacers through the cell membrane and therefore absorption is much slower compared with common sweeteners such as sucrose and glucose. Sorbitol, mannitol and xylitol are absorbed by diffusion and except for mannitol are directly converted and contribute energy to the body. Maltitol, isomalt, lactitol, the disaccharide polyols and polydextrose are partially hydrolysed to their sorbitol, mannitol and glucose moieties which in turn can be absorbed.

Because these processes take place at a slow rate, the small intestine is exposed for a longer period to a high osmotic load, causing water influx from the mucosal wall into the intestinal lumen, resulting in physical pain

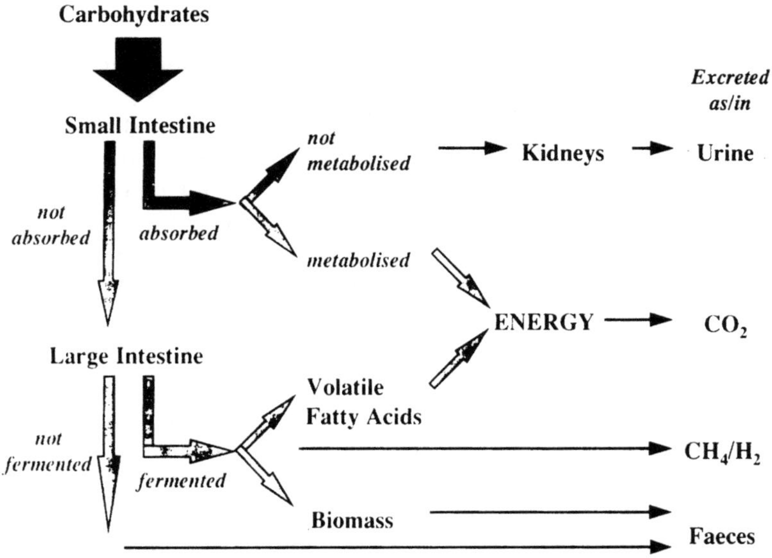

Figure 8.5 Schematic presentation of the digestive routes for carbohydrates in the human body.

and diarrhoea. This also means that a substantial length of time is taken to pass through the small intestine without hydrolysis or absorption. In the large intestine, these materials serve as good substrates for gut colonic microflora which ferment them to yield volatile fatty acids (VFA) and gases (H_2; CH_4). They also renew microbial biomass. The VFAs are absorbed and further metabolised in the liver, whereas the remaining energy in the gases and the biomass are excreted as flatus and faeces.

8.5.2 The metabolic fate of erythritol

In the search for new substances providing reduced calories and with minimal potential side-effects, initial efforts were focused on molecules completely inert to the digestive system, passing through the gut unabsorbed and unfermented. Erythritol however offers a surprising solution, as it is absorbed quickly and completely but is not subject to metabolic conversion in the body. A great number of animal and human studies have been carried out to elucidate its metabolism.

Most of the details on the metabolic fate of erythritol have been obtained from feeding experiments with rats using the ^{14}C-labelling technique. After oral or intravenous administration, the fate of the ^{14}C from labelled erythritol was followed in the breath, the urine and the faeces (Noda and Oku 1990a,b, 1992). When male rats were fed a 0.1 g/kg dose of labelled erythritol, 6% of the total radioactivity was exhaled as $^{14}CO_2$ in the breath and 88% was excreted unchanged in the urine within 24 h. Intravenous administration reduced these levels to 1% in the breath and 94% in the urine, indicating that some of the $^{14}CO_2$ recovered after oral feeding must have originated from microbial fermentation. As adaptation of the colonic flora to substrates is a well known phenomenon, rats were fed a diet containing 10% erythritol for 2 weeks, after which $^{14}CO_2$ in the breath increased to 10%. This adaptation was further confirmed by *in vitro* fermentation assays on ^{14}C-erythritol: examination of the faeces of adapted rats showed that over 20% was converted to $^{14}CO_2$ and 60% to volatile fatty acids.

Similar experiments (unpublished) with normal and germ-free rats confirmed that the gut microflora is almost solely responsible for the CO_2 production from erythritol. Ultra-pure ^{14}C-erythritol was prepared for feeding trial by removing the 2% ^{14}C-erythrose contamination. The results demonstrated that 98% of the ^{14}C radioactivity was recovered in the urine within 24 h, indicating that erythritol must be absorbed extremely quickly from the small intestine and excreted in the urine, and that only the minor amounts reaching the colon are fermented.

A similar pattern of metabolic behaviour was observed in the human digestive tract. In this case, the increased level of the natural ^{13}C-isotope occurring in erythritol was used to trace its metabolism (Hiele *et al.*, 1989, 1993). Six human volunteers were given a 25 g oral dose of erythritol. There was no increase in $^{13}CO_2$ in the breath in the following 6 h sampling period, whereas similar doses of glucose, xylose or lactitol resulted in ^{13}C (as CO_2) recovery levels of 28, 23 and 14% respectively (see Figure 8.6).

The experiment also showed a slower $^{13}CO_2$ response from lactitol than from glucose, clearly indicating that its metabolism is delayed until it is fermented in the colon. This fermentative activity could be confirmed by measuring the hydrogen which appears in the breath in the same tests. Both xylose and (especially) lactitol are known to undergo colonic fermentation, and produce hydrogen in the breath, whereas glucose and erythritol gave no response at all (Figure 8.7).

The absence of any metabolism of erythritol was confirmed by examination of the urinary excretion pattern. Over 50% of the ingested erythritol dose was recovered in the urine after 6 h, rising to an average of 84% after a 24 h period. Other experiments (unpublished) have independently confirmed the excretion pattern for erythritol, showing that after oral doses erythritol concentration in the blood-stream rises sharply, followed by a steady decline reflecting the high renal filtration rate of erythritol from the blood into the urine (Figure 8.8).

It is clear from these results that erythritol is absorbed from the intestinal lumen extremely quickly. Since there are no specific membrane transport mechanisms known to enhance erythritol uptake, erythritol must be

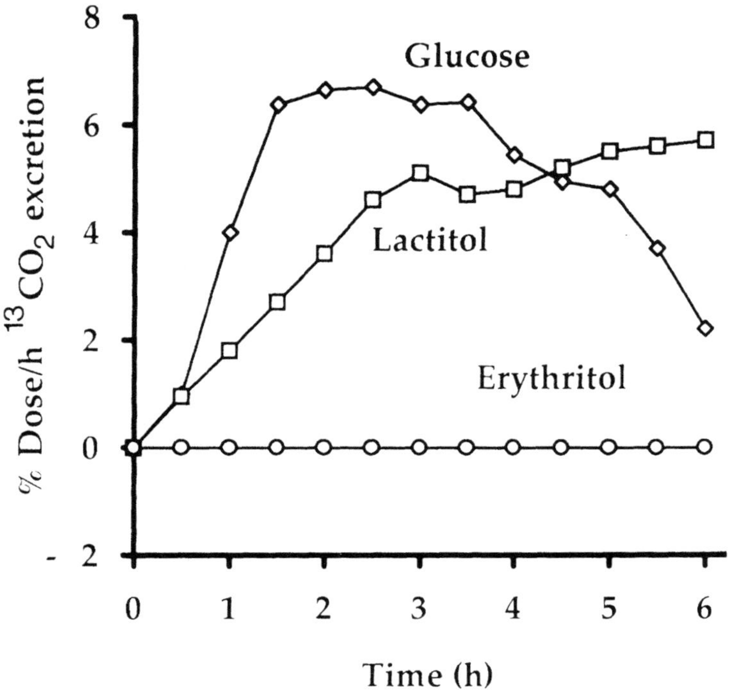

Figure 8.6 Excretion of $^{13}CO_2$ in the breath expressed as the percentage of the administered dose after a 25 g single oral intake of ^{13}C-labelled glucose, lactitol or erythritol. Data represent the average of 6 healthy human volunteers measured over a 6 h period following intake.

absorbed by passive diffusion at very high speed. It is known that the permeability coefficient of sugars and sugar alcohols is related to their carbon number (Burkitt *et al.*, 1972). Only substances with a molecular size in the range of the membrane pore diameter may penetrate the cell membrane without apparent resistance, which is apparently the case for the small, uncharged molecule of erythritol. Driven by the osmotic pressure it causes in the small intestine, erythritol quickly penetrates the mucosal cells and passes into the blood, rapidly restoring the osmotic equilibrium between intestinal lumen and tissues.

It can be concluded that erythritol behaves completely differently from other known sugar replacers and is not comparable to the other polyols or polydextrose. Because of this, an extensive study programme has been undertaken to establish beyond any doubt that erythritol is also a completely safe substance for the replacement of other sweeteners in reduced-calorie and sugar-free foods.

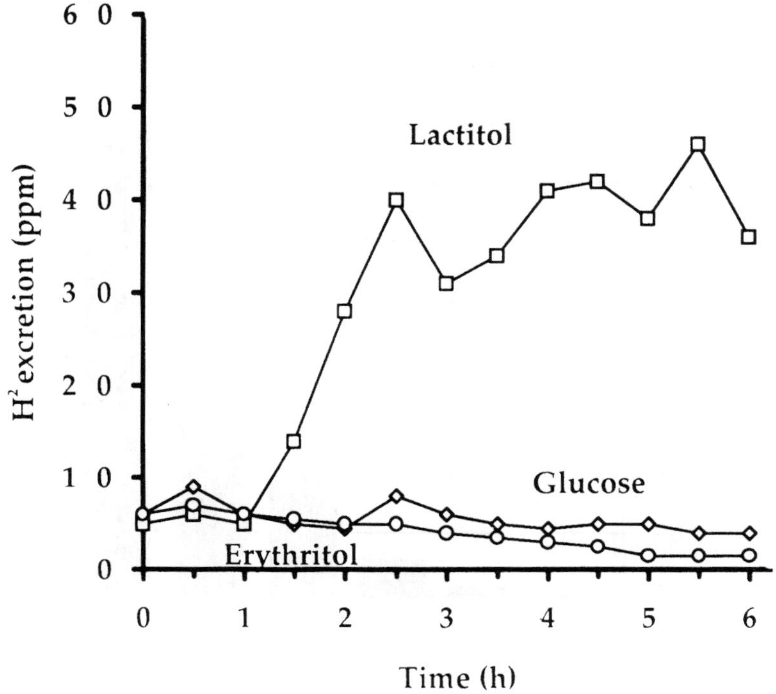

Figure 8.7 Hydrogen concentration excreted in the breath after a 25 g single oral intake of glucose, lactitol or erythritol. Data represent the average of 6 healthy human volunteers measured over a 6 h period following intake.

8.5.3 Caloric value of erythritol

The caloric value of a food material is very difficult to measure directly, and in many cases the theoretical value calculated on the basis of the substance's metabolic behaviour, if sufficiently understood, is often the only reliable measure (Anon, 1987; Grossklaus, 1987; Würsch and Schweizer, 1987). For erythritol it becomes a very easy exercise, as it can be assumed that from a normal average dose (20–50 g in a single intake) at least 80% is absorbed and excreted in the urine. If we assume the remaining 20% is completely fermented in the colon to volatile fatty acids, only half of this energy will be available to the body (van Es, 1987). This brings the energy availability to less than 10% of its initial value of 4 kcal/g, thus giving a significant reduction in caloric value compared to other low-calorie bulk sweeteners which release on average 50 to 60% of their initial energy value to the body. Indeed the European Nutrition Labelling Directive attributed a value of 2.4 kcal/g to polyols which reflects a

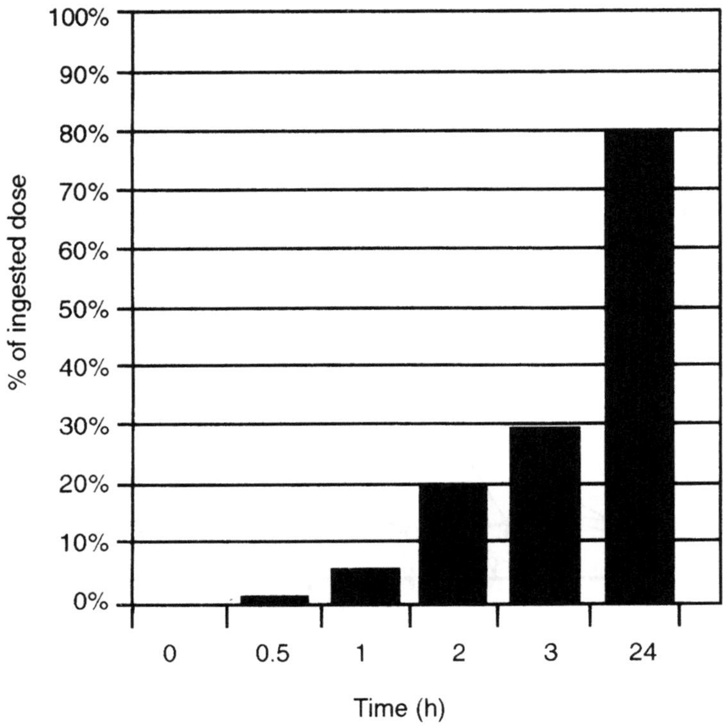

Figure 8.8 Average cumulative amount of erythritol excreted in the urine expressed as the percentage of the administered dose after a 25 g single oral intake of erythritol. Data represent the average of 6 healthy human volunteers measured over a 24 h period following intake.

reasonable average to simplify the caloric value calculation of consumer goods, whereas polydextrose was proposed to have a value of 1 kcal/g. Figure 8.9 is a schematic representation of this energy availability/distribution after ingestion of erythritol compared with other low-calorie bulk sweeteners.

8.5.4 High digestive tolerance

The relatively rapid and complete ($\geqslant$ 80%) absorption of erythritol from the small intestine means that unabsorbed products with their potential side effects are absent. A high proportion of the breakdown products of other low-calorie bulking agents remain unabsorbed and are subsequently fermented to give two adverse side effects: diarrhoea and flatulence. It is therefore recommended that their use be restricted to avoid an average daily intake in excess of 20 g as a single dose, although up to 50 g is normally tolerated when taken over a period or on a regular basis. Both

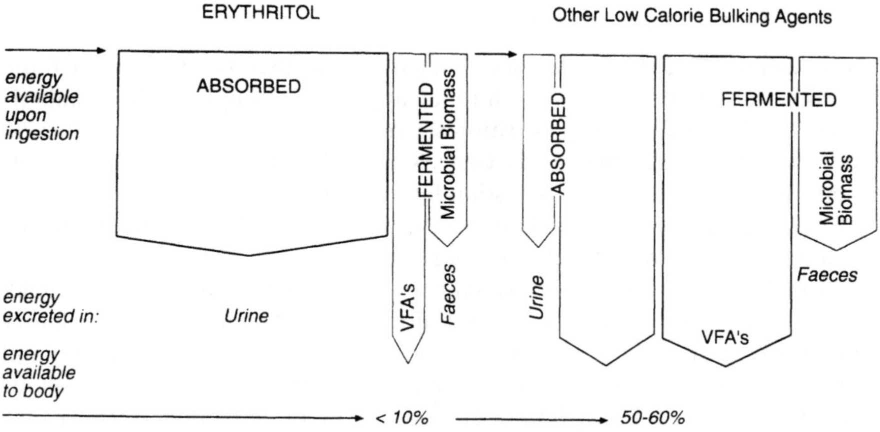

Figure 8.9 Schematic representation of the energy flow in the digestive system after oral administration of erythritol compared to the average behaviour of other low calorie bulk sweeteners.

effects depend very much on the individual's condition. In some cases these effects may be accompanied by stomach cramps and rumbling, even at lower dose levels. The frequency of intake as well as the type of food containing the sugar-replacers have an influence on the amount of inconvenience; regular intake may result in increased tolerance.

Diarrhoea may be the result of a high concentration of unabsorbed carbohydrates in the small intestine, resulting in a high local osmotic pressure. This causes water influx from the mucosal wall, and if the excess water cannot be reabsorbed sufficiently diarrhoea will quickly arise. Increased flatulence comes from the extensive fermentation of these unabsorbed carbohydrates, giving volatile substances in excess of what can be absorbed through the gut wall and blood stream. This coincides very often with a dilatation of the caecum, a physiological effect that results from the increased microbial biomass and gas formation (Bär, 1984; Leegwater, 1974; Hodgkinson *et al.*, 1982).

The high digestive tolerance of erythritol was shown in long-term experiments with rats, in which a diet containing 10% erythritol caused no side effects, whereas other sugar replacers such as sorbitol induced a permanent laxative condition (Oku and Noda, 1990). In the past this laxative effect proved to be an obstacle in assessing the safety of other sugar replacers at higher dose levels. All the studies which we have performed to assess the safety of erythritol show that erythritol does not cause a significant enlargement of the caecum even in the long term, indicating that the gut microflora does not easily adapt to ferment erythritol.

8.5.5 Suitability for diabetics

The usual objective of replacing sugar in the diet is to reduce calorie intake, but there may be additional appeal for people with a diabetic condition, although today nutritionists favour the opinion that diabetics do not need to avoid sugar at all costs. Controlled intake of small amounts may even be advisable to stimulate metabolic control mechanisms. However, the craving of diabetics for sweets would be helped by the availability of a wider range of sugar-replacers. Erythritol, with its low metabolic profile, is absolutely safe for diabetics because it is a virtually inert substance in the body and does not interfere with regulatory processes such as those for glucose and insulin. Blood plasma glucose and insulin levels in human volunteers remained perfectly stable in a 3-hour period following a 1 g/kg body weight oral dose of erythritol (Bornet *et al.*, 1992), as can be seen from Figure 8.10.

8.5.6 Non cariogenicity

Caries is a destructive dental process in which the tooth enamel deteriorates as the result of a complex mechanism initiated by bacteria in the oral cavity. These ferment dietary carbohydrates to organic acids like acetic, propionic and lactic acid and produce polymeric components of the dental plaque, which adheres to the tooth surface. In the absence of thorough dental hygiene, the plaque persists and provides an ideal environment for oral bacteria to continue their acid production. The combination of both processes over time disturbs the equilibrium of the distribution of calcium and phosphate between enamel hydroxyapatite and the dental plaque, resulting finally in enamel demineralisation, or caries (Kleinberg, 1970).

Sugar has been identified as one of the main factors causing caries, as it is a good substrate for oral bacteria, especially *Streptococcus mutans*, which is one of the main culprits in dental caries (Loesche, 1986). The caries-promoting effect of sugar, together with the fact that sweets eaten between meals are particularly harmful, have been the driving forces in the development of sugar-free confectionery. Sugar replacers have provided valid alternatives to sucrose (Linke, 1987; Loesche, 1985) and non-cariogenicity has therefore become a *sine qua non* for sugar-replacers.

The non-cariogenicity of erythritol has been demonstrated using various methods. Incubation *in vitro* with a range of *Streptococci* species has shown that neither lactic acid (Kawanabe *et al.*, 1992) nor other organic acids are produced from erythritol. Under the same test conditions, xylitol gave a similar response whereas maltitol, sorbitol, mannitol, lactitol and isomalt showed minor acid production (unpublished results; see Figure 8.11).

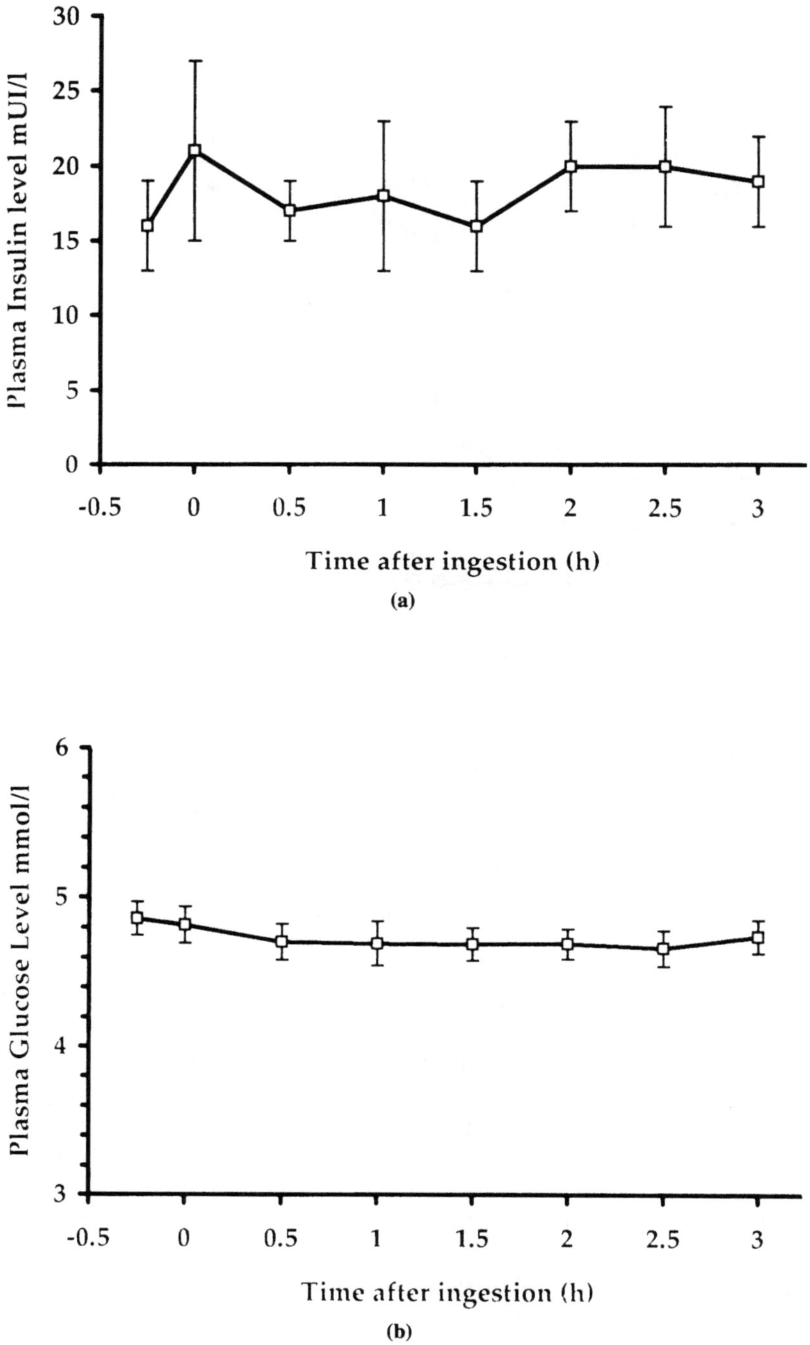

Figure 8.10 Average glycaemic and insulinaemic plasma responses after a 1 g/kg bodyweight single oral intake of erythritol by 6 healthy human volunteers.

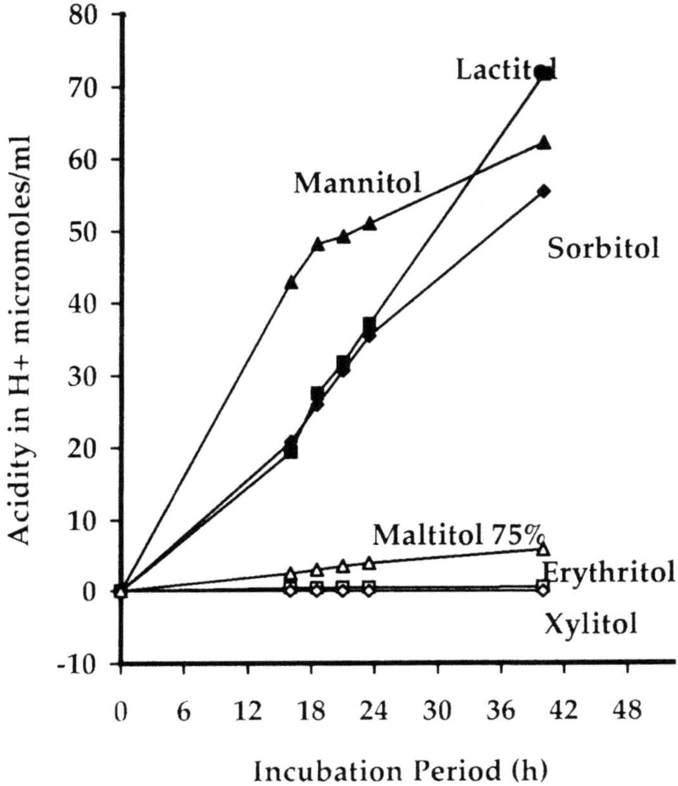

Figure 8.11 Development of the acidity by *Streptococcus mutans* strains growing on a 3% solution of erythritol or another polyol under controlled *in vitro* conditions. Cultures with an initial bacteria count of 3×10^9 were incubated in test tubes at 37°C in a biological buffering system. Acidity is expressed as the total H^+ concentration in micromoles/ml accumulating over the time period.

Streptococci were also unable to grow on erythritol and therefore could not produce the glucosyltransferase which enables them to synthesise their insoluble glucan plaque material. This was confirmed in separate tests *in vitro* showing that *Streptococci* did not produce any polymeric plaque material and subsequently could not adhere to a glass surface when incubated with erythritol, in contrast to the control groups incubated with sucrose.

A study *in vivo* in pathogen-free rats artificially infected with *S. mutans* to evaluate the caries-inducing potential of diets based on starch alone or in combination with erythritol or sucrose indicated clearly that after 70 days, the erythritol-fed rats were virtually caries-free compared to the sucrose-fed group which had caries scores showing more than 60% of teeth affected. This was confirmed in a similar feeding experiment with chocolate in addition to a standard starch diet. The rats receiving sucrose-based chocolate had 82% of teeth caries, whereas those on erythritol chocolate reached only 7%. It was not clear why, in both experiments, the scores of erythritol-fed rats were always lower than the control group receiving chocolate with only starch as carbohydrate (Kawanabe, 1992).

With regard to cariogenicity in humans, plaque pH telemetry has been used to measure the cariogenic potential of foods. This is based on the formation of organic acids in dental plaque after exposure to fermentable dietary carbohydrates, and records the fall in plaque-pH. The potential of a food to reduce plaque-pH values, in other words the plaque acidogenicity of a food, is generally agreed to be closely associated with the initiation of dental decay (Loesche, 1995).

Plaque-pH telemetry measures continuously the acidity (H^+ concentration) under an undisturbed layer of plaque at the tooth enamel surface following the intake, chewing and sucking of foods (Graf and Mühlemann, 1965; Imfeld, 1982, 1983). It does not disturb the normal diffusion of substrates into the interdental areas and the neutralising effect of salivary buffers, and the influence of acids or buffers in the food is assessed in a natural environment. Although the frequency of food consumption needs to be considered to predict cariogenicity, it can be disregarded if *in vivo* intra-plaque acid formation does not produce pH values below 5.7. Therefore, if the telemetrically-recorded plaque pH values after ingestion of a food are above this arbitrary limit of 5.7 the food can be regarded low in cariogenic potential (Muhlemann, 1979). Erythritol was submitted to this test in the form of a crystallised tablet (or lozenge). The pH curve from the intra-plaque measurements during chewing of erythritol tablets is represented in Figure 8.12.

It is clear that with an average minimum pH value of 6.05, erythritol tablets are of low cariogenic potential and can be assigned tooth-friendly status.

Although this point is proved, it is still necessary to formulate foods in which it can be used to take advantage of this property.

8.6 Functional properties of erythritol

Erythritol has properties similar to those of other polyols at present used as food ingredients, such as xylitol, sorbitol, mannitol or the disaccharide polyols maltitol, lactitol and isomalt.

It is a moderately sweet bulking agent with 60–70% of the sweetness of sucrose in a 10% solution. As it has a taste profile very close to that of sucrose with no bitter aftertaste, it is ideal for improving the taste in combination with intense sweeteners like aspartame or acesulfame-K.

Erythritol like other polyols, has excellent heat- and acid stability, but

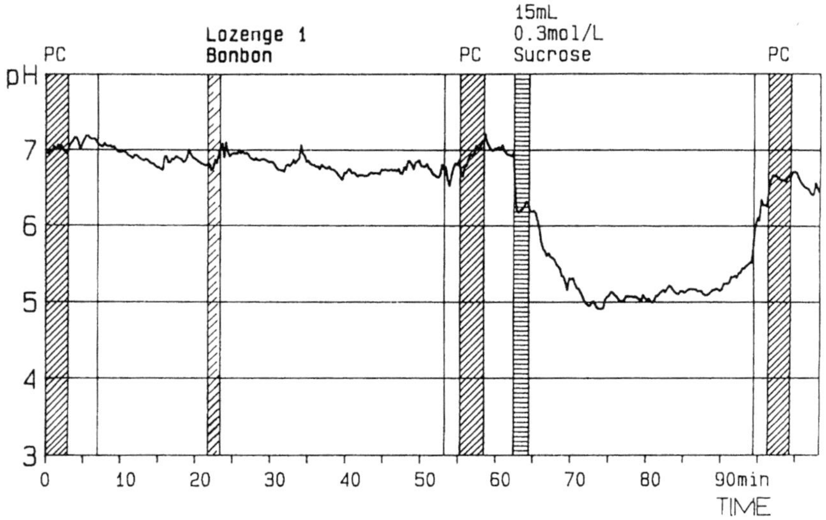

Figure 8.12 Example of a telemetrically recorded pH of interdental plaque in one volunteer during the 30 minutes after consuming an erythritol lozenge. Two-minute 0.3 mol/l (10%) sucrose rinses were administered as positive controls. (PC) = Paraffin chewing. (id) = age of interdental plaque (I) in days (d).

differs in having low solubility, a property also of mannitol. It also has a very low heat of solution, like xylitol.

When erythritol is dissolved in water the crystals need energy to dissolve. This high negative heat of solution provides the crystalline material with a strong endothermic cooling effect. The heat of solution of erythritol is −23.3 kcal/kg. This property is particularly important in applications such as chewing gum, fondant or fat cream.

The excellent heat stability of erythritol ensures that there will be no product decomposition and/or discoloration from it at temperatures up to 170°C (Figure 8.13). Erythritol resists decomposition in both acidic and alkaline media, and remains stable under prolonged exposure to pHs in the range of 2 to 12.

Its low solubility, only 37% at 25°C, provides it with excellent crystalline and powder properties. These can be exploited in replacement applications where a crystalline structure previously provided by sucrose is essential.

The solubility of erythritol rises with increasing temperature. Figure 8.14 shows the solubility of erythritol at different temperatures. Because of its excellent crystallisation behaviour, high viscosity syrups such as maltitol syrup can be added to prevent unwanted crystallisation, depending on the application. This is similar to the use of glucose syrup, as so-called 'doctor sugar' in traditional confectionery.

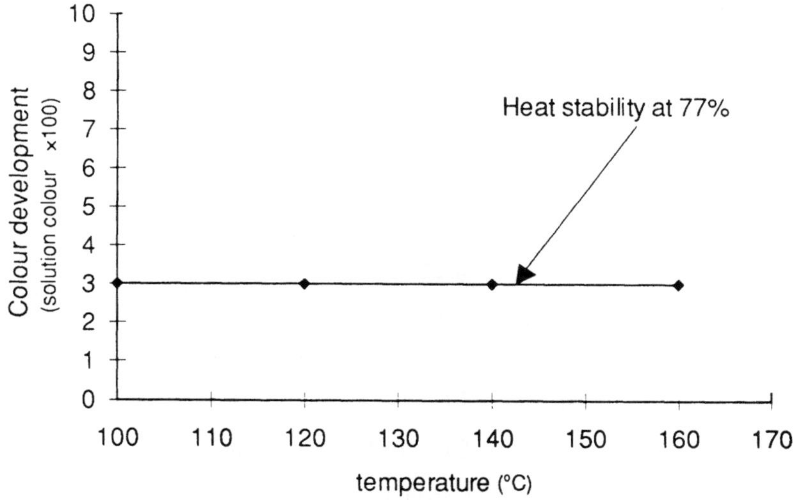

Figure 8.13 Optical density of an erythritol solution after exposure during 3 minutes to increasing temperatures.

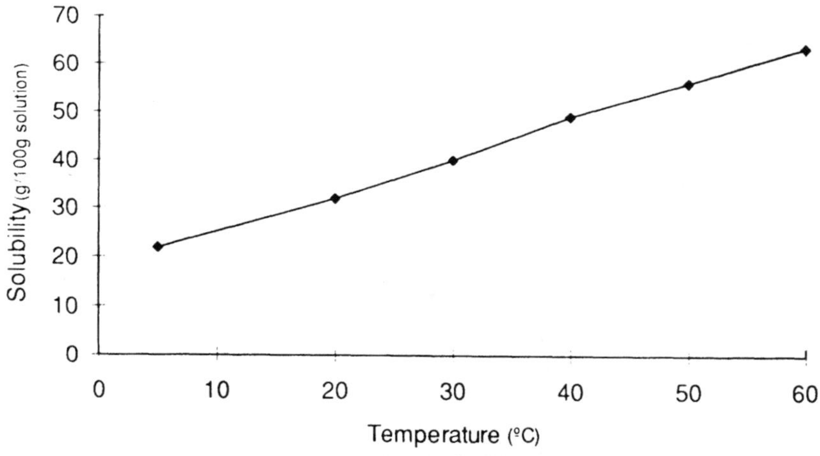

Figure 8.14 Solubility of erythritol in water as percentage weight by weight at different temperatures.

Due to its low solubility, only relatively low concentrations in solution can be achieved. These solutions have very low viscosities. (Viscosities at different concentrations and temperatures are shown in Figure 8.15.)

As erythritol is a small molecule, it has strong colligative properties, i.e. strong freezing point depression and boiling point elevation effects as well as high osmotic pressure. Erythritol has a much more pronounced effect on

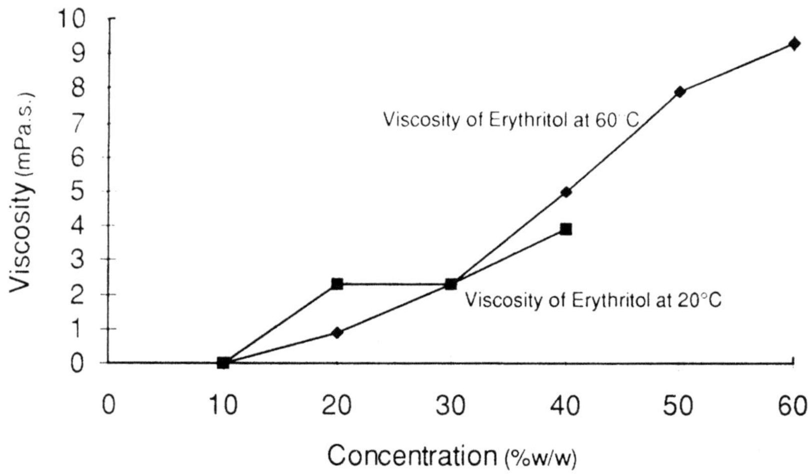

Figure 8.15 Viscosity (in mPa.s) of erythritol solutions at different concentrations and temperatures using Brookfield viscosimeter.

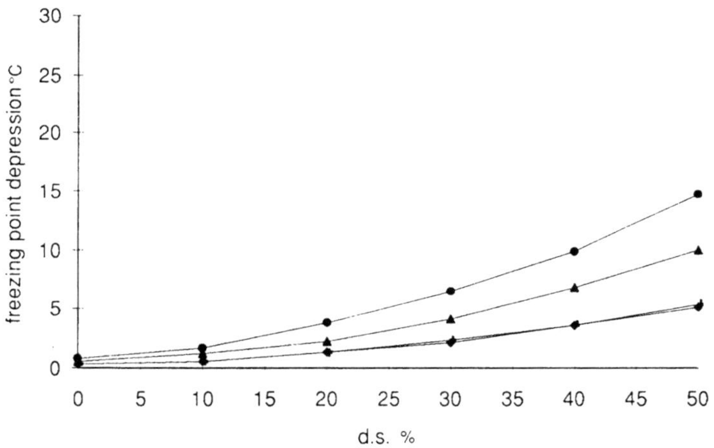

Figure 8.16 Freezing point depression of erythritol, sorbitol, maltitol and sucrose in solution. ●, erythritol; ▲, sorbitol; +, maltitol; ♦, sucrose.

freezing point depression than sorbitol and sucrose for example (see Figure 8.16). This means that ice cream based on 100% erythritol will be very soft with fast melt-down characteristics. It also means that in solutions of equal dry substance, erythritol provides a much higher osmotic pressure and therefore offers better preservative properties.

Unlike sorbitol, erythritol crystals are non-hygroscopic (see Figure 8.17). Hygroscopicity gives an indication of the ability of the substance to

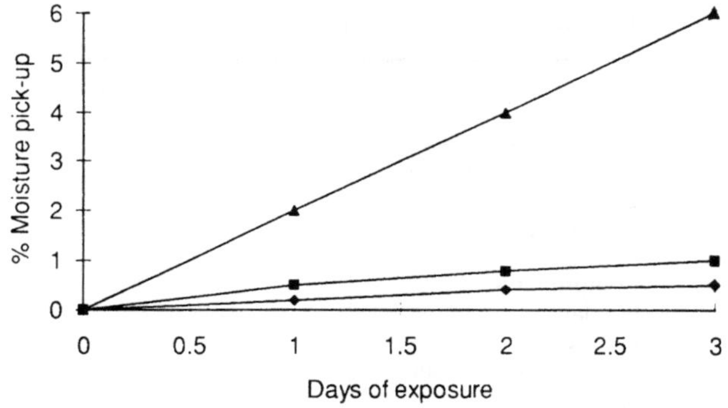

Figure 8.17 Water absorption of erythritol compared to sorbitol. ▲, sorbitol at 75% R.H.; ■, erythritol coarse at 75% R.H.; ♦, erythritol fine at 75% R.H.

absorb water, as it is closely related to water activity (a_w) and equilibrium relative humidity. Transfer of moisture between a product and the surrounding atmosphere or between parts of a product is determined by reference to the water activity of each component.

The water activity is the ratio of the water vapour pressure around a substance to that above pure water at the same temperature. It is expressed on a scale of 0–1. The water activity is important in the regulation of moisture pick-up and release. It also influences enzyme activity, the Maillard reaction, fat oxidation, microbiological stability and texture. Erythritol is very useful for reducing and controlling the water activity of foodstuffs.

8.7 Food applications of erythritol

The combined nutritional and functional properties of erythritol make it a unique low-calorie bulk sweetener offering new possibilities for foods. It is especially suitable for applications where a high level of incorporation in powder or crystalline forms is required, which is impossible with the other low-calorie bulking agents currently available. Using erythritol, good quality products can be produced while achieving considerable reductions in their caloric value (see Table 8.1).

To summarise, erythritol can significantly reduce the caloric value of foodstuffs mainly based on carbohydrates which can be substituted with erythritol. It can improve the tolerance of reduced-calorie foods by lessening the risk of side effects caused by other 'light' food ingredients, and it can offer ways of improving taste, texture quality and storage stability.

Table 8.1 Maximum sugar replacement

Applications	% reduction in calories
Table top sweeteners	90
Chocolate	34
Chewing gum	85
Lozenges	90
Comprimates	85
Fondant	65
Fat cream	36
Instant bakery cream	31
Sponge cake	25

Some examples of the possible food applications of erythritol are given below.

8.7.1 Sweet taste improvement

Today a wide range of non-sugar sweeteners such as stevioside, glycyrrhizin, neohesperidine dihydrochalcone, aspartame, cyclamate, saccharin and acesulfame-K are available. These intense sweeteners are far sweeter than sugar, and have therefore been incorporated in foods and drugs to replace sugar, mainly to reduce calorie levels. As they have such high sweetening properties they are used in foods at an extremely low level. They must therefore be combined with a bulking agent in foods such as candies or tablets.

Although erythritol alone is inferior to sugar in the quality of its sweetness and its sweetness level, formulations have been worked out in which erythritol is blended with saccharin (0.075–4.4% by weight, based on the erythritol) to give a sweetener combination suitable for toothpaste, fondant, fudge bars, marshmallows, diet ice creams, sugar-free chewing gum or low-calorie soda pop which is superior in taste to that obtained using erythritol alone. Furthermore, it has been shown that a sweetness very similar to that of sugar, which is considered ideal, can be achieved by blending erythritol with other intense sweeteners such as stevioside, thaumatin, aspartame or glycyrrhizin, with the advantage that the combination is low-calorie. A typical application would be low-calorie soft drinks.

One problem when making a blend of erythritol and an intense sweetener is to ensure that the intense sweetener, which is present in only very small quantity, is evenly distributed throughout the mass of the erythritol. Erythritol is crystalline, which is an advantage in this respect.

Spray-drying (and spray-drying with agglomeration) ensures an even level of sweetness in an erythritol/intense sweetener combination, useful in specialised applications such as tableting, chewing gum, chocolate, crystallised tablets, fondants and especially table-top sweeteners. By spray-drying a solution of erythritol and an intense sweetener, a blend is obtained in which the intense sweetener is evenly distributed throughout the erythritol at any desired level. Spray-dried erythritol, although still crystalline, is a light, free-flowing powder which can be produced in a range of particle sizes and densities by including an agglomeration step in the process, so as to suit the requirements of each application.

8.7.2 *Tableting*

Although erythritol is a highly crystalline solid, it is unsuitable for direct compression. Even with the addition of a high level of microcrystalline cellulose as a binding agent, erythritol used as an excipient gives capping and low tablet hardness problems.

These are resolved by using spray-dried erythritol with or without the addition of an intense sweetener. The erythritol should preferably be spray-dried with a suitable binding agent having a low DE value (3–4) such as potato maltodextrin, in order to obtain sufficiently hard tablets.

8.7.3 *Chewing gum*

Increasing interest in health foods has caused major changes in chewing gum formulations, since a product which contains up to 75% by weight of sucrose or dextrose has a high calorie count and is cariogenic. In consequence, gums have been developed with sweeteners which are either low-calorie or non-cariogenic, or which possess both of these attributes.

Today, sugar-free chewing gum production technology is well established. It is based principally on a solid phase, i.e. sorbitol powder, and a liquid phase, usually a non-crystalling sorbitol syrup, to control the plasticity and chewability. Maltitol syrups can also be used as the liquid phase to achieve the desired balance of relative humidity which prevents drying out and give the right viscosity to ensure the correct plasticity. The gum may still show stickiness and excessive water activity which disturbs production, but this can be reduced by dusting with fine mannitol or maltitol powder. Xylitol is often added to provide a cooler taste, with the additional advantage of its low cariogenicity.

When erythritol is used as the solid phase in a chewing gum greater flexibility, softer texture and improved shelf life are obtained. Since pure erythritol induces too much crystallisation, maltitol syrup with a 75% maltitol content is used to control crystallisation by reducing the water activity. This erythritol/maltitol chewing gum has outstanding shelf-life.

A typical erythritol-based chewing gum formulation giving a soft product is shown in Table 8.2.

Table 8.2 A typical erythritol-based chewing gum formulation

Ingredients	%
Gum base	30
Erythritol	55
Maltitol syrup	13
Glycerine	2

Table 8.3 Formulation for a harder chewing gum

Formulation	%
Gum base	38
Erythritol	45.5
Maltitol (75%)	2.0
Mannitol	10.0
Glycerol	3.0
Mint flavour	1.5

In order to achieve a product with good appearance, chewability, texture and stability, the following method is recommended. The temperature is a critical factor:

Manufacturing process

- Heat the gum base to 60°C and knead.
- While still kneading add the hydrogenated starch hydrolysate:maltitol syrup.
- Add one third of the erythritol and knead for 6 min.
- Add one half of the glycerine and knead for 1 min.
- Add the second third of erythritol and knead for 1 min.
- Add the second half of the glycerine and knead for 1 min.
- Add the remaining third of erythritol and knead for 6 min.
- Discharge the chewing gum and cool.

A gum with a harder texture can be produced using the formulation in Table 8.3.

In both formulations erythritol particle sizes of up to 300 microns may be used although it is generally preferred that the particle size be less than 300 microns to avoid a grainy texture.

Due to the negative heat of solution, erythritol-based chewing gum has a very pleasant cooling effect in the mouth. However, for sugar-free chewing gums its non-cariogenic properties are also very important.

8.7.4 Chocolate

Traditional chocolate based on cocoa mass, cocoa butter and sucrose is a high-calorie product owing to its sugar and fat content. Its fine texture makes it difficult to develop alternatives that are acceptable in both taste and texture.

This problem has been particularly evident when replacing sucrose by one of the many sugar replacers (e.g. sorbitol, maltitol, lactitol and isomalt) to provide calorie-reduced, sugar-free chocolate for diabetics and slimmers. The use of sugar alcohols (except crystalline maltitol) makes chocolate processing very difficult. With erythritol, however, there are virtually no differences in processing from the normal.

Chocolate is typically prepared in four stages:

(i) mixing or refining;
(ii) conching;
(iii) tempering;
(iv) moulding.

In stage (i), the ingredients are mixed together in a process which also involves grinding or rubbing, e.g. on a multiple roll press, to provide a smooth, fluid paste. The ingredients may be added sequentially, and in particular the cocoa butter may be added step-wise to control the viscosity of the composition. The sugar may also be pre-ground to a smaller particle size to reduce the length of time required in the grinding/rubbing of the chocolate mixture. At this stage, erythritol with a standard S2 sugar granulometry behaves like sugar. Finer erythritol crystals have a tendency to cause a lumpier chocolate mass. Energy savings could be achieved during refining since erythritol crystals are more fragile than sucrose or sorbitol and therefore the required particle size of 50–60 μm is obtained after just one passage through the rollers (for sucrose two passes are generally required).

Most types of chocolate, and certainly all good quality products, undergo the process of 'conching' after the mixing stage, in which the mixture is subjected to mechanical working to give it a fuller and more homogeneous flavour. Other ingredients such as flavours e.g. vanilla, and extra cocoa butter may be added at this stage if desired. Lecithin (or another emulsifier) is frequently added to improve the flow properties of the chocolate, thereby enabling a reduction in the amount of cocoa butter required. Chocolate makers have found that the relatively low temperatures (55°C for lactitol or isomalt, and a maximum of 40°C for sorbitol) required for the conching of chocolate compositions containing a sugar alcohol sweetener usually necessitate lengthy wet conching, and the flavour development is not as satisfactory as when higher conching temperatures are used.

Erythritol chocolate however, like sucrose chocolate, can be dry-conched at an elevated temperature. Under standard conching conditions, the viscosity of an erythritol chocolate mass is only slightly higher than that obtained with sugar, but the good heat stability and low moisture pick-up of erythritol makes it possible to work at temperatures up to 80°C. This results in enhanced flavour development with normal processing times. It should even be possible to use conching temperatures far above 80°C since erythritol is perfectly stable and non-hygroscopic.

The third stage of chocolate preparation is called 'tempering', in which nuclei are provided in the liquid chocolate to facilitate the rapid crystallisation of its fat on cooling. The final appearance of the chocolate, its texture and its shelf-life depend upon correct tempering conditions. After tempering, the erythritol chocolate may finally be put into moulds to set or may be used in an enrobing process to produce chocolate-coated confectionery, similarly to traditional chocolate.

A typical high quality product is obtained with the formulation in Table 8.4.

Table 8.4 Comparative caloric content of chocolate containing erythritol or sucrose

Formulation		EROH Chocolate		Sucrose Chocolate	
Ingredients	kcal/g	%	kcal/100 g	%	kcal/100 g
Cocoa mass	6.1	39	237.9	42	256.2
Cocoa butter	9.3	13	120.9	13.5	125.5
Erythritol	0.4	47.7	19	–	–
Sucrose	4.0	–	–	44	176
Lecithin	9.3	0.48	4.5	0.48	4.5
Vanillin	–	0.02	–	0.02	–
Aspartame	–	0.03	–	–	–
Caloric value	in kcal		382.3		562.2
Reduction	%		68		100

Erythritol-based chocolate is not only easy to produce, but it can also be considered a true 'reduced-calorie' product providing less calories. Other polyols with a calorific value of 2.4 kcal g do not achieve the 30% calorie reduction required for a 'light' chocolate by the simple substitution of sucrose, but need major reformulation, leading to inferior quality. The chocolate obtained with erythritol is non-hygroscopic with good gloss, breaking characteristics and melting properties in the mouth (Figure 8.18).

The cooling effect of erythritol-based products could be considered as either a special asset or a disadvantage, depending on application. However, trials have revealed that using blends of erythritol with crystalline maltitol i.e. 70/30, the cooling effect is dramatically reduced, while maintaining the workability advantages of both materials (i.e.

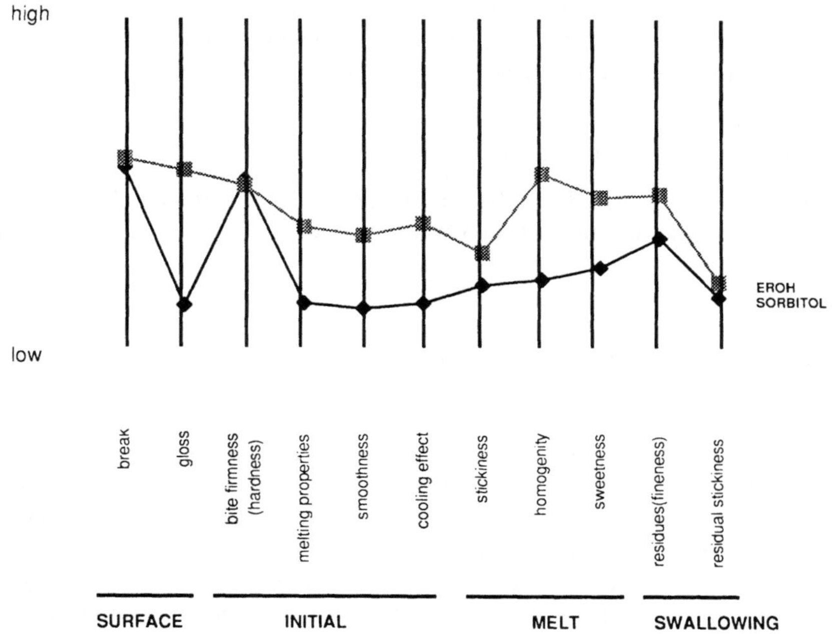

Figure 8.18 Sensorial profile of chocolate based on erythritol in comparison with sorbitol based product. EROH = erythritol.

conching at high temperatures up to 80°C). It is therefore possible by using these blends, to produce 'light' chocolate – 30% calorie reduction – with sensorial properties similar to those of standard products made with sucrose.

A typical erythritol-based low-calorie/sugar-free chocolate manufacturing process could be:

Kneading
- Insert erythritol, cocoa mass (liquid) and 5–10% cocoa butter into the mixer.
- Kneading time: 10–15 min.
- Kneading temperature: 30–40°C

Refining
- The mass is milled on a cooled 5-roll refiner.

Conching
- Temperature: up to 80°C
- Total conching time: 16–22 h
- Conching conditions: according to equipment

Near the end of the conching period the remaining cocoa butter and the lecithin are added, e.g. for a 16 h conching time; addition of the remaining cocoa butter after 14 h; addition of lecithin after 15 h

Tempering

- Tempering temperature: 28–31°C

Should the sweetness level be unsatisfactory this can be adjusted, for example, by the addition of 0.03% aspartame.

8.7.5 Lozenges

Lozengers are usually made from finely-milled sugar which is kneaded into a doughy consistency with water and a binding agent. A gum such as gum arabic or gelatin is usually used to help in retaining tablet shape. The still malleable dough is cut into suitable shapes which are dried and hardened.

Lozenges may of course contain suitable flavours, colours and/or acids, and should have a hard, brittle consistency. Here again polyols such as xylitol, sorbitol, mannitol maltitol, isomalt and lactitol have been suggested in order to produce low calorie/non-cariogenic lozenges. Although all of the sugar alcohols mentioned above find many applications in confectionery, they are far from equivalent in all applications. In the confectionery field the contrasting demands of the various types of products require careful selection of ingredients to find the correct combination of the properties to replace part or all of the sugar.

Research has shown that lactitol, crystalline maltitol and erythritol are the only suitable materials to produce a lozenge with a lower caloric value than conventional sugar-based lozenges, but which appear otherwise identical.

In addition to the lower calorie content and pronounced cooling effect, erythritol has the added advantage over lactitol and crystalline maltitol, of requiring a shorter drying time to obtain the required texture at low residual moisture content. Erythritol-based lozenges have excellent shelf life when stored under high humidity conditions.

A typical manufacturing process for erythritol based lozenges (see Table 8.5) could be that shown below:

Table 8.5 A typical erythritol-based lozenge formulation

Formulation	Content
Erythritol	1.5 kg
Gelatine solution (10% d.s.)	230 ml

Manufacturing process

- Prepare 10% gelatin (170 bloom) solution by slowly adding the gelatin to warm water (±50°C).
- Put erythritol into the Z-blade mixer preheated to 40–45°C.
- Slowly add the warm gelatin solution while continuing to mix.
- Mix for ±10 min to obtain a smooth, homogeneous paste.
- Remove the paste from the kneader, roll out and cut to shape.
- Stove at 45°C for 8 h.

8.7.6 Fondant

Fondants and creams are sugar confectionery products which contain mixed sugars in two phases. The sugar crystals constituting the solid phase are dispersed in a syrup containing higher sugars. Fondants and creams are similar in composition, though creams contain slightly higher residual water and a greater proportion of doctor solids.

Using erythritol, it is possible to obtain sugar-free fondant with identical properties to classical products, which was impossible with the other low-calorie bulking agents. Pure erythritol induces too high a crystallisation, but the use of maltitol syrup as a liquid phase helps to control this.

The preparation method for an erythritol-containing fondant is similar to that for a sucrose-based product. The erythritol crystals are dissolved in a specified amount of syrup phase, then the mixture is cooked in an open system to the required dry-substance point. After cooling the mass to 40°C, it is beaten to achieve the correct consistency and crystal size. The fondant is discharged into small containers and matured for 1 day before use. Fondants with slightly different textures and fluidities can be obtained, depending on the beating temperature and time. Examination under the microscope indicates that a good quality fondant has a majority of crystals between 5 and 10 μm. An erythritol fondant moisture content of ±6% and the low water activity (in the range of ±0.5) prevent microbiological spoilage.

A typical erythritol based fondant manufacturing process recipe is given overleaf.

Table 8.6 A typical erythritol-based fondant formulation

Formulation	%
Erythritol	50
Maltitol syrup (75% maltitol content)	50

Manufacturing process

- Dissolve the erythritol in the maltitol syrup.
- Cook the mix to ±140°C.
- After cooling to 40–45°C, the mass is beaten for 5–10 min to obtain the required consistency and crystal size.
- The fondant is placed in containers and matured for 1 day.

8.7.7 *Fudge*

Fudge is a soft sweet with a texture between that of a caramel and a fondant. It is a grained medium-boiled confection which contains both milk solids and a high fat level and both a solid and a liquid phase. Sugar-free fudge with texture and shelf-life properties equivalent to those of conventionally sweetened fudge can be produced using erythritol in combination with maltitol syrup (75% maltitol) to control undesirable crystallisation. Depending on the cooking temperature and seeding level, the texture can be manipulated.

A typical formulation (Table 8.7) and manufacturing process for fudge containing erythritol is shown below.

Manufacturing process

- Mix erythritol, maltitol syrup and water and heat to ±70°C.
- Add the milk, melted fat and GMS and homogenise for ±2 min.
- The mix is further cooked to ±140°C.
- Cool the mass to ±100°C and seed with 5–10% of seeding blend.
- Cool and cut.

Erythritol can also be used in a range of other confectionery applications such as candy or gums, but with limitations owing to its crystallisation behaviour.

Table 8.7 A typical erythritol-based fudge formulation

Formulation	%
Erythritol	22.5
Maltitol syrup (75% maltitol content)	45.5
Water	5.5
Unsweetened whole concentrated milk	18.7
Hydrogenated cocoa fat GMS	7.5
Sodium bicarbonate	0.2
Seeding blend	
Erythritol 81%	5–10
Water	19

8.7.8 Bakery

Bakery products, with their high proportion of flour, butter, sugar and other ingredients, are difficult to make in calorie-reduced versions. However, due to the very low caloric value of erythritol, sugar replacement gives some interesting results.

8.7.8.1 Fat composition. Conventional 'fat plus sweetener' compositions contain sucrose as the sweetener and are especially used in butter cream fillings for cake and biscuits. Such compositions, besides having a substantial caloric value, have a typical fatty mouthfeel.

Although these characteristics are accepted by the consumer, it would be acceptable to have a similar cream filling with a less fatty mouthfeel, provided the composition also had a reduced calorie level. Many ideas have been put forward to reduce the caloric value by replacing part or all of the sugar with a substitute sweetener.

Two sugar alcohols, xylitol and erythritol, are known to give a cooling sensation in the mouth, on account of their negative heat of solution. This has the effect of masking the fatty mouthfeel, giving a more refreshing and attractive product. In addition, the caloric value of the composition is also reduced (particularly in the case of erythritol).

The composition may contain up to 75% of erythritol to take advantage of its sweetening effect. The best products are obtained with particle sizes below 300 μm. The manufacturing process and shelf life properties of the erythritol-based fat cream are similar to those for conventional fat/sucrose compositions.

Over-all reduction of calories in bakery products is difficult if only the sugar they contain is replaced. Erythritol can reduce the caloric value by more than 30% depending upon the application, without introducing undesirable side effects, and additionally it improves the shelf-life.

Erythritol can also be applied successfully in the following bakery applications:

- Instant pie filling or instant bakery cream, in combination with spray-dried glucose syrup to avoid recrystallisation upon cooling.
- Sponge cake in combination with maltitol powder, providing improved shelf life and cake volume in comparison with classical (sucrose-based) product.
- Hard biscuits to improve baking stability and shelf life but also in biscuit, pound cake, shortbreak, almond dough etc.

Generally, compared to sucrose, erythritol gives a different melting behaviour, a more compact dough, softer end products, less colour formation and better water-binding capacity.

Typical formulations and manufacturing process of bakery products using erythritol are shown in Tables 8.8–8.14.

8.7.8.2 *Fat cream*

Table 8.8 A typical erythritol-based fat cream formulation in bakery

Formulation	%
Erythritol < 300 microns	60
Shortening	40

Manufacturing process

Gently mix all the ingredients for 5 min at full speed in a Hobart mixer using the harp beater configuration.

8.7.8.3 *Instant pie filling*

Table 8.9 A typical erythritol-based instant pie filling formulation

Formulation	%
Sorbitol powder	10.3
Erythritol	6.2
Fructose	1.4
Spray dried glucose syrup 21 DE	2.7
Cold soluble modified waxy maize starch	10.3
Citric acid	0.7
Water	68.4

Manufacturing process

- Dissolve the citric acid in the water.
- Add the mixture of the other ingredients to the solution and mix for 30 sec at low speed and 30 sec at higher speed in the Hobart mixer.

8.7.8.4 *Instant bakery cream*

Table 8.10 A typical erythritol-based instant bakery cream formulation

Formulation	%
Sorbitol powder	8
Erythritol < 300 microns	4.65
Fructose	1.1
Spray dried glucose syrup 21 DE	1.4
Cold soluble modified potato starch	7.1
Whole milk solids	5.7
Lacticol D336	0.5
Colour/flavour	0.05
Water	71.5

Manufacturing process

- Blend the dry ingredients.
- Add the mix to the water and blend in a Hobart mixer for 30 sec at low speed and for 3 min at higher speed (whisk).
- Deposit the cream.

8.7.8.5 Pound cake

Table 8.11 A typical erythritol-based pound cake formulation

Formulation	%
Flour	19.38
Native wheat starch	9.68
Maltitol powder	18.6
Erythritol	4.64
Aspartame	0.15
Eggs	24.22
Cake margarine	17.42
Sodium bicarbonate	0.48
Fat emulsion	2.91
Sodium acid pyrophosphate	0.58
Water	1.94

Manufacturing process

- Measure all the ingredients in a Hobart bowl.
- Mix for 20 sec at speed I and 5 min at speed II.
- Pour 400 g of creamed mix into a rectangular baking pan (240 mm × 90 mm).
- Bake at 180°C for 45 min.
- Wrap the cake with cling film after 1 h at room temperature.

8.7.8.6 Hard biscuit

Table 8.12 A typical erythritol-based hard biscuit formulation

Formulation	%
Margarine	10
Sucrose	7.65
Erythritol	7.65
Whole milk solids	1.23
Salt	0.46
Pyrophosphate	0.15
Vanillin	0.04
Biscuit flour	61.5
Sodium bicarbonate	0.3
Ammonium bicarbonate	0.15
Sulphite	0.07
Water	10.8

Manufacturing process

- Cream margarine, sucrose and erythritol in the Hobart mixer for 5 min at high speed.
- Dissolve the baking agents in a part of the water.
- Sieve the flour in the Z-blender.
- Add the cream to the flour/sulphite and mix during 30 min at high speed; dough temp: 32–36°C.
- Add the dissolved products and the water during the mixing process.
- Let the dough rest for 30 min.
- Take out the dough with a biscuits-seal or with a biscuit machine.
- Bake the biscuit for 7 min at 250°C.

8.7.8.7 *Sponge cake*

Table 8.13 A typical erythritol-based sponge cake formulation

Formulation	%
Flour	11.44
Native wheat starch	14.64
Emulsifier	5.08
Glucono-δ-lactone	0.57
Trisodium phosphate	0.286
Colouring	0.011
Vanillin	0.034
Sodium bicarbonate	0.286
Aspartame	0.023
Eggs	0.023
Maltitol powder	14.85
Erythritol	9.9
Water	8.58

Manufacturing process

- Measure the liquid phase into a bowl.
- Blend the dry ingredients.
- Mix at high speed for 6 min in the Hobart mixer.
- Measure 300 g of batter in a biscuit tin.
- Bake for 20 min at 190/200°C.

8.7.8.8 *Shortbread*

See Table 8.14 on the top of page 183.

Manufacturing process

- Cream all the ingredients except the flour for 2 min at low speed and for 5 min at high speed with the harp beater in the Hobart mixer.

Table 8.14 A typical erythritol-based shortbread formulation

Formulation	%
Flour	37.4
Native wheat starch	4.2
Margarine	29.8
Maltitol powder	10.8
Erythritol < 300 microns	7.2
Fructose	1.8
Maltodextrin 18 DE	1.8
Egg powder	0.7
Whole milk solids	0.4
Salt	0.4
Sodium bicarbonate	0.2
Lemon	1.1
Water	4.2

- Add the flour and blend for 1 min at high speed.
- Bake for 15–16 min at 180°C.

8.8 Conclusions

The object has been to demonstrate the unique properties of erythritol in comparison to other low-calorie and non-cariogenic bulk sweeteners. In respect of its nutritional and metabolic characteristics, erythritol displays a number of interesting properties, creating new possibilities for calorie-reduced and toothfriendly products by the replacement of sugar. The wide range of applications described here also show the functional and technical advantages of erythritol in specific formulations.

On the basis of its presence in nature and in foods, erythritol has been regarded a normal and safe type of food by the Japanese authorities and has been accepted for general use in food formulations there. Because of its low caloric value and non-cariogenicity, erythritol continues to be more and more used in a wide range of foods. In other cases the authorities of course require the usual thorough safety evaluation before approving a new substance as a food ingredient or food additive. Extensive safety studies have been conducted to establish beyond any doubt that erythritol can be safely consumed, and the necessary documents are being prepared for submission to the relevant authorities.

References

Anon (1985) Scientific Consensus Conference on Methods for Assessment of the Cariogenic Potential of Foods. San Antonio, Texas, November 17–21. *J. Dent. Res.* **65**, 1473–1543.
Anon (1987) The energy value of sugar alcohols. *Voeding* **48**(12), 357–365.

Aoli, M.A.Y., Pastore, G.M. and Park, Y.K. (1993) Microbial transformation of sucrose and glucose to erythritol. *Biotechnology Letters* **15**(4), 383–388.

Bär, A. (1984) Safety assessment of polyol sweeteners; some aspects of toxicity. *Food Chemistry* **16**, 231–241.

Bär, A. (1988) Caries prevention with xylitol. In *World Review Nutrition and Dietetics*, Vol. 55, G.H. Bourne (ed.), S. Karger, Basel. pp. 183–209.

Bornet, F., Dauchy, F., Chevalier, A. and Slama, G. (1992) Etude de devenir metabolique, après ingestion chez l'homme sain, d'un nouvel edulcorant de charge basse calorie: l'érythritol. *Gastroenterol. Clin. Biol.* **16**, A169.

Burkitt, D.P., Walker, A.R.P. and Painter, N.S. (1972) Effect of dietary fiber on stool and transit time, and its role in the causation of disease. *Lancet* **30**, 1408–1411.

Cunha, M., Firme, P., San Romao, M.V. and Santos, H. (1992) Application of nuclear magnetic resonance to elucidate the unexpected biosynthesis of erythritol by *Leuconostoc oenos*. *Appl. Env. Microbiol.* **58**(7), 2271–2279.

Estrella, M.I., Hernandez, M.T. and Olano, A. (1986) Changes in polyalcohol and phenol compound contents in the ageing of sherry wines. *Food Chemistry* **20**(2), 137–152.

Firme, M.P., Leitao, M.C. and San Romao, M.V. (1994) The metabolism of sugar and malic acid by *Leuconostoc oenos*: effect of malic acid, pH and aeration conditions. *J. Appl. Bact.* **76**(2), 173–181.

Flourié, B. (1989) The digestion of starches and sugars present in the diet. In *Dietary Starches and Sugars in Man: a Comparison*. J. Dobbing (ed.), Springer-Verlag, Berlin, Heidelberg, pp. 49–66.

Gomez-Cordoves, M.C. and Hernandez, L. (1987) Polyalcohols and sugars in wines with different degrees of pressing. *Alimentaria* **183**, 109–111.

Graf, H. and Mühlemann, H.R. (1965) Glass electrode telemetry of pH-changes of interdental human plaque. *J. Dent. Res.* **44**, 1039.

Grossklaus, R. (1987) Dosisabhängigkeit der energetischen Nutzung von Zuckeraustauschstoffen. *Dtsch. Zahnärztl.* **42**, S154–S158.

Hattori, K. and Suzuki, T. (1974) Production of erythritol by *n*-alkene grown yeasts. *Agric. Biol. Chem.* **38**(3), 581–586.

Hiele, M., Ghoos, Y., Rutgeers, P. and Vantrappen, G. (1989) How to determine the absorptive and metabolic characteristic of new sugars and sugar alcohols. *Gastroenterology* **96**, A208.

Hiele, M., Ghoos, Y., Rutgeers, P. and Vantrappen, G. (1993) Metabolism of erythritol in humans: comparison with glucose and lactitol. *British Journal of Nutrition* **69**.

Hodgkinson, A., Davis, D., Fourman, F., Robertson, W.G. and Roe, F.J.C. (1982) A comparison of the effects of lactose and of two chemically modified waxy maize starches on mineral metabolism in the rat. *Fd Cosmet. Toxicol.* **20**, 371–382.

Imfeld, T. (1982) Interdental plaque pH telemetry. In *Surface and Colloid Phenomena in the Oral Cavity: Methodological Aspects*. Proceedings of a workshop in saliva-dental plaque and enamel surface interactions, Frank and Leach (eds), IRL Press Ltd., London. pp. 143–156.

Imfeld, T. (1983) *Identification of Low Caries Risk Dietary Components*. Monographs in Oral Science, Vol. 11, S. Karger, Basel.

Ishizuka, H., Wako, K., Kasumi, T. and Sasaki, T. (1989) Breeding of a mutant of *Aureobasidium* sp. with high erythritol production. *Journal of Fermentation and Bioengineering* **68**(5), 310–314.

Kawanabe, J., Hirasawa, M., Takeuchi, T., Oda, T. and Ikeda, T. (1992) Noncariogenicity of erythritol as a substrate. *Caries Research* **26**, 258–362.

Kleinberg, I. (1970) Regulation of the acid–base metabolism of the dento-gingival plaque and its relation to dental caries and periodontal disease. *International Dental Journal* **20**, 451–465.

Leegwater, D.C., De Groot, A.P. and van Kalmthout-Kutper, M. (1974) The etiology of caecal enlargement in the rat. *Fd Cosmet. Toxicol.* **12**, 687–697.

Leutner, C. (1981) (ed.), *Geigy Scientific Tables*, Vol. 1. Ciba Geigy Ltd., Basle, Switzerland, pp. 84–85.

Levin, R.J. (1989) Dietary carbohydrate and the kinetics of intestinal functions in relation to

hexose absorption. In *Dietary Starches and Sugars in Man: a Comparison*. J. Dobbing (ed.), Springer-Verlag, Berlin, Heidelberg. pp. 89–118.
Linke, H.A.B. (1987) Sweeteners and dental health: the influence of sugar substitutes on oral microorganisms. In *Development in Sweeteners – 3*. Grenby T.H. (ed.); Elsevier. Amsterdam. pp. 151–188.
Loesche, W.H. (1986) *Role of Streptococcus mutans in human dental decay. Microbiology Review* **50**, 353–380.
Loesche, W.H. (1995) The rationale for caries prevention through the use of sugar substitutes. *Int. Dent. J.* 35, 1–8.
Meyrial, V., Delgenes, J.P., Moletta, R. and Navarro, J.M. (1991) Xylitol production from D-xylose by *Candida Guilmlermondi*: Fermentation behaviour. *Biotechnology Letters* **13**(4), 281–286.
Mühlemann, H.R. (1979) Sugar substitutes and plaque pH telemetry in caries prevention. *J. Clin. Periodontol.* extra issue **7**, 47.
Noda, K. and Oku, T. (1990a) The fate and availability of erythritol in rats. *International Symposium of Caloric Evaluation of Carbohydrates – Book of Abstracts*. Kyoto 11–12 Jan.
Noda, K. and Oku, T. (1990b) The erythritol balance study and evaluation of its metabolisable energy. *International Symposium of Caloric Evaluation of Carbohydrates – Book of Abstracts*. Kyoto 11–12 Jan.
Noda, K. and Oku, T. (1992) Metabolism and distribution of erythritol after oral administration to rats. *Journal of Nutrition* **122**, 1266–1272.
O'Connor, M. (1992) Europe and nutrition: prospects for public health. *Brit. Med. J.* **304**, 178–180.
Oku, T. and Noda, K. (1990) Influence of chronic ingestion of newly developed sweetener, erythritol on growth and gastrointestinal function of the rats. *Nutrition Research* **10**, 987–996.
Ribereau-Gayon, P. and Bertrand, A. (1972) New applications of gas chromatography for analysis and quality control of wines. *Vitis* **10**(4), 318–322.
Roberts, G.P., MacDiarmid, A. and Gleed, P. (1976) The present of erythritol in the fetal fluid of fallow deer. *Res. Vet. Science* **20**, 154–256.
Roboz, J., Kappatos, D.C., Greaves, J. and Holland, J.F. (1984) Determination of polyols in serum by selected ion monitoring. *Clin. Chem.* **30**(10), 1611–1615.
Sasaki, T. (1989) Production and properties of erythritol obtained by fermentation of glucose by *Aureobasidium. Journal of the Agricultural Chemistry Society of Japan* **63**(6), 1130–1132.
Servo, C., Pabo, J. and Pitkaenen, E. (1977) Gas chromatographic separation and mass spectrometric identification of polyols in human cerebrospinal fluid and plasma. *Acta Neurol. Scand.* **56**, 104–110.
Shindou, T., Sasaki, Y., Miki, H., Eguchi, T., Hagiwara, K and Ichikawa, T. (1988) Determination of erythritol in fermented foods by HPLC. *Journal of the Food Hygienic Society of Japan* **29**(6), 419–422.
Shindou, T., Sasaki, Y., Miki, H., Eguchi, T., Hagiwara, K. and Ichikawa, T. (1989) Identification of erythritol by HPLC and GC–MS and quantitative measurement in pulps of various fruits. *J. Agric. Food Chem.* **37**, 1474–1476.
Southgate, D.A.T. (1989) The role of the gut microflora in the digestion of starches and sugars: with special reference to their role in the metabolism of the host, including energy and vitamin metabolism. In *Dietary Starches and Sugars in Man: a Comparison*. J. Dobbing (ed.), Springer-Verlag, Berlin, Heidelberg, pp. 67–88.
Spencer, J.F.T. (1968) Production of polyhydric alcohols by yeast. *Progress in Industrial Microbiology* **7**, 1–42.
Sponholz, W.R. and Dittrich, H.H. (1985) Sugar alcohols and myoinositol in wines and sherries. *Vitis* **24**(2), 97–105.
Sponholz, W.R., Lachev, M. and Dittrich, H.H. (1986) The formation of alditols by the yeasts of wine. *Chemie Mikrobiologie Technologie der Lebensmittel.* **10**(1–2), 19–24.
Størset, P., Stooke, O. and Jellum, E. (1978) Monosaccharides and monosaccharide derivatives in human seminal plasma. *J. Chrom.* **145**, 351–357.

Tomana, M., Prchal, J.T., Cooper Garner, L., Skala, H.W. and Barker, S.A. (1984) Gas chromatographic analysis of lens monosaccharides. *J. Lab. Clin. Med.* **103**(1), 137–142.

van Es, A.J.H. (1987) Energy utilisation of low digestibility carbohydrates. In Leegwater D.C. *et al.* (eds.) Wageningen, Pudoc. 121–127.

Würsch, P. and Schweizer, T. (1987) Sugar substitutes and their energy value for the human body. *Dtsch. Zahnärztl.* **42**, S151–S153.

Yoshida, H., Sugahara, J. and Hayashi, J. (1984) Studies on free sugars and free sugar alcohols of mushrooms. *Journal of the Japanese Society of Food Science and Technology* **31**(12), 765–771.

9 Sweet taste and solution properties of α,α-trehalose as a new cryoprotectant sugar

M.-O. PORTMANN and G.G. BIRCH

9.1 Introduction

Cryprotectants include mono-, oligo- or polysaccharides, di- and poly-alcohols, hydroxymonocarboxylic acids, di- and tricarboxylic acids (Matsumoto, 1980). A cryoprotective effect has also been attributed to triglycerides (Wessel *et al.*, 1981), amino acids and quaternary amines (Jiang *et al.*, 1987a; Loomis *et al.*, 1988, 1989). The nucleotides ATP, ADP and IMP, have been shown to have a protective effect on fish actomyosin (Jiang *et al.*, 1987b). The first mixture of cryoprotectants, found to be successful in a commercial application was a combination of sucrose (10%) and polyphosphate (0.2–0.5%) which prevented denaturation of muscle proteins of Alaska pollack, a fish very sensitive to frozen storage. Cryoprotective agents have been studied not only for the preservation of foods, but also for storing microorganisms and biological materials such as enzymes, vaccines, blood and organs (Matsumoto, 1980).

Recently the UK government has permitted the use of α,α-trehalose as a novel cryoprotectant for freeze-dried foods. This sugar has for a long time been known for its unusual ability to protect organisms against complete dehydration during drought, protecting both the structural and functional integrity of membrane vesicles against dehydration during the freeze-drying process (Crowe and Crowe, 1984; Crowe *et al.*, 1984a). When membranes lyophilized in the presence of trehalose were rehydrated they were seen to have a morphology and an activity similar to fresh membranes (Crowe *et al.*, 1983; Payen, 1949). This protective effect can be applied in food manufacturing to preserve cell structure and minimize damage caused by desiccation. The interaction of the sugar molecules with water molecules is an essential process in the membrane preservation mechanism but no real evidence of this hydration process has yet been produced. Consequently interest is now particularly focused on the interactions between α,α-trehalose and water in order to explain its efficiency as a cryoprotectant. Information on hydration is gathered by investigating solution properties, including intrinsic viscosity [η], apparent molar and specific volumes and NMR analysis. The parameters indicating water/solute interactions are prime determinants of taste qualities. Sweetness intensity and persistence of α,α-trehalose are therefore interpreted here as functions of concentration based on its solution properties and its effect on water structure. This theoretical approach leads to a real food technology application when the differences of texture, colour and sweetness of fruit pulps freeze-dried with maltose, sucrose and α,α-trehalose can be compared. In every case it was found that samples freeze-dried with α,α-trehalose had the characteristics closest to those of fresh pulps.

9.2 Carbohydrates as cryoprotectants

9.2.1 Low-MW and high-MW sugars

The most common cryoprotectants in the food industry have been low-MW sugars and polyols. Sucrose and sorbitol are commonly chosen because of their low cost, ready availability, good safety records, permitted status, good solubility, beneficial functional effects and low tendency to cause Maillard browning (McDonald and Lanier, 1991). However, sugars impart a sweet taste to foods, and their use as cryoprotectant additives can be objectionable in certain applications. It is in this context that less sweet additives such as polydextrose (Lanier and Akahane, 1986), dextrose or maltodextrins have been investigated as alternatives, but less so for dextrose on account of Maillard browning reactions which may limit applications in light or white-coloured fish and poultry products (McDonald and Lanier, 1991). Lactulose, lactitol, maltitol, isomalt or hydrogenated glucose syrups could be considered for special application (Sych *et al.*, 1990a,b). Edible gums have also been proposed as effective cryoprotectants, but tests have failed to demonstrate their effectiveness (Daponte *et al.*, 1985a–c).

9.2.2 α,α-Trehalose, a new cryoprotectant for freeze-dried foods

Known as mycose or mushroom sugar, α,α-trehalose (α-D-glucopyranosyl-1,1-α-D-glucopyranoside) is a non-reducing disaccharide of glucose. It was discovered in 1832 in the fungal disease of rye, ergot, and has since been found in many other organisms, especially in mushrooms; hence it has acquired the name of 'mushroom' sugar. It has been found at relatively high levels (as much 20% of dry weight) in many anhydrobiotic organisms (Tadford, 1980; Crowe *et al.*, 1984b): brine shrimp embryos (Crowe *et al.*, 1982; Clegg, 1965), dry active baker's yeast (Crowe *et al.*, 1983; Payen, 1949), and spores of certain fungi (Crowe *et al.*, 1988; Sussman, 1959). Some insects contain considerable amounts of trehalose in their haemolymph, and in many cases trehalose is the major blood sugar (Lee, 1980). Survival of dehydration by some of these organisms is connected with synthesis of trehalose during dehydration or its degradation to glucose during the breaking of dormance (Crowe *et al.*, 1984a,b). In the D-glucosyl D-glucosides three linkages are possible (α,α; α,β and β,β) but only the α,α isomer is widespread in nature (Birch, 1963; Lee, 1980).

There are benefits from adding trehalose to foodstuffs to induce long term stability in the dry state and rapid rehydration, but apart from protecting the structure and function of food molecules from damage caused by desiccation, this sugar possesses other properties of use in the food industry (Roser, 1991; Green and Angell, 1989), (e.g. confectionery, soft drink and ice-cream):

(i) It is sweet, but only slightly, and it does not alter the flavour of foods.
(ii) It is safe and non-toxic.
(iii) The trehalose-water system has a significantly higher glass transition temperature than other disaccharides.
(iv) The glossy structure of α,α-trehalose can trap volatile aromas.

9.3 Mechanism of cryoprotection

The physical mechanisms of cryoprotection by carbohydrates are not fully understood. It is still unclear, for example, whether trehalose stabilizes biological membranes through direct binding or indirectly through interaction with water (Crowe *et al.*, 1987; Chandrasekhar and Gaber, 1988; Johnston *et al.*, 1984). Matsumoto (1979) envisioned a protective 'coating' of the protein by cryoprotectant molecules. Some studies (Crowe *et al.*, 1983, 1984b, 1988; Crowe, 1971; Clegg, 1986) suggest that cryoprotectant activity seems to be related to the replacement of the main hydration shell of the polar head group of the membrane phospholipids by OH groups of sugars. In other words, sugars such as trehalose serve as 'water substitutes' when the hydration shell of the proteins is removed. In order of efficiency, α,α-trehalose, maltose and sucrose are followed by lactose, fructose, glucose, cellobiose, glycerol, raffinose and myo-inositol (Crowe *et al.*, 1985). All these sugars interact strongly with water and participate in the water lattice through hydrogen bonds. Agents protecting cells during freezing probably act by inducing additional lattice or clathrate formation, either intracellularly or extracellularly (Karow and Webb, 1965). This presumably leads to the formation of extended regions of hydrogen bonded (structured) water in the vicinity of the cryoprotective molecule, which in turn may stabilize the surface hydration of cell membranes and protect them from freezing injury. According to Doebler and Rinfret (1962), the capacity of a compound to form multiple hydrogen bonds with water is the best measure of its effectiveness as a cryoprotectant. There is considerable evidence (Kabayama and Patterson, 1958; Tait *et al.*, 1972; Franks, 1975) that the most favourable water–carbohydrate interactions involve equatorial –OH groups, but from the studies of Crowe *et al.* (1984a,b), it appears that the primary structure of a sugar (i.e. number and position of OH groups) is not involved in membrane preservation. Their stereochemistry seems more important in this. For example, trehalose and cellobiose have the same numbers of OH groups available for intermolecular bonding and all the OH groups are equatorial, yet trehalose is twice as effective as cellobiose in preserving membrane function (Crowe, 1984a).

9.4 Physico-chemical properties of α,α-trehalose in aqueous solution

Physico-chemical parameters give a direct measurement of the degree of displacement and disturbance of water by solutes. Intrinsic viscosity [η] reflects a dynamic tumbling effect of hydrated solute molecules within the structure of water. Viscosity constants in aqueous solution were derived from the time necessary for a given volume to flow through a capillary at a constant temperature, and intrinsic viscosity was obtained from a triple extrapolation procedure (Meffroy-Biguet, 1975a,b, 1978; Seuvres, 1988) which permits an accuracy determination of three viscosimetric functions:

$$\text{reduced specific viscosity } \eta_{sp}/c = [(\eta d - \eta_0 d_0 / \eta_0 d_0 c] \Leftrightarrow \eta_{sp}c$$
$$= [(td - t_0 d_0) / t_0 d_0 c]$$

$$\text{inherent viscosity } \eta_{inh} = [1/c \ln(\eta d / \eta_0 d_0)] \Leftrightarrow \eta_{inh}$$
$$= [1/c \ln(td / t_0 d_0)]$$

$$\text{reduced differential viscosity } \eta_{diff} = [(\eta d - \eta_0 d_0) / \eta dc] \Leftrightarrow \eta_{diff}$$
$$= [(td - t_0 d_0) / tdc]$$

[η] is defined as the limit of η_{sp}/c, η_{inh} and η_{diff} towards $c = 0$ (Figure 9.1) where η and η_0 are viscosities of solution and solvent respectively and *c* the concentration in g/dl, *d* density of solute and d_0 density of solvent, *t* and t_0 times of flow for the solute and the solvent in the Ubbelhode viscosimeter. The second viscometric constant is the Huggins constant *k'*. It is representative of the reactivity between the solute and the solvent (Kemp *et al.*, 1990; Shamil *et al.*, 1987), i.e. the ease of rupturing and establishing of intermolecular hydrogen bonds between water and solute, and is derived from Huggins' relationship (Huggin, 1942)

$$\eta_{sp}/c = [\eta] + k'[\eta]^2 c + \ldots \Leftrightarrow k' = (\eta_{sp}/c - [\eta]) / [\eta]^2$$

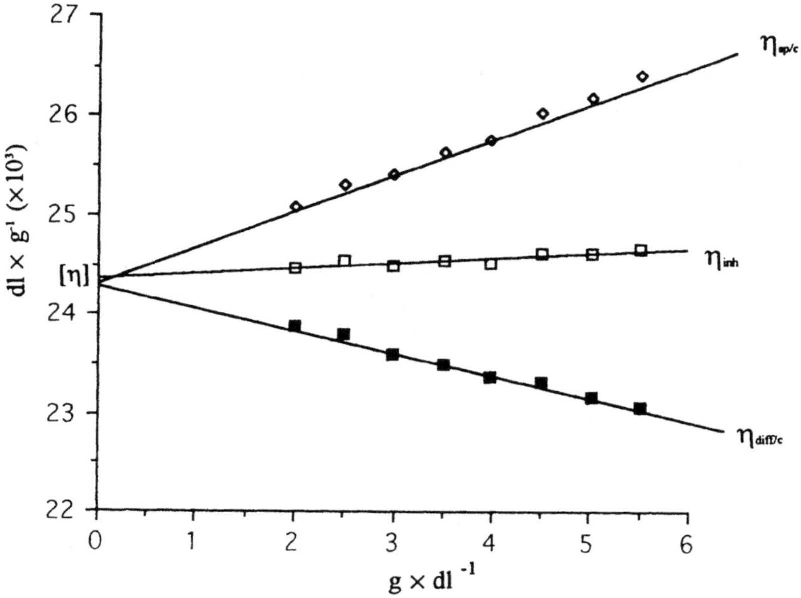

Figure 9.1 Triple extrapolation method of determination of intrinsic viscosity [η].

The Jones and Dole *B*-viscosity coefficient is another way to express the overall hydrodynamic volume of the solute. It is derived from the Jones and Dole equation (Jones and Dole, 1929)

$$\eta_{rel} = \eta/\eta_0 = 1 + Bc + Dc^2$$

where B is comparable to [η] and is the sum of B_{size} and $B_{structure}$, respectively accounting for the size of the solute and its effect on solvent structure (Miyajima *et al.*, 1983; Kemp *et al.*, 1990) with

$$B_{size} = 2.5\ \phi(V_{O_2}) / 1000$$

and

$$B_{structure} = B - B_{size}$$

All of these parameters are essential indicators of the dynamic interaction of trehalose molecules with water molecules, and they serve to interpret its unique effect.

Apparent specific volume reflects a static packing of solute molecules within the structure of water (Kemp *et al.*, 1990) and is indicative of the degree of compatibility between the solute and the water structure. Apparent molar volume $\phi(V_{O_2})$ is determined from density values using the following formulae

$$\phi(V_{O_2}) = Mw^* \left[(1/s) - (W_1/\varrho)\right] / W_2$$

where ϱ is the density of the solution, s is the density of water at 20°C ($0.9982\ g\ cm^{-3}$), W_1 is the mass fraction of water and W_2 is the mass fraction of the solute. From the B-viscosity coefficient and the apparent molar volume the hydration numbers h are calculated for solute molecules supposed to have a spherical shape, according to the equation (Herkovitz and Kelly, 1973)

$$h = [(1000B / 2.5 - (M_2V_2)] / M_1V_1$$

with M_1V_1 apparent molar volume of the solvent ($18\ ml\ mol^{-1}$ for water) and M_2V_2 apparent molar volume of the solute. The hydration number derives from the water molecules with a residence time in the vicinity of sugars longer than in that of other water molecules (Harvey and Symons, 1978).

^{1}H-NMR pulse relaxation studies represent another way to explore the state of order of sugar solutions. Spin–lattice (T_1) and spin–spin (T_2) relaxation times are determined by the tumbling action of the water and solute molecules and reflect the dynamic hydration state of the solute. When a sample is placed in an external stationary magnetic field, absorption of electromagnetic radiation will occur. This energy is then dissipated by two processes: spin–spin relaxation time, T_2; and spin–lattice relaxation time, T_1. The longitudinal relaxation time T_1 serves as a measure of the rate at which the spin system comes into equilibrium with environment. The transverse relaxation time, T_2, characterises the rate at which the transverse component decays to zero and determines the width of the line of the nuclear magnetic resonance. T_2 is, in addition, determined by the exchange of protons, and is thus relatable to the hydronium ion concentration. The net water mobility measured by low resolution NMR is usually a combined contribution of more than one motion of water molecules. Because water exchanges very rapidly, the results of such NMR experiments reflect the average of motion of all water molecules. Again, these effects aid in the explanation of the sweet properties of α,α-trehalose and its unique effects in a water medium.

9.5 Interpretation of interactions of α,α-trehalose with water

Intrinsic viscosity [η] is strongly dependent on the conformation, size and state of hydration of the molecule. The values of [η] for the disaccharides are close to each other but the hydration shell surrounding α,α-trehalose is slightly larger than other disaccharides (Table 9.1). If it is assumed that [η] is a shape factor of the hydrated molecule, sugars have similar intrinsic viscosity values because of the essentially hydrophilic nature of their hydration and the quasi-spherical shape of the hydrated molecule. k' values for sucrose and trehalose are similar and greater than those of

Table 9.1 Physico-chemical parameters of α,α-trehalose, maltose, lactose and sucrose in aqueous solution, at 20°C

Parameters	α,α-Trehalose	Maltose	Lactose	Sucrose
[η] ($cm^3 g^{-1}$)	2.58	2.55	2.50	2.45
k'	1.28	1.0	0.85	1.27
B(1 mol^{-1})	0.88	0.87	0.89	0.83
B_{sise} (1 mol^{-1})	0.52	0.54	0.52	0.53
$B_{structure}$ (1 mol^{-1})	0.36	0.33	0.37	0.30
B_{size}/B (%)	59	62	58	64
$B_{structure}/B$(%)	41	38	42	36
V_{O_2} (ml g^{-1})	0.61	0.63	0.61	0.62
$\phi(V_{O_2})$ (ml mol^{-1})	208	216	209	211
h	8.0	7.3	8.3	6.8

maltose and lactose, indicating a better exchange of hydration water with the surrounding water. Low values of Huggins' constant may originate from the fact that the exchange of water molecules between the hydration shell and the surrounding water is prevented by its greater hydrophobic character.

The contributions of B_{size} and $B_{structure}$ to the hydrodynamic molar volume B are respectively about 1/3 and 2/3, which correspond with values previously found for sugars (Mathlouthi *et al.*, 1993). Hydrophilic hydration increases the hydrodynamic volume of α,α-trehalose. However, with α,α-trehalose or lactose, the hydration layer is more disturbed than with sucrose or maltose (Galema and Høiland, 1991).

Molecules of similar constitution have close apparent molal volumes. Obviously, a more hydrophilic molecule is more likely to interact with water. If it packs well within the water structure it will exhibit a low apparent molar volume. Overall, carbohydrates have small apparent specific volumes in relation to their molecular weights as a result of extensive solute–solvent interaction (Barone *et al.*, 1983). The apparent specific volumes of sugar at normal tasting levels usually lie between 0.60 and 0.63 $cm^3 g^{-1}$. α,α-Trehalose and maltose are at these two extremes. The value for α,α-trehalose is the lowest for the disaccharides. α,α-Trehalose is well hydrated, and packs better among water molecules than the other cryoprotectant sugars, with maltose the least compatible. The relative position of the hydroxyl groups $(OH)_4$ and $(OH)_2$ in the carbohydrate molecule is important in determining the overall compatibility of the carbohydrate with the three-dimensional hydrogen-bonded structure of water.

One pecularity of α,α-trehalose is its high ratio of hydrogen bond numbers, with 11 donors and 3 acceptors. Maltose has 8 donors and 4 acceptors whereas lactose has 7 donors and 4 acceptors (Gaffney *et al.*, 1988). Hydration numbers of α,α-trehalose and lactose are comparable, but the significance of these numbers is not the same. α,α-Trehalose attracts water molecules whereas lactose, being less soluble in water tends to repulse them and to immobilize them in a 'clathrate' structure. Difference between α-1,1 and α-1,4 linkages might also make different numbers of hydration sites available. The α-1,4 linkages cause structural incompatibility between maltose and the ice-like structure of water (Miyajima *et al.*, 1983). The hydration number depends on many factors, such as the number or position (axial or equatorial) of each OH and the balance between hydrophilicity and hydrophobicity of the solute. α,α-Trehalose is a glucose dimer. Its OH groups are therefore in equatorial positions, which improves the establishment of linear hydrogen bonds between the solute and the solvent, leading to the stabilization of their hydration shells (Barone *et al.*, 1983; Kabayama and Patterson, 1958). α,α-Trehalose is a flexible molecule which predominantly adopts a 'folded' conformation (Rees and Thom, 1977), in which adjacent sugar rings shield the C–H groups from water and the hydroxyl groups are exposed to water molecules. This conformation increases the availability of the two hydroxyl groups $(OH)_6$ and $(OH)_2$ to establish some hydrogen bonds with water (Rees and Thom, 1977). This particular arrangement of a hydrophobic area packed inside the molecule and a hydrophilic area exposed to water, may explain the low value of V_{O_2}, showing a high compatibility between the sugar and the structure of water.

Spin–lattice (T_1) and spin–spin (T_2) relaxation times values are presented in Tables 9.2 and 9.3. ^{1}H-NMR pulse reflects the overall perturbation of water molecules because the proton NMR relaxation of dilute aqueous solutions is dominated by that of the solvent and observation of the solute relaxation can be difficult (Birch *et al.*, 1989). Within any particular disaccharide, increasing the concentration of the solute causes a decrease in T_2 value, as the proportion of ordered protons in solution is increased. With increasing concentration of a solute in water, the amount of 'free' water decreases whereas the relative amount of 'bonded' water increases (Mora-Gutierrez and Baianu, 1989). The molecular tumbling is slower and

Table 9.2 Spin–spin relaxation times (T_2) in aqueous solution, at 20°C for glucose, fructose, α,α-rehalose and sucrose

Components	Spin–spin relaxation times T_2 (sec)		
	Concentration (g/100 ml)		
	5	10	1
α,α-Trehalose	3.2	2.5	2.2
Sucrose	2.7	2.1	2.0
Glucose	1.8	1.1	0.9
Fructose	2.9	2.3	1.6

Table 9.3 Spin–lattice relaxation times (T_1) in aqueous solution, at 20°C for glucose, fructose, α,α-trehalose and sucrose

Components	Spin–lattice relaxation times T_1 (sec)		
	Concentration (g/100 ml)		
	5	10	15
α,α-Trehalose	4.2	3.6	3.3
Sucrose	4.0	3.6	3.0
Glucose	4.1	3.4	3.3
Fructose	3.8	3.5	3.3

therefore the relaxation time T_2 is shorter. Pulse NMR analysis (Birch *et al.*, 1989) has shown that sucrose and maltose solutions probably contain only hydration water. When concentration is increased, sucrose molecules form between them intermolecular hydrogen bonds, resulting in the formation of clusters of sucrose molecules in water (Mora-Gutierrez and Baianu, 1989). Such clusters containing bound water molecules would have longer relaxation times than unassociated sucrose molecules. Similar phenomena can be suggested to explain the highest values of T_2 observed for α,α-trehalose. T_2 value for glucose is lower than other sugars. The OH groups of glucose in the equatorial configuration fit within the quasi-tetrahedral arrangement of water molecules in the liquid state (Tait *et al.*, 1972). Solutions with greater degrees of order may be expected to exchange energy more easily by proton–proton spin relaxation (Birch and Karim, 1992). As a result, binding of water to glucose is stronger causing T_2 values to be shorter than other mono/disaccharide solutions at the same concentration. The T_2 value for pure water is about 4 sec. Values slightly below indicate that of bulk water is an indication that water is not tightly bound. Such behaviour has been observed with solutions of glucose, fructose, ribose, maltose and lactose (Mora-Gutierrez and Baianu, 1989; Lai and Schmidt, 1990, 1991; Richardson *et al.*, 1987). All these sugars share with α,α-trehalose the characteristic of establishing strong hydrogen bonds with water molecules.

α,α-Trehalose is subject to 'preferential hydration' due to its equatorial OH groups. This type of specific hydration structure may account for the superior cryoprotective effect of ring polyhydroxy compounds (Doebbler, 1966) because equatorially substituted sugars may prevent ice nucleation due to the incompatibility between the ice lattice geometry and the carbohydrate hydration structure (Franks, 1975).

9.6 Intensity/time relationships of α,α-trehalose: relation of its physico-chemical properties to its sweet taste

9.6.1 Relationships between the solution properties and sweet taste of α,α-trehalose

Before access to the sweet receptor the stimulus passes through layers of saliva as the aqueous medium for propagating the sweet taste perception. Sapid molecules must be water-soluble so their solution properties affect their taste. A sapid molecule must arrive at an appropriate receptor region of the taste epithelium before binding and activation of the receptor take place. Previous work (Mathlouthi and Luu, 1980; Mathlouthi, 1983; Mathlouthi *et al.*, 1986; Birch and Shamil, 1986) has demonstrated that an understanding of mechanism of the taste chemoreception might be achieved by studying water/solute interactions such as viscometric constants, intrinsic viscosity [η], Huggins coefficient k', apparent molar and apparent specific volumes. Apparent specific volume may be the prime determinant of taste quality (Birch and Kemp, 1989). This measurement of the direct displacement of the water by solute, has already been said to influence the sweetness of sugars (Shallenberger and Acree, 1967, 1969; Birch, 1976, 1983), as well as the arrival of sweet molecules at receptor sites (Birch, 1981). The mutual interaction of water and the stimulus molecule is therefore likely to affect the occupation and the activitation of receptors, detected by the quality, intensity and persistence of response (Shamil *et al.*, 1987).

9.6.2 Intensity and persistence of sweetness as a function of concentration

A time–intensity test measures the rate, duration and intensity of a single stimulus (Amerine *et al.*, 1965). The values of sweetness intensity and persistence obtained with SMURF (Sensory Measurement Unit Recording Flux) (Birch and Munton, 1981) plots are reported in Tables 9.4 and 9.5 for α,α-trehalose solutions, and also for glucose, fructose and sucrose, lactose and maltose solutions. The values represent the mean panel response with standard deviations given in parentheses. The variation of sweetness intensity as function of concentration is illustrated in Figure 9.2. Persistence data are reported in Figure 9.3. Increasing concentration leads to a gain in perceived sweetness of α,α-trehalose, maltose and lactose, with an expansion trend exactly like that of D-glucose, whereas fructose and sucrose exhibit a compression near 7–9% (w/v). Such observations are confirmed when the ANOVA test at a significance level of 0.05–0.01 is applied (Table 9.6).

Compared with other sugars, the sweetness of α,α-trehalose is lower but its persistence is longer. Analysis of variance reveals statistically

Table 9.4 Sweetness intensity[a] of α,α-trehalose, glucose, fructose, sucrose, maltose and lactose at 20°C

Sugars	Concentration (w/v) 2.3%	4.6%	6.9%	9.2%
α,α-Trehalose	4.3 (0.61)	10.9 (1.12)	21.2 (1.73)	35.7 (1.97)
Glucose	14.5 (3.55)	25.9 (4.38)	39.5 (5.32)	61.7 (4.64)
Fructose	28.7 (3.36)	61.7 (7.10)	84.9 (3.13)	89.24 (3.29)
Sucrose	27.8 (3.52)	54.8 (6.71)	75.0 (5.04)	90.2 (2.96)
Maltose	6.1 (0.91)	13.3 (1.46)	23.3 (1.79)	38.9 (2.93)
Lactose	5.9 (0.92)	13.6 (1.55)	26.6 (2.52)	40.8 (3.64)

[a] Intensity mean values are SMURF values and (SD).

Table 9.5 Sweetness persistence[a] of α,α-trehalose, glucose, fructose, sucrose, maltose and lactose at 20°C

Sugars	Concentration (w/v) 2.3%	4.6%	6.9%	9.2%
α,α-Trehalose	4.3 (0.71)	10.6 (0.89)	17.2 (1.35)	23.9 (1.71)
Glucose	3.3 (0.45)	5.8 (1.04)	6.9 (1.42)	8.5 (1.43)
Fructose	5.6 (0.98)	9.5 (1.86)	10.0 (2.09)	9.7 (1.36)
Sucrose	6.5 (1.20)	8.2 (1.58)	10.0 (1.43)	12.2 (2.48)
Maltose	7.9 (1.40)	12.1 (1.34)	16.3 (1.48)	22.6 (2.27)
Lactose	7.0 (1.19)	11.7 (1.26)	16.1 (1.73)	18.7 (2.27)

[a] Persistence values are in seconds and (SD).

significant differences in persistence only for α,α-trehalose when the concentration increases (Table 9.6). The order of taste intensity is: sucrose > maltose = lactose > trehalose (Diamant *et al.*, 1963, 1965; Jakinovich, 1976). Whether a sugar is reducing or non-reducing does not affect the intensity of the stimulation, but for the reducing disaccharides, it is the non-reducing glycosyl residue that binds to the sweet receptor (Jakinovich, 1976). α,α-Trehalose is a non-reducing disaccharide, a poor stimulus compared to sucrose, and is no sweeter than its reducing counterparts, lactose and maltose. α,α-Trehalose is a dimer of glucose, and it has been

Table 9.6 Analysis of variance on sensory results of α,α-trehalose, glucose, fructose, sucrose, maltose and lactose

	Range of concentration variation (g/100 ml)						
	2.3		4.6		6.9		9.2
α, α-Trehalose							
Intensity		**		**		**	
Persistence		**		**		**	
Glucose							
Intensity		ns		ns		**	
Persistence		ns		ns		ns	
Sucrose							
Intensity		**		*		*	
Persistence		ns		ns		ns	
Fructose							
Intensity		**		**		ns	
Persistence		ns		ns		ns	
Maltose							
Intensity		**		**		**	
Persistence		ns		ns		ns	
Lactose							
Intensity		**		**		**	
Persistence		ns		ns		ns	

Key: **, $p < 0.01$; **, $p < 0.05$; ns, not significant.

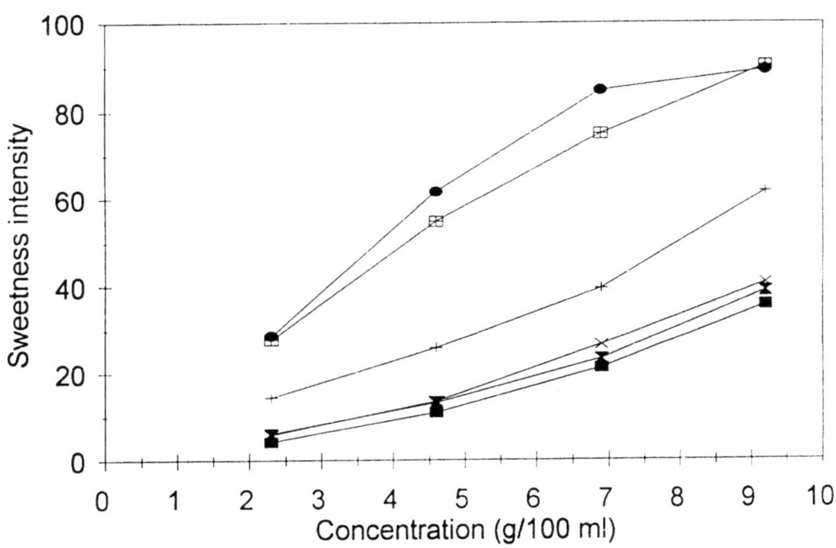

Figure 9.2 Variation of the sweetness intensity in function of concentration. ■, Trehalose; +, glucose; ●, fructose; □, sucrose; ⧗, maltose, ×, lactose.

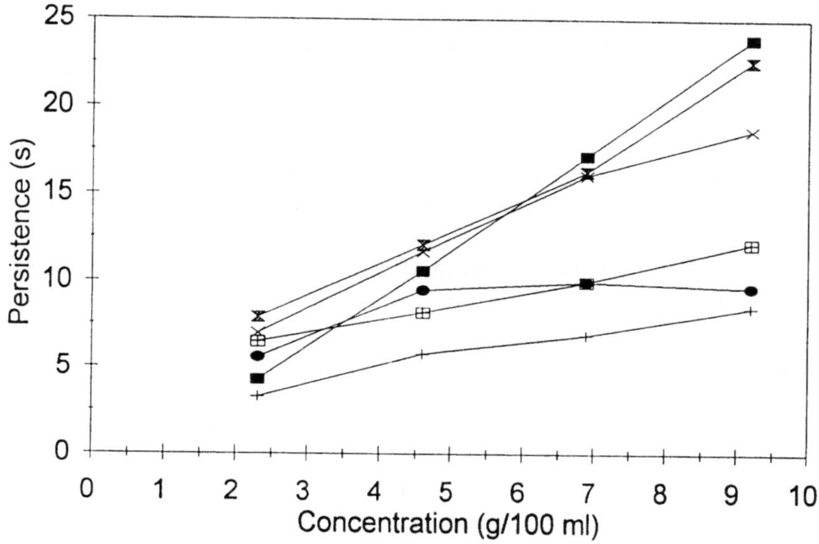

Figure 9.3 Variation of the sweetness persistence in function of concentration. ■, Trelahose; +, glucose; ●, fructose; □, sucrose; ⧗, maltose; X, lactose.

suggested as a general characteristic (Birch *et al.*, 1970 that dissacharides (probably with the exception of sucrose) are less sweet than their parent monosaccharides. In α,α-trehalose only one of the glucose residues binds to the taste receptor, the other being excluded, presumably because of steric hindrance involving the substituents at position C-1 of the glucopyranose ring. When Beidler (1954) formulated his taste theory, he envisioned a single stimulus molecule binding to a single receptor site to form a stimulus–receptor complex. Jakinovich (1976) explained the dominant effectiveness of sucrose compared to disaccharides by the presence of a sucrose-receptor site. The other disaccharides may fit into the same site but not perfectly. However, Lee and Birch (1975) suggested that it is the fructofuranosyl residue of sucrose which elicits sweetness, glucose being half as sweet as sucrose. For access to the receptor site, macroscopic and physicochemical data suggest that there is an optimal volume corresponding to a certain packing of water molecules around sweet molecules (Mathlouthi *et al.*, 1993). From the solution properties of α,α-trehalose, it is concluded that, due to its peculiar conformation in aqueous solution, α,α-trehalose is surrounded by a large and stable hydration shell (high intrinsic viscosity value [η]) and it is highly compatible with the water structure (low apparent specific volume V_{O_2}). Molecules with low apparent specific volumes may be conveyed by water to deep layers of the taste epithelium. Water acts as a vehicle to transport solutes deep into the receptor epithelium, greater depth being achieved by those solutes which interact with water (Shamil *et al.*, 1987). α,α-Trehalose molecules are efficient at packing in the localized receptor environment. In terms of the 'orderly queue' proposed by Birch *et al.* (1980), this means a more efficient localized concentration with high persistence. An alternative explanation is that the intensity of the sweet taste is related to the degree of mobility of the water molecules around the sugar. According to Mathlouthi (1984), when free water molecules are in the vicinity of the sweet receptor, the transport of Na^+/K^+ ions across the membrane of the tongue is facilitated. The more the water is dissociated, the greater the permeability of the membrane to Na^+. This leads to a high value of the membrane potential (Sybesma, 1977) and could enhance the intensity of the sweet taste. But around α,α-trehalose the translational motions of water molecules in solution are not very strong (low value for k') due to the good fit between the equatorially disposed hydroxyl groups and the structure of water. This stereochemical compatibility (water–trehalose) prevents AH–B bonding to the receptor, or at least weakens it, and confers lower sweetness on α,α-trehalose than on D-glucose, D-fructose or other disaccharides.

9.6.3 Power functions of intensity and persistence of α,α-trehalose

Intensity and persistence may be related to concentration by a power function of the type

$$S = kC^{ns}, \ P = kC^{np}$$

with S intensity, P persistence, n exponent and k constant (Moskowitz, 1977; Stevens, 1957). Constants k and n are determined from plotting log S and log P as functions of concentrations as the ordinate at the origin and the slope of the line, respectively. k represents the ability of the sweet molecules to reach the receptor whereas n reflects a more profound interaction between stimulus and receptor depending on chemical structure (Moskowitz, 1977). Their values are listed in Table 9.7 giving both intensity and persistence. The higher n value for α,α-trehalose may be

Table 9.7 k_s, n_s, k_p and n_p coefficients, correlation coefficients r_s and r_p for sucrose, glucose, fructose, α,α-trehalose, maltose and lactose, at 20°C

Sugars	k_s	n_s	r_s	k_p	n_p	r_p
Sucrose	14.1	0.83	0.99	4.46	0.46	0.98
Glucose	5.7	1.12	0.99	2.0	0.87	0.99
Fructose	15.24	0.85	0.97	4.16	0.45	0.89
Trehalose	1.15	1.51	0.99	2.88	0.95	0.99
Maltose	1.83	1.34	0.99	4.05	0.75	0.99
Lactose	1.70	1.42	0.99	3.78	0.72	0.99

interpreted as indicating that α,α-trehalose is as sweet or even sweeter than other mono- and disaccharides when the concentration is high enough. The ease of access to the sites is low for α,α-trehalose, as also is its efficiency of interaction with the sweet receptor. It seems that the linkage between constituents of the disaccharides partially governs sweetness (Lee and Birch, 1975). In general, persistence exponents are always lower than intensity exponents (Munton and Birch, 1985). For a given sugar, persistence of sweetness will be maximal at a higher concentration than that at which the sweetness intensity is maximal (Birch and Ogunmoyela, 1980). This gives a clue to localized concentration effects on the taste receptor and the relative accessibilities of these sites (Munton and Birch, 1985). These results show that intensity and persistence are very likely to proceed from different and independent mechanisms of taste perception (Birch *et al.*, 1980).

9.7 Protective effect of α,α-trehalose in freeze-dried fruit purées

The colour retention of freeze-dried fruit purée is improved by the presence of trehalose. α,α-Trehalose gives a lower brown score and a higher yellow score than mannitol or no treatment (Figure 9.4). However, it is important to note that freeze-dried samples are very different from fresh samples, and much colour change still takes place. Trehalose had

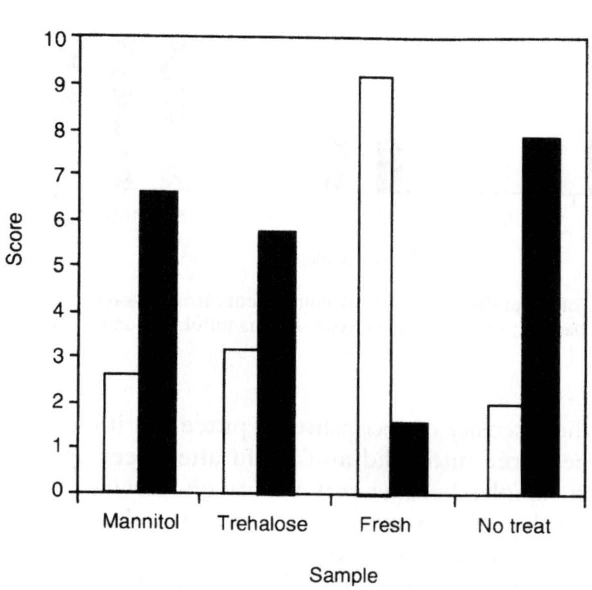

Figure 9.4 Colour assessment of freeze-dried banana purée. □, Yellow; ■, brown.

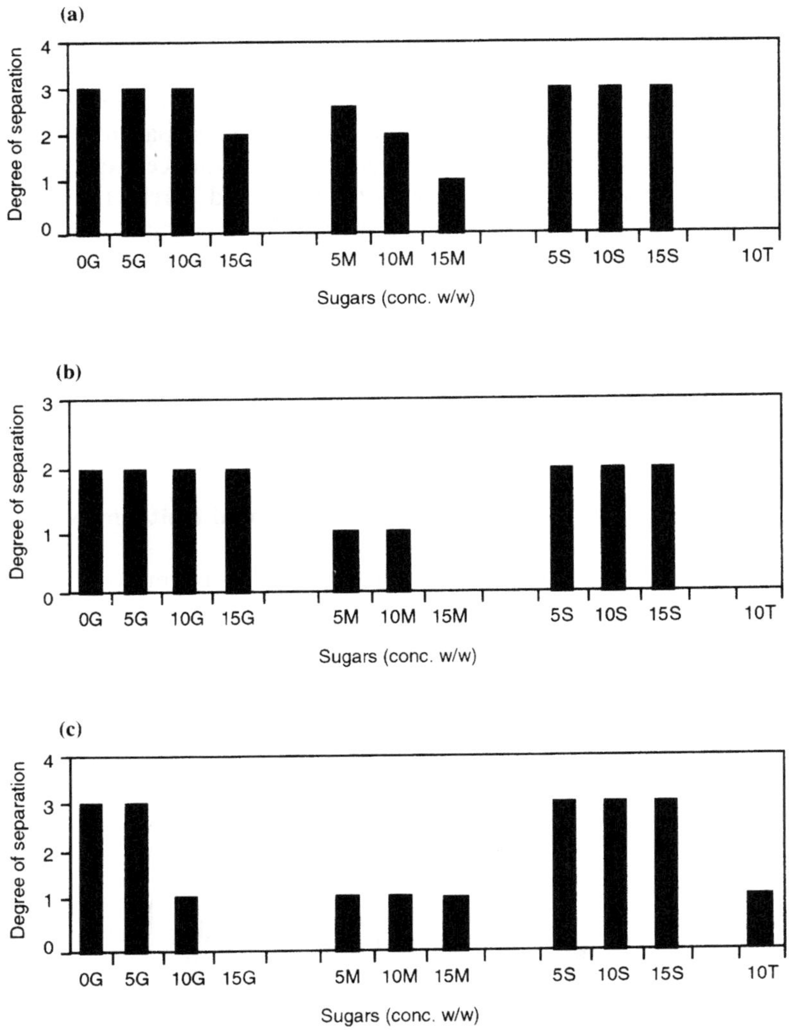

Figure 9.5 Degree of separation of aqueous components from thawed-frozen fruit slurry. (a) Apple; (b) banana; (c) pear. G, glucose; M, mannitol; S, sucrose; T, trehalose.

little effect on the viscosity of reconstituted purée but it did minimize the separation of the purée into solid and liquid after freezing (Figure 9.5); untreated frozen samples became very watery on thawing. It was noticed by some panellists that off-flavours are associated with banana freeze-dried in the absence of trehalose. These flavours are described as straw-like and woody and appear to linger in the mouth for some time. The addition of trehalose eliminates this problem and gives a more 'rounded' flavour. It is suspected that this is attributable to the loss of certain volatiles from the untreated material prior to reconstitution. Trehalose may have the effect of trapping these volatiles in food and conferring a more smooth and fresher flavour.

Physico-chemical and sensory analysis measurements confirm that cryoprotective effects of α,α-trehalose may be related to its ability to structure the surrounding water. Greater interaction between a cryoprotectant and the water lattice leads to greater protein–protein repulsion, resulting in a reduction in the quantity of frozen water in the protein matrix (Noguchi, 1974). Ice crystal growth is inhibited, as well as the migration of bonded water molecules from the protein which stabilize in their native form during the freeze-drying process. α,α-Trehalose has no direct internal hydrogen bond (Rosser, 1991). This gives it an unusual flexibility so that the molecule conforms more closely to the irregular surface of macromolecules. The disaccharide molecules could create a stable, immobile network forming a sort of 'coat' around the proteins, enzymes or indeed all biomolecules whose structural and functional integrity is preserved from damage during freeze-drying.

9.8 Conclusion

α,α-Trehalose undergoes a hydrophilic hydration. The peculiar conformation it adopts in aqueous solution gives it good compatibility with water structure. Physico-chemical and sensory analyses have been used to explain its particular taste properties, as well as its unique cryoprotectant properties. Its long taste persistence may be related to its capacity to fit in an orderly manner to the irregular protein surfaces of a taste receptor. This accompanies its freeze-protective effects on isolated proteins and phospholipid bilayers (Crowe *et al.*, 1984).

Acknowledgement

The authors thank the European Community for a grant in support of this work and Mr Ian Hunt for his work on food technology applications.

References

Amerine, M.A., Pangborn, R.M. and Roessler, E.B. (1965) *Principles of Sensory Evaluation of Food*, Academic Press, New York.

Barone, G., Castronuovo, G., Doucas, D., Ella, V. and Mattia, C.A. (1983) Excess of enthalpies of aqueous solutions of monosaccharides at 298.15 K: Pentoses and deoxysugars. *J. Phys. Chem.* **87**, 1931–37.

Beidler, L.M. (1954) A theory of taste stimulation. *J. Physiol.* **38**, 133–9.

Birch, G.G. (1963) Trehalose. In *Advances in Carbohydrate Chemistry* (eds. M.L. Wolfrom and S.R. Tipson), Academic Press, London, pp. 201–25.

Birch, G.G. (1976) Structural relationships of sugars to taste. *CRC Critical Reviews of Food Science and Nutrition* **8**, 57–91.

Birch, G.G. (1981) A molecular approach to intensity/time phenomena in sugar sweetness. In *Biochemistry of Taste and Olfaction* (eds. R.H. Cagnan and M.R. Kare), Academic Press, pp. 163–73.

Birch, G.G. (1983) QSAR of carbohydrate sweeteners and their role in time-intensity studies. *Chem. & Ind.* **3**, 13–15.

Birch, G.G. and Karim, R. (1992) Apparent molar volumes and ^{1}H-NMR relaxation values of glucose syrups. *J. Sci. Food Agric.* **58**, 563–68.

Birch, G.G. and Kemp, S.E. (1989) Apparent specific volumes and taste of amino acids. *Chem. Senses* **14**(2), 249–58.

Birch, G.G. and Munton, S.L. (1981) Use of the SMURF in the taste analysis chemical. *Chemical Senses* **6**(1), 45–52.

Birch, G.G. and Ogunmoeyela, G. (1980) Effect of surfactants on the taste and flavour of drinking chocolate. *J. Food. Sci.* **45**(4), 981–4.

Birch, G.G. and Shamil, S. (1986) Structure-activity relationships in sweetness. *Food Chem.* **21**, 245–58.

Birch, G.G., Cowell, N.D. and Eyton, D. (1970) A quantitative investigation of Shallenberger's sweetness hypothesis. *J. Food. Technol.* **5**, 277–80.

Birch, G.G., Latymer, Z. and Holloway, M. (1980) Intensity-time relationships in sweetness. *Chem. Senses* **5**(1), 63–78.

Birch, G.G., Grigor, J. and Derbyshire, W. (1989) Identification of proton type in concentrated sweet solutions by pulsed NMR analysis. *J. Sol. Chem.* **18**(8), 795–801.

Chandrasekhar, L. and Gaber, B.P. (1988) *Journal of Biomolecular Structure and Dynamics* **5**, 1163–1171.

Clegg, J.S. (1965) *Comparative Biochemistry and Physiology* **14**, 135.

Clegg, J.S. (1986) The physical properties and metabolic status of Artemia cyst at low water contents. In *Membranes, metabolism and dry organisms* (ed. A.C. Leopold), Cornell University Press, Ithaca, New York, pp. 169–87.

Crowe, J.H. (1971) Anhydrobiosis: An unsolved problem. *Am. Naturalist* **105**, 563–74.

Crowe, L.M. and Crowe, J.H. (1982) Hydration dependent hexagonal phase lipid in a biological membrane. *Arch. Biochem. Biophys.* **217**(2), 582–87.

Crowe, J.H. and Crowe, L.M. (1984) *Biological Membrane* **5**, 57.

Crowe, J.H., Crowe, L.M. and Jackson, S.A. (1983a) Preservation of structural and functional activity in lyophilized sarcoplasmic reticulum. *Arch. Biochem. Biophys.* **220**(2), 477–84.

Crowe, J.H., Crowe, L.M. and Mouradian, R. (1983b) *Cryobiology* **20**, 346–356.

Crowe, L.M., Mouradian, R., Crowe, J.H., Jackson, S.A. and Womersley, C. (1984a) Effects of carbohydrates on membrane stability at low water activity. *Biochem. Biophys. Acta* **769**, 141–150.

Crowe, J.H., Crowe, L.M. and Chapman, D. (1984b) Preservation of membranes in anhydrobiotic organism: Role of water. *Science* **223**, 701–3.

Crowe, J.H., Crowe, L.M., Rudolph, A.S., Womersley, C. and Appel, L. (1985) *Archives of Biochemistry and Biophysics* **232**, 240–7.

Crowe, J.H., Spargo, B.J. and Crowe, L.M. (1987) Preservation of dry liposomes does not require retention of residual water. *Proc. Natl. Acad. Sci.* **83**, 1537–40.

Crowe, J.H., Crowe, L.M., Carpenter, J.F., Rudolph, A.S., Aurell Wistrom, C., Spargo, B.J. and Anchordoguy, T.J. (1988) Interactions of sugars with membranes. *Biochem. Biophys. Acta* **947**, 367–84.

DaPonte, D.J.B., Herfst, J.M., Roozen, J.P. and Pilnik, W. (1985a) Effects of different types of carrageenans and carboxymethylcellulose on stability of frozen stores minced fillets of cods, whiting. *J. Food Technol.***20**, 587–598.

DaPonte, D.J.B., Herfst, J.M., Roozen, J.P. and Pilnik, W. (1985b) Effects of additions on stability of frozen stored minced fillets of whiting: I. Various anionic hydrocolloids. *J. Food Quali.* **8**, 51–69.

DaPonte, D.J.B., Herfst, J.M., Roozen, J.P. and Pilnik, W. (1985c) Effects of additions on stability of frozen stored minced fillets of whiting: I. Various anionic and neutral hydrocolloids. *J. Food Quali.* **8**, 183–198.

Diamant, H., Funakoshi, M., Strom, L. and Zotterman, Y. (1963) *Olfaction and Taste*, Vol. 1, Pergamon Press, Oxford, pp. 193–203.

Diamant, H., Oakley, B., Strom, L., Wells, C. and Zotterman, Y. (1965) *Acta Physiologica Scandinavia* **64**, 67–74.

Doebbler, G.F. (1966) *Cryobiology,* **3**, 1.

Doebbler, G.F. and Rinfret, A.P. (1962) *Biochimica et Biophysica Acta* **58** 449.

Franks, F. (1975) *Water Relation in Foods* (ed. R.B. Duckworth), Academic Press, London.

Gaffney, S.H., Haslam, E., Lilley, T.H. and Ward, T.R. (1988) Homeotactic and heterotactic interactions in aqueous solutions containing some disaccharides. *J. Chem. Soc. Faraday Trans 1* **84**(8), 2545–52.

Galema, S.A. and Høiland, H. (1991) Stereochemical aspects of hydration of carbohydrates in aqueous solutions – 3. Density and ultrasound measurements. *J. Phys. Chem.* **95**, 5321–26.

Green, J.L. and Angell, C.A. (1989) Phase relation and vitrification in saccharide/water solutions and trehalose anomaly. *J. Phys. Chem.* **93**, 2880–2882.

Harvey, J.M. and Symons, M.C.R. (1978) The hydration of monosaccharides. A NMR study. *J. Sol. Chem.* **7**, 571–586.

Herkovitz, T.T. and Kelley, M.M. (1973) Viscosity studies of aqueous solutions of alcohols, ureas and amides. *J. Phys. Chem.* **77**, 381–8.

Huggins, M. (1942) Viscosity of dilute solutions of long chain molecules – Dependence of concentration. *J. Am. Chem. Soc.* **64**, 2716–18.

Jakinovich, W. (1976) *Brain Research* **110**, 481–90.

Jiang, S.T., Hwang, B.O. and Tsao, C.T. (1987a) Protein denaturation and changes in nucleotides of fish muscle during frozen storage. *J. Agric. Fd Chem.* **35**, 22–27.

Jiang, S.T., Hwang, B.O. and Tsao, C.T. (1987b) Effect of adenosine nucleotides and their derivatives on the denaturation of myofibrillar protein in vitro during frozen storage at −20°C. *J. Fd Sci.* **52**(1), 117–123.

Jones, G. and Dole, M. (1929) The viscosity of aqueous solutions of strong electrolytes with special reference to barium chloride. *J. Am. Chem. Soc.* **51**, 2950–64.

Johnston, D.S., Copperd, E., Parera, G.V. and Chapman, D. (1984) Langmuir film balance study of the interactions between carbohydrates and phospholipid monolayers. *Biochemistry* **23**, 6912.

Kabayama, M.A. and Patterson, D. (1958) The thermodynamics of mutarotation for some sugars. *Can. J. Chem.* **36**, 563–573.

Karow, A.M. Jr. and Webb, W.R. (1965) *Cryobiology* **2**, 99.

Kemp, S.E., Birch, G.G., Portmann, M.O. and Mathlouthi, M. (1990) Intrinsic viscosity and apparent specific volumes of amino acids and sugars: 'Effective size' and taste of sapid molecules. *J. Sci. Food Agr.* **51**, 57–107.

Lai, H.S. and Schmidt, S.J. (1990) Water mobility and crystallization action of lactose–water systems by oxygen-17 and carbon13 NMR. *J. Food Sci.* **55**, 1345–40.

Lai, H.S. and Schmidt, S.J. (1991) Proton, deuterium and oxygen-17 nuclear magnetic resonance relaxation studies of lactose and sucrose–water system. *J. Agr. Food Chem.* **39**, 1921–26.

Lanier, T.C. and Akahane, T. (1986) Method of retarding denaturation of meat products. US patent 4 572 838.

Lee, C.K. (1980) Trehalose. In *Developments in Food Carbohydrate – 2* (ed. C.K. Lee), Applied Science Publisher Ltd, London, pp. 1–89.

Lee, C.K. and Birch, G.G. (1975) Structural function of taste in the sugar series: Binding characteristics of disaccharides. *J. Sci. Food Agr.* **26**, 1513–21.

Loomis, S.H., Carpenter, T.J. and Crowe, J.H. (1988) *Biochimica et Biophysica Act* **943**, 113.

Loomis, S.H., Carpenter, T.J., Achordoguy, T.J., Crowe, J.H. and Branchini, B.R. (1989) *Journal of Experimental Zoology* **252**, 9.

MacDonald, G.A. and Lanier, T. (1991) Carbohydrates as cryoprotectants for meat and surumi. *Food Technology* **March**, 150–159.

Mathlouthi, M. (1984) Relationships between the structure and the properties of carbohydrates in aqueous solution: Solute–solvent interactions and sweet taste of D-fructose, D-glucose and sucrose. *Food Chem.* **13**, 1–16.
Mathlouthi, M. and Luu, D.V. (1980) Laser-Raman spectra of D-glucose and sucrose in aqueous solution. *Carbohy. Res.* **81**, 203–13.
Mathlouthi, M., Seuvre, A.M. and Birch, G.G. (1986) Relationships between the structure and the properties of carbohydrates in aqueous solution. *Carbohy. Res.* **152**, 47–61.
Mathlouthi, M., Bressan, C., Portmann, M.O. and Serghat, S. (1993) Role of water structure in the sweet taste chemoreception. In *Sweet-Taste Chemoreception* (eds. M. Mathlouthi, J.A. Kanters and G.G. Birch), Elsevier Applied Science, London, pp. 141–74.
Matsumoto, J.J. (1979) Denaturation of fish muscle proteins during frozen storage. In *Proteins at Low Temperatures* (ed. O. Fennema), ACS Advances in Chemistry Series 180, American Chemical Society, Washington DC, pp. 206–224.
Matsumoto, J.J. (1980) Chemical deterioration of muscle proteins during frozen storage. In *Deterioration during frozen storage* (eds. J.R. Withaker and M. Fujimak), ACS Symposium Series 123, American Chemical Society, Washington DC, pp. 96–124.
Meffroy-Biget, A.M. (1975a) Flexible transcendental equation generalizing. *Comptes Rendus de l'Academie des Sciences, Serie C* **280**, 183–185.
Meffroy-Biget, A.M. (1975b) Solving of transcendental equation relating the relative viscosity to the related concentration variable. *Comptes Rendus de l'Academie des Sciences, Serie C* **280**, 263–265.
Meffroy-Biget, A.M. (1978) On two transcendental equations generalizing the relations between the viscosity of polymer solution on its concentration. *J. Chem. Phys.* **4**, 2646.
Miyajima, K., Sawada, M. and Nakagaki, M. (1983) Studies of aqueous solutions of disaccharides. II: Viscosity B-coefficients, apparent molar volumes and activity coefficients of D-glucose, maltose, maltotriose in aqueous solution. *Bul. Chem. Soc. Japan* **56**, 1954–7.
Mora-Gutierrez, A. and Baianu, I.C. (1989) ^{1}H-NMR relaxation and viscosity measurements on solutions and suspensions of carbohydrates and starch from corn: The investigation of carbohydrate hydration and sterochemical aggregation effects in relation to ^{17}O and ^{13}C NMR data for carbohydrate solutions *J. Agr. Food Chem.* **37**(6), 1459–65.
Moskowitz, H.R. (1973) Effects of solution temperature on taste intensity in humans. *Physiol. Behav.* **10**, 289.
Munton, S.L. and Birch G.G. (1985) Accession on sweet stimuli to receptors. I: Absolute dominance of one molecular species in binary mixtures. *J. Theor. Biol.* **112**, 539–551.
Noguchi, S. (1974) Studies on the control of denaturation of fish muscle proteins during frozen storage. *Sc.D. Thesis*, Sophia University, Tokyo.
Rees, D.A. and Thom, D. (1977) Polysaccharides conformation. Part 10: Solvent effects on the optical rotation and conformation of model carbohydrates. *J. Chem. Soc. Perkins II*, 191–201.
Richarson, S.J., Baianu, I.C. and Steinberg, M.P. (1987) Mobility of water in sucrose solution determined by deuterium and oxygen-17 nuclear magnetic resonance measurements. *J. Food Sci.* **52**(3), 806–9.
Roser, B. (1991) Trehalose, a new approach to premium dried foods. *Trends Food Sci. Technol.* **July**, 166–169.
Seuvre, A.M. (1988) Contribution à l'étude des interactions solvant–soluté en solution aqueuse de sucres et de sels. *Ph.D Thesis*, Rouen.
Shallenberger, R.S. and Acree, T.E. (1967) Molecular theory of sweet taste. *Nature* **216**, 480–2.
Shallenberger, R.S. and Acree, T.E. (1969) Molecular structure and sweet taste. *J. Agr. Food Chem.* **17**, 701–3.
Shamil, S., Birch, G.G., Mathlouthi, M and Clifford, M.N. (1987) Apparent molar volumes and taste of molecules with more than one sapophore. *Chem. Senses* **12**(2), 397–409.
Stevens, S.S. (ed.) (1957) *Pychophysics*, J. Wiley, New York.
Sussman, A.S. (1959) *Canadian Journal of Research* **27**, 749.
Sybesma, C. (1977) *Introduction to Biophysics*, Academic Press, New York, pp. 180–185.
Sych, J., Lacroix, C., Adambounou, L.T. and Castaigne, F., (1990a) Cryoprotective effects of Lactitol, Palatinit and Polydextrose on cod-surimi proteins during frozen storage. *J. Fd Sci.* **55**, 356.

Synch, J., Lacroix, C., Adambounou, L.T. and Castaigne, F. (1990b) Cryoprotective effects of some material on cod-surimi proteins during frozen storage. *J. Fd Sci.* **55**, 1222.
Tait, M.J., Suggett, A., Franks, F., Ablett, S. and Quickenden, P.A. (1972) *Journal of Solution Chemistry* **1**(2), 131–151.
Tandford, C. (ed.) (1980) *The hydrophobic effect.* J. Wiley, New York.
Wessel, J.P.H., Simmonds, C.K., Seaman, P.D. and Avery, L.W.J. (1981) *Annual Report No. 35*, Fishing Industry Research Institute, Cape Town, South Africa, 32 pp.

10 Intense sweeteners and calorie control: the weight of a body of evidence

A.G. RENWICK

10.1 Introduction

There is a wide range of intense sweeteners currently approved for use in human food and drink by national and international advisory bodies such as the European Scientific Committee for Food (SCF) and the Joint WHO/FAO Expert Committee on Food Additives (JECFA). Intense sweeteners currently used include acesulfame-K, alitame, aspartame, cyclamate, saccharin, sucralose and thaumatin, although different countries permit different sweeteners. Intense sweeteners are used in a wide range of products to provide sweetness without unnecessary calories as well as for technological reasons related to stability, texture, etc. The past 15 years have seen a rapid growth in the sales of foods and drinks containing intense sweeteners (Currie, 1994) which has coincided with an increased awareness of the importance of consuming a balanced, healthy diet. Obesity and overweight are associated with increased health problems, particularly cardiovascular disease, and current advice includes not eating too much, reducing fat intake and taking exercise (DoH, 1992). Intense sweeteners have been accepted by consumers as a part of a healthy diet and lifestyle (Currie, 1994) and as a way of enjoying sweet foods and drinks without the intake of unnecessary calories.

Against this background, it is not surprising that considerable interest was generated by research published in 1986 which suggested that intense sweeteners increased appetite and caused rises in calorie intake and in weight gain. These observations (see below) were the exact opposite of what would intuitively be expected, and they overturned 'perceived wisdom' in this area. The possibility that so-called diet foods could make you put on weight attracted considerable media attention at the time. The results of these published studies were both unexpected and fascinating. Since then a large body of research has been performed to investigate the findings in greater detail. This subject has been reviewed previously (Rolls, 1991) and this chapter serves to update a review of the human data undertaken two years ago (Renwick, 1994).

10.2 The original observations

The suggestion that intense sweeteners could increase rather than decrease body weight arose in 1986 from two different areas of research, epidemiology and controlled experimental observations. Although both studies are open to criticisms (see below) the combination of two different types of study leading to the same unexpected conclusion was interesting and appeared credible.

10.2.1 Epidemiological evidence

The epidemiology evidence (Stellman and Garfinkel, 1986) arose from an analysis of a group selected from the massive American Cancer Society database developed for a prospective analysis of mortality in one million subjects in the USA. The study group analysed in relation to body weight and intense sweeteners was a narrowly defined cohort in order to minimise variability. The cohort consisted of 78 694 white women aged 50–69 years old with a high school education and no history which would be likely to affect body weight, e.g. changes in diet, diabetes, changes in smoking habits. Such a narrow selection group would exclude people using intense sweeteners as part of a calorie-controlled diet. The measurement of change of body weight was the difference in response to questions on what do you weigh now (in 1982)? – and what did you weigh a year ago? – which is open to major recall bias. The changes in body weight were related to the use of artificial sweeteners (saccharin and cyclamates). The results showed that the percentage of women who used intense sweeteners rose with increase in body weight. Surprisingly, a greater proportion of women who used artificial sweeteners had gained weight over the preceding 12 months, compared with non-users. Although there were minor indications of possible benefits of sweeteners for women who lost the most weight, the overall conclusion of the study reported by the authors was:

> *"these data do not support the hypothesis that long-term AS [artificial sweetener] use either helps in losing weight or prevents weight gain."*

This carefully worded conclusion is consistent with the study data but the authors did not discuss the potential for bias in their study design. The selection of the study group and stratification based on outcome have been identified as methodological deficiencies (Lavin *et al.*, 1994) but the major problem with the study is that the division into users and non-users was not randomised but by self-selection. In consequence, the groups were not equally matched, and motivation to use intense sweeteners represents a major basis which is insurmountable by any subsequent data analysis. Common sense suggests that women who have difficulty controlling their weight are more likely to use intense sweeteners, which is an alternative way of interpreting the study findings. The exclusion of those actively modifying their diet to reduce weight may have restricted the study user group to those who were half-hearted about tackling a weight problem, i.e. the 'cream cake plus coffee with intense sweetener' type. The study design can give no information on the weight gain the users would have had if they had not used intense sweeteners since the non-users are not an appropriate control group.

10.2.2 Experimental evidence

In May 1986, Blundell and Hill published a letter in the *Lancet* which reported the data on a controlled experiment with 95 young (18–22 years old) volunteers. The study was part of a programme of research on appetite, food taste and the intake of calories and of food. The study involved a comparison of sweet aqueous solutions of glucose (50 g in 200 ml) or aspartame (162 mg in 200 ml) with water (200 ml). Measurements of the sweetness intensity and pleasantness of a sugar solution and of appetite motivation (e.g. hunger, desire to eat and fullness) were made subsequently using visual analogue scales every 10 min over the following 60 min. The study showed that the glucose gave a decline in motivation to eat and an increase in ratings of fullness compared with baseline assessments. Unexpectedly, the aspartame solution resulted in an increased rated motivation to eat and decreased ratings of fullness. The appetite rating after aspartame was significantly increased from baseline "on certain occasions in the second half of the test period". The study is difficult to assess since important information was not reported. The authors claimed that the 'residual hunger' left after aspartame resulted in increased food intake, but no data were provided; given that the aspartame dose was equivalent to 188 kcal, a key question is whether the increased food intake exceeded 185 kcal (the difference between glucose and aspartame doses). The main problem for interpretation is the exact nature of the study design.

Important questions include:

- Were the 95 subjects given all three treatments? The text implies that different individuals had different treatments. For example ". . . *did the volunteers who consumed aspartame . . .?*" and "*For volunteers tested shortly after eating the aspartame load*". If the study was an inter-group comparison then matching of subjects based on body weight and obesity would be crucial. Also all measurements would have been against baseline rather than comparison with placebo (water).
- Was the study double-blind with respect to the sweet solutions since this could bias the results for subjective ratings?

10.3 Recent studies on the effects of intense sweeteners

Clearly neither study by itself provides definitive evidence that intense sweeteners increase food intake and body weight, but the observations pointed in the same direction at the same time and stimulated a number of subsequent studies. These new studies have investigated a range of observations from simple perceptions of appetite through biochemical changes to actual measures of calorie intake and body weight change. The recent data are complicated by the presence of numerous possible variables:

- sweetener(s) studied
- vehicle used and placebo/control
- measurements made
- time between sweetener administration and measurements
- status of subjects prior to study, e.g. time since last meal
- population studied, e.g. age, weight, sex, etc.

In the following review the publications are assessed based on the characteristics measured.

10.3.1 Intense sweeteners and motivation to eat

The original study design of Blundell and Hill has been adopted or extended in a number of subsequent studies (Table 10.1). A common feature to most was the assessment of appetite or motivation to eat at different times after treatment using a visual analogue scale. Despite the similarity of many protocols there was not a clear or consistent indication that intense sweeteners increased hunger compared with a control. Only four subsequent studies have reported a subjective increase in hunger ratings; two by Blundell and colleagues (one with saccharin in yoghurt and one essentially a repeat of their earlier study – see below), one by Tordoff and Alleva (1990a) who incorporated different concentrations of aspartame into chewing gum and one by Black *et al.* (1993), which gave an inconsistent effect.

The follow-up study by Blundell and his colleagues (Rogers *et al.*, 1988) was reported as a full publication with more details of the study design. The study group comprised 12 young normal-weight subjects, each of whom was given water, glucose (50 g), aspartame (as before), acesulfame-K or saccharin as equisweet aqueous solutions under single blind conditions on separate days in a counterbalanced design. The test solutions were administerd at noon, following which a series of motivational ratings were made for 1 h, after which a meal was offered. (The food intake data are discussed later.) Although glucose suppressed ratings of hunger and desire to eat, the intense sweeteners were associated with increased ratings during

Table 10.1 Hunger ratings following single doses of intense sweeteners

Sweetener[a]	Vehicle	n[b]	Time after dose	Hunger rating[c]	Reference
AceK	water	12	60 min	⟷	Rogers *et al.* (1988)
ASP	water	95	40–60 min	↑	Blundell and Hill (1986)
ASP	water	12	60 min	↑	Rogers *et al.* (1988)
ASP	water	27	60 min	⟷	Rogers *et al.* (1990)
ASP	drink	20	90 min	⟷	Anderson *et al.* (1989)[d]
ASP	drink	42	30, 60 min	⟷	Rolls *et al.* (1990)
ASP	drink	20	60 min	⟷	Black *et al.* (1991)
ASP	drink/water	18	60 min	⟷ ↑	Black *et al.* (1993)
ASP	dessert	32	up to 120 min	⟷	Rolls *et al.* (1989)
ASP	cheese	24	30, 60, 120 min	⟷	Drewnowski *et al.* (1994)
ASP	gum	120	30–90 min	↑	Tordoff and Alleva (1990a)
ASP	capsules	26	105 min	⟷	Ryan-Harshman *et al.* (1987)
ASP	capsules	27	60 min	↓	Rogers *et al.* (1990)
ASP	capsules	16	60 min	⟷	Rogers *et al.* (1991)
ASP	capsules	18	60 min	⟷	Black *et al.* (1993)
SACC	water	12	60 min	⟷	Rogers *et al.* (1988)
SACC + ASP	drink	20	30–60 min	⟷	Canty *et al.* (1991)
SACC	yoghurt	24	60 min	↑	Rogers and Blundell (1989)

[a] ASP = aspartame; SACC = saccharin; AceK = acesulfame-K. [b] n = number of subjects. [c] ⟷ = no difference compared with placebo; ↓ = decreased compared with placebo; ↑ = increased compared with placebo. [d] This study compared aspartame with cyclamate not with an unsweetened drink.

the 1 h period. The data for aspartame, but not those for acesulfame-K or saccharin, showed a statistically significant difference compared with the water solution. Ratings of fullness and prospective consumption showed smaller treatment-related differences that were not statistically significant. Therefore this study confirmed the effect for aspartame on appetite reported previously, but this did not translate into an increase in food intake (see Table 10.3 below). An increase in motivational ratings compared to placebo was also found when a saccharin-sweetened yoghurt was consumed (Rogers and Blundell, 1989).

In contrast, no significant effect was found with many other similar studies (Table 10.1), suggesting that if there is an effect of intense sweeteners on appetite, it is 'fragile' and easily perturbed by other variables. The data of Tordoff and Alleva (1990a) are particularly interesting in this regard since an effect was found when stimulation of sweet taste receptors was maximised by incorporation of the sweetener into chewing gum base. However, even this conclusion should be interpreted with caution because the effect was significant only in females and because of the possible influence of the unsweetened gum base which had a significantly less pleasant taste after chewing.

An important variable identified in the study of Black *et al.* (1993) was the volume of solution given, since a significant increase in subjective appetite ratings was produced by aspartame (340 mg) in 280 ml carbonated mineral water but not by aspartame (340 mg) in 560 ml of carbonated mineral water. However, this increased motivation to eat was not consistent across different measures of appetite, since the consumption of aspartame-sweetened mineral water resulted in subjects finding foods less appealing. Importantly, none of the treatments resulted in an increase in energy intake at a lunchtime meal given 65 min after the test solutions or after encapsulated aspartame (see Table 10.3 below).

The data from studies with aspartame in capsules showed that there was the possibility of a decrease in appetite following the ingestion of aspartame without tasting it (see below).

An interesting study on sweetness, but which did not use intense sweeteners, compared the satiating effects of meals providing the same number of calories but containing either sweet or non-sweet carbohydrates (De Graaf *et al.*, 1993). The sweet meals were found to suppress appetite for sweet foods more than for savoury foods, while the non-sweet meals suppressed appetite for savoury more than for sweet foods. Two studies were performed. One showed that the sweet carbohydrate meal was less satiating overall than the non-sweet carbohydrate meal, but the second study showed no difference and a suppression of food intake two hours after the sweet meal.

10.3.2 Intense sweeteners and changes in glucose and/or insulin

The original observations of Blundell and Hill (1986) were consistent with the hypothesis that stimulation of sweet taste receptors caused insulin release which in turn resulted in a lowering of blood sugar and a heightened motivation to eat. There have been a number of studies of insulin release following ingestion of intense sweeteners.

The ingestion of food causes the release of insulin which results in a decrease in blood glucose due to its uptake from the blood by tissues. The release of insulin takes place in two phases:

(a) *cephalic or preabsorption phase*, about 5 min after the ingestion of food. This precedes the absorption of glucose, is a minor and variable response (Teff *et al.*, 1991) and is caused by the organoleptic properties of the food i.e. its taste, smell and texture.
(b) *post-prandial phase*, after about 15 min, is the major phase of insulin release, and is related to the effects of absorbed glucose and/or amino acids.

The question of a possible effect of intense sweeteners on this system has been raised because it could provide a scientifically credible rationale for the proposed effects on hunger. If stimulation of sweet taste receptors was a major component of the cephalic phase of insulin release then the taste of intense sweeteners could stimulate this process, causing a decrease in blood glucose concentrations and thereby stimulating appetite. This hypothesis is supported by data from animal experiments (Ionescu *et al.*, 1988; Tordoff and Friedman, 1989).

Studies on the effects of intense sweeteners on blood insulin and/or glucose levels are summarised in Table 10.2. A decrease in blood glucose compared with baseline was reported following aspartame ingestion in both normal and diabetic subjects (Okuno *et al.*, 1986). In this study there was no placebo dose and the blood levels during the 3 hours after aspartame showed a steady decrease whereas the insulin levels did not change, which indicates that this was a simple fasting effect rather than a cephalic phase insulin response.

A problem with a number of the studies is that the timing of blood sample collection would not have allowed detection of the cephalic phase insulin response because it occurs very early, i.e. at about 4 min after administration (Teff *et al.*, 1991). The only study to demonstrate both an increase in insulin and a decrease in blood glucose (Teff *et al.*, 1991) used an aspartame-sweetened strawberry flavoured gelatin dessert combined with dairy fat to produce a mousse. The study was repeated on three separate days and the response was highly reproducible within subjects but showed wide inter-individual differences. Although the study demonstrated very clearly the presence of a cephalic phase insulin response, the contribution of the sweet taste itself was not investigated.

Table 10.2 Effects of intense sweeteners on insulin and blood glucose concentrations

(a) Healthy subjects

Sweetener[a]	Vehicle	n[b]	Time	Insulin concn.[c]	Glucose concn.[c]	Reference
AceK	water	14	5	⟷	⟷	Härtel *et al.* (1993)
ASP	water	17		⟷	⟷	Bruce *et al.* (1987)
ASP	water	–	15	⟷	⟷	Carlson and Shah (1989)
ASP	water	7	30	⟷	↓	Okuno *et al.* (1986)
ASP	water	24	10	⟷	⟷	Rodin *et al.* (1990)
ASP	water	14	5	⟷	⟷	Härtel *et al.* (1993)
ASP	drink	12		↑	⟷	Horwitz *et al.* (1988)
ASP	dessert	20	2	↑	↓	Teff *et al.* (1991)
ASP	capsule	–	15	⟷	⟷	Carlson and Shah (1989)
CHS	water	14	5	⟷	⟷	Härtel *et al.* (1993)
SACC	water	14	5	⟷	⟷	Härtel *et al.* (1993)
SACC	drink	12		⟷	⟷	Horwitz *et al.* (1988)

(b) Diabetic patients

Sweetener[a]	Vehicle	n[b]	Time	Insulin concn.[c]	Glucose concn.[c]	Reference
ASP	water	22	30	⟷	↓	Okuno *et al.* (1986)
ASP	water	26		⟷	⟷	Shigeta *et al.* (1985)
ASP	drink	10		⟷	⟷	Horwitz *et al.* (1988)
SACC	drink	10		⟷	⟷	Horwitz *et al.* (1988)

[a] ASP = aspartame; SACC = saccharin; CHS = cyclamate; Time = time of first blood sample. [b] n = number of subjects. [c] ⟷ = no difference compared with placebo; ↓ = decreased compared with placebo; ↑ = increased compared with placebo.

The most comprehensive study of the role of sweet taste *per se* in the cephalic phase insulin response was a study by Härtel *et al.* (1993) which compared the effects of solutions of different sweeteners on blood insulin and blood glucose. The study was a cross-over design in 14 subjects with blood samples taken prior to drinking the solution and at 5 min intervals up to 15 min. The blood insulin data (Figure 10.1) showed no statistically significant difference between the intense sweeteners and water, but sucrose ingestion had a major effect. There was no indication of a decrease in blood glucose in the first 30 min (Figure 10.1) which could have accounted for the increased appetite discerned in some studies in Table 10.1.

10.3.3 Intense sweeteners and food intake

A number of studies outlined in Table 10.1 extended the protocol to determine food consumed at a meal offered at a fixed time after the intake of intense sweetener. It is clear from the data summarised in Table 10.3 that there was a significant increase in food consumed in only one study, which used saccharin with yoghurt. None of the studies comparable to the original Blundell and Hill (1986) observation, i.e. an aqueous solution of

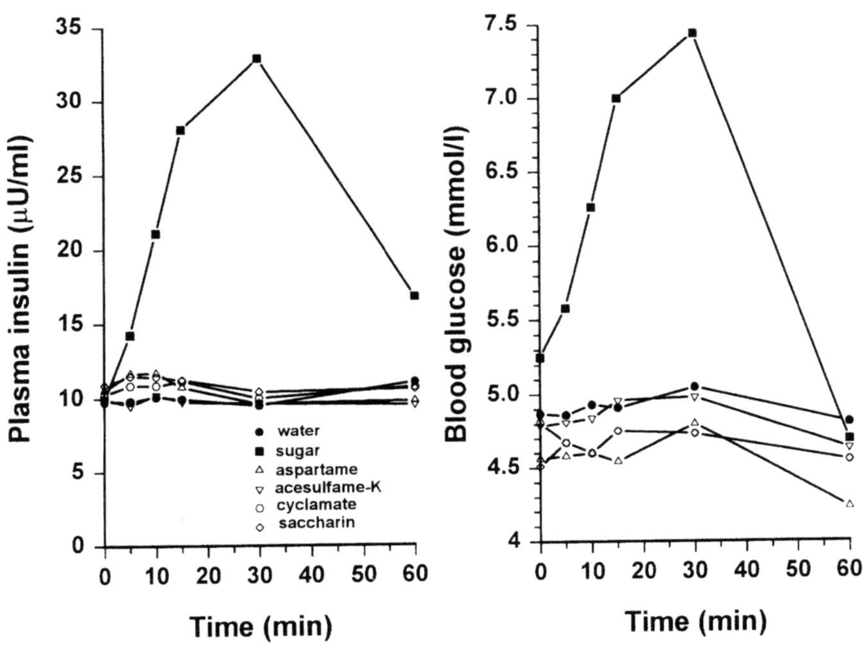

Figure 10.1 The effects of intense sweeteners and sucrose on insulin release and blood glucose in healthy volunteers. Data adapted from Härtel *et al.* (1993).

the sweetener with increased motivation at about 1 h, showed evidence to support the concept of an increase in appetite which translated into a functional rise in food intake.

Monneuse *et al.* (1991) gave volunteers yoghurt containing five different concentrations of aspartame and reported that the two lowest concentrations tasted more pleasant than the others and were associated with higher calorie intakes over the next 14 h. The comparisons were made with the higher concentrations of aspartame, not with a placebo, and this difference was found for men but not for women. It is difficult to interpret the findings in comparison with other studies quoted in Table 10.3. Although this study was interpreted in the French media as proving that intense sweeteners increase appetite, the study actually showed a fall in food intake with a rise in aspartame concentration, and therefore it is equally consistent with the capsule data in Table 10.3 which indicate that aspartame may decrease food intake.

When groups of obese and non-obese women were given a breakfast preload of white cheese, unsweetened or sweetened with sucrose or aspartame, ingestion of the low-calorie preload was associated with increased motivation to eat at lunch (Drewnowski *et al.*, 1994b). However, actual energy intakes at the lunch were not dependent on the calorie content of the preload and there was no indication that aspartame ingestion stimulated appetite.

An interesting and relatively consistent finding is that aspartame administered in capsule form was associated with a decrease in food intake. The study by Ryan-Harshman *et al.* (1987) is not consistent with this trend but a reanalysis of the study data led Rogers and Blundell (1994) to conclude that encapsulated aspartame does suppress food intake. Rogers *et al.* (1991) suggested that the anorexic action of encapsulated aspartame should be evaluated for a possible therapeutic benefit in obesity. Mechanisms which have been proposed to explain the decrease in appetite when aspartame is given as capsules are (i) an effect on cholecystokinin release due to phenylalanine or the dipeptide, which would produce a feeling of satiety (Smith *et al.*, 1973), and (ii) effects on satiety via central catecholamines produced following metabolism of aspartame to phenylalanine, although the clinical evidence does not support this mechanism (Ryan-Harshmann *et al.*, 1987; Karstaedt and Pincus, 1993). The claimed paradoxical effects of aspartame on appetite, depending on whether it is tasted or not (Figure 10.2), represent an interesting extension of the debate which is based on an unsubstantiated increase and an unproven decrease in appetite. Saccharin and acesulfame-K would produce only the postulated sweet-taste-mediated increase in appetite (Figure 10.2) but these two sweeteners showed little or no effect on appetite in the comparative study of Rogers *et al.* (1988), which reported an increase in appetite with aspartame. It is unlikely therefore that the proposed 'dual effect' of

Table 10.3 Food intake following single doses of intense sweeteners

Sweetener[a]	Vehicle	n[b]	Time after dose	Food intake[c]	Reference
AceK	water	12	65 min	⟷	Rogers *et al.* (1988)
ASP	water	12	65 min	⟷	Rogers *et al.* (1988)
ASP	water	27	60 min	⟷	Rogers *et al.* (1990)
ASP	water	24	38 min	⟷	Rodin (1990)
ASP	drink	20	90 min	⟷	Anderson *et al.* (1989)[d]
ASP	drink	20	60 min	⟷	Black *et al.* (1991)
ASP	drink	24	0, 30, 60 min	↓ or ⟷	Birch *et al.* (1989)
ASP	drink	106	60 min	⟷	Brala and Hagen (1983)
ASP	drink	42	30, 60 min	⟷	Rolls *et al.* (1990)
ASP	drink/water	18	65 min	⟷	Black *et al.* (1993)
ASP	dessert	32	120 min	⟷	Rolls *et al.* (1989)
ASP	cheese	24	3, 5.5, 9.5 h	⟷	Drewnowski *et al.* (1992, 1994a)
ASP	capsules	27	60 min	↓	Rogers *et al.* (1990)
ASP	capsules	16	60 min	↓	Rogers *et al.* (1991)
ASP	capsules	13	1.75 h	⟷ or ↓	Ryan-Harshman *et al.* (1987)
ASP	capsules	18	65 min	⟷	Black *et al.* (1993)
SACC	water	12	65 min	⟷	Rogers *et al.* (1988)
SACC+ASP	drink	20	60 min	⟷	Canty and Chan (1991)
SACC	yoghurt	24	60 min	↑	Rogers and Blundell (1989)

[a] ASP = aspartame; SACC = saccharin; AceK = acesulfame-K. [b] n = number of subjects. [c] ⟷ = no difference compared with placebo; ↓ = decreased compared with placebo; ↑ = increased compared with placebo. [d] This study compared aspartame with cyclamate not with an unsweetened drink.

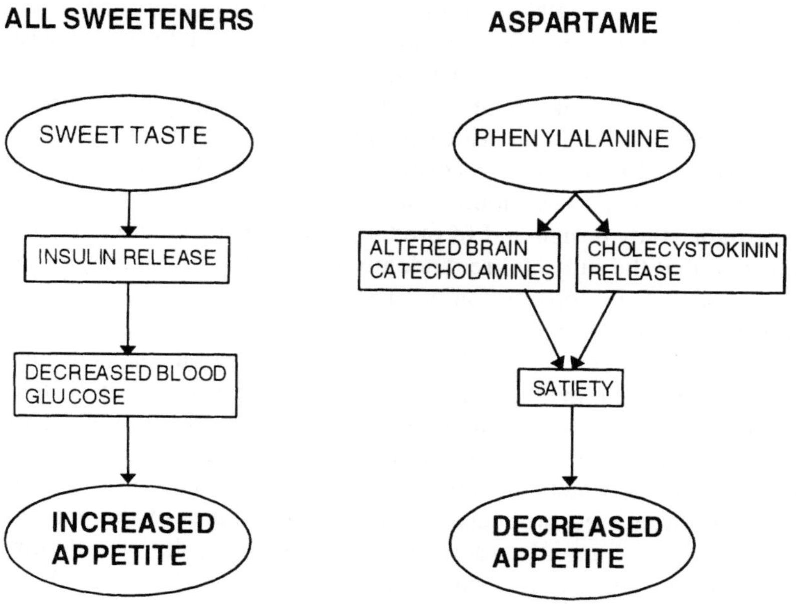

Figure 10.2 Mechanisms proposed to explain the claimed increase or decrease in appetite due to intense sweeteners.

aspartame (Figure 10.2) would have contributed to the differences in study outcome in Tables 10.1 and 10.3. It is interesting that almost no media attention has been given to the more credible effect of untasted aspartame to diminish food intake.

10.3.4 Intense sweeteners and body weight and weight control

From the review above it is clear that intense sweeteners are essentially neutral and produce neither an increase nor a decrease in appetite, and do not significantly alter blood composition or food intake. In consequence, the potential effect of intense sweetener intake on body weight will be entirely dependent on the circumstances and form in which the sweetener is consumed.

There are major problems in undertaking human studies on the effects of intense sweeteners on body weight. Studies must be of sufficient duration to affect body weight, the diets must be able to produce sufficient differences in calorie intake to affect body weight, and most importantly the subjects should not be aware of the nature of the diet, i.e. the presence of intense or caloric sweeteners. The results of studies in this area relate more closely to the epidemiological findings of Stellman and Garfinkel (1986) than to the mechanistic research of Blundell and Hill (1986).

Chen and Parham (1991) compared food intakes in users and non-users of intense sweeteners in a 24 h dietary recall study. The results suggested a reduced intake of sugars in women but an increase in men; but this trial has the same basic flaw of self-selected groups as the study of Stellman and Garfinkel (1986). A similar criticism applies to the earlier study by Parham and Parham (1980) which recorded reduced calorie intake in college students who chose to use intense sweeteners. In order to avoid such a bias in the basic study design and to produce scientifically credible results it is important that the subjects are randomly allocated to treatment groups and are unaware of the substitution of sucrose by an intense sweetener.

Intense sweetener consumption by men and women during a weight loss programme has been reported (Morris *et al.*, 1993) but the self-selection basis prevents meaningful analysis of any benefit. Another study investigated the possible advantage of intense sweeteners for subjects who had lost weight in a weight reduction programme (Kanders *et al.*, 1990b); although subjects were allocated to one of two groups either using or not using aspartame-sweetened products, the aspartame intakes of the groups had converged by 12 months, demonstrating the difficulty of attempts to produce even simple long-term dietary modifications. Interestingly, there was a positive association between aspartame intake and the maintenance of weight loss, and a negative correlation between aspartame intake and weight gain. However, both the aspartame intake data and the weight change may simply reflect individual long-term motivation.

During continuous studies in humans it is possible to replace the sugar in soft drinks with an intense sweetener. A study has been reported in which volunteers were given a large amount of soda sweetened with either high-fructose corn syrup or aspartame (Tordoff and Alleva, 1990b), or no experimental drinks for periods of 3 weeks (Table 10.4). Because the subjects drank the sugar-based soda on top of their normal diet there were significant increases in both the total calories consumed and in their body weight. In contrast, there were significant decreases in calorie intake and in body weight when the subjects drank the aspartame soda compared with when they had no extra drink. This is the exact opposite of what would be expected in intense sweeteners increased appetite.

Obviously it is not possible to replace food continuously with non-caloric products, but in a number of studies some items of the diet have been replaced by low-calorie equivalents without the individual being aware of the substitution (Table 10.4). Under such circumstances there was usually a decrease in total calorie intake although this may not have been sustained. Again the decrease in calorie consumption is the exact opposite of what would be expected if intense sweeteners enhanced appetite and food intake. The calorie intakes in these covert replacement studies sometimes showed an initial rapid fall which tended to revert towards normal later. This indicates that people may be unlikely to lose weight just

Table 10.4 Food intake and body weights following chronic intakes of intense sweeteners

Sweetener[a]	Vehicle	n[b]	Duration of study[c]	Effect[d]	Reference
ASP	sodas	30	3 weeks	↓ kcals	Tordoff and Alleva (1990b)
ASP	sodas	30	3 weeks	↓ weight	Tordoff and Alleva (1990b)
ASP	cereal	24	4 weeks	⟷ weight	Mattes (1990)
ASP	diet	8	6 days	↓ kcals	Porikos *et al.* (1977)
ASP	diet	6	12 days	↓ kcals	Porikos *et al.* (1982)
ASP	diet	13	12 days	↓ kcals	Porikos and Pi-Sunyer (1984)
ASP	capsules	108	24 weeks	⟷ weight	Leon *et al.* (1989)
CHS–SACC	diet	25	41 days	↓ weight	Berryman *et al.* (1968)

[a] ASP = aspartame; SACC = saccharin; CHS = cyclamate. [b] n = number of subjects. [c] The studies were run under blind or covert conditions so that subjects were unaware of the dietary manipulation or which product was sweetened with an intense sweetener or a sugar. [d] ⟷ = no difference compared with placebo; ↓ = decreased compared with placebo; ↑ = increased compared with placebo.

by using intense sweeteners without intentional control of their total calorie intake.

The long-term study on aspartame given in capsules (Table 10.4) did not support the suggestion that encapsulated aspartame could be useful as an appetite suppressant (see above).

Studies in subjects on weight control programmes have shown that intense sweeteners can be helpful in making the regime both more acceptable and more successful (Evans, 1989; Kanders *et al.*, 1988, 1990a and 1991).

10.4 Conclusions

It is clear that recent comprehensive studies do not support the hypothesis developed in the late 1980s that intense sweeteners can enhance appetite and increase food intake and body weight. Despite the convincing weight of evidence now available, the media – particularly in Germany – continue to give space to this classic 'man bites dog' story.

The findings of the original studies were interesting and worthy of further exploration. The epidemiological data of Stellman and Garfinkel could never prove a cause and effect relationship, but such computerised data bases are valuable in hypothesis generation. The basic flaw in the study design effectively prevents a clear conclusion, but long-term studies have shown that there is no evidence that intense sweeteners are associated with increased body weight. However, the carefully worded conclusion of the authors (see above) remains true and is possibly a reflection of the way some people use intense sweeteners. Intense sweeteners are not appetite suppressants, and unless there is an intentional limiting of calorie intake, there will not be a reduction in body weight. Studies indicate that subjectively intense sweeteners can help to make dieting more 'palatable' – both literally and metaphorically. Also, many people use intense sweeteners to allow themselves to enjoy some other item of food without increasing total calories, although a shift from carbohydrate to fat intake should be avoided (Naismith and Rhodes, 1995).

The experimental data of Blundell and Hill (1986) showed an effect that did not stand the basic test of scientific validity, i.e. the observation was not reproducible under apparently similar conditions. This could be because the effect was critically dependent on the precise circumstances, and because some unrecognised variable in the original study design was not controlled adequately in subsequent investigations. Rogers and Blundell (1993) conclude that the apparent discrepancies in the data from different studies may arise from differences in protocol, e.g. the time elapsing between ingestion of the sweet-tasting material and the test meal to assess effect on food intake. Discrepancies could also arise because some studies reported as showing no effect gave findings that approached statistical significance. In other cases prior knowledge about substitution of an intense sweetener may have biased the outcome (Rogers and Blundell, 1993). The chewing gum data of Tordoff and Alleva (1990a) support the concept of a very weak or 'fragile' effect which is easily perturbed by other variables. Even if there is a weak and fragile effect, it is clear that there is no evidence for the proposed mechanism involving cephalic phase insulin release and no consistent effect on food intake or body weight.

An alternative interpretation of the original studies would be that they were either flawed (Stellman and Garfinkel, 1986) and/or gave erroneous results (Blundell and Hill, 1986). If this is the case, then the 'intense sweetener – food intake' story is a good example of the fact that a false negative (type II error) can be corrected by a single subsequent study of greater power and/or better design, but that a false positive (type I error) cannot be laid to rest until numerous possible variables have been investigated and the weight of the body of evidence is sufficient to negate the positive result.

References

Anderson, G.H., Saravis, S., Schacher, R., Zlotkin, S. and Leiter, L.A. (1989) Aspartame: Effect on lunch-time food intake, appetite and hedonic response in children. *Appetite* **13**, 93–103.

Berryman, G.H., Hazel, G.R., Taylor, J.D., Sanders, P.G. and Weinberg, M.S. (1968) A case for safety of cyclamate and cyclamate–saccharin combinations. *American Journal of Clinical Nutrition* **21**, 673–686.

Birch, L.L., McPhee, L. and Sullivan, S. (1989) Children's food intake following drinks sweetened with sucrose or aspartame: Time course effects. *Physiology and Behavior* **45** 387–395.

Black, R., Tanaka, P., Leiter, L. and Anderson, G. (1991) Soft drinks with aspartame: Effect on subjective hunger, food selection, and food intake of young adult males. *Physiology and Behavior* **49**, 803–810.

Black, R.M., Leiter, L.A. and Anderson, G.H. (1993) Consuming aspartame with and without taste: differential effects on appetite and food intake of young adult males. *Physiology and Behavior* **53**, 459–466.

Blundell, J.E. and Hill, A.J. (1986) Paradoxical effects of an intense sweetener (aspartame) on appetite. *Lancet* May 10th, 1092–1093.

Brala, P.M. and Hagen, R.L. (1983) Effects of sweetness perception and calorie value of a preload on short term intake. *Physiology and Behavior* **30**, 1–9.

Bruce, D.G., Storlien, L.H., Furler, S.M. and Chisholm, D.J. (1987) Cephalic phase metabolism responses in normal weight adults. *Metabolism* **36**, 721–725.

Canty, D.J. and Chan, M.M. (1991) Effects of consumption of calorie versus non calorie sweet drinks on indices of hunger and food consumption in normal adults. *American Journal of Clinical Nutrition* **53**, 1159–1164.

Carlson, H.E. and Shah, J.H. (1989) Aspartame and its constituent amino acids: Effects on prolactin, cortisol, growth hormone, insulin and glucose in normal humans. *American Journal of Clinical Nutrition* **49**, 427–432.

Chen, L.N.A. and Parham, E.S. (1991) College students' use of high-intensity sweeteners is not consistently associated with sugar consumption. *Journal of the American Dietetic Association* **91**, 686–690.

Currie, E. (1994) A view of the world from low-calorie sweeteners, in *Health, Nutrition, Low-Calorie Sweeteners: Common Themes in a Changing Europe* (eds S.G. Lisansky and K.E. Hall), CPL Press, Newbury, pp. 21–27.

De Graaf, C., Schreurs, A. and Blauw, Y.H. (1993) Short-term effects of different amounts of sweet and nonsweet carbohydrates on satiety and energy intake. *Physiology and Behavior* **54**, 833–843.

DoH (1992) *The Health of the Nation: A Strategy for Health in England*, ISBN 0–10–119862–0, Her Majesty's Stationery Office, London.

Drewnowski, A., Louise-Sylvestre, J., Massien, C., Fricker, J. Chapelot, D. and Apfelbaum, M. (1992) Effects of sucrose and aspartame on hunger, energy intake and taste responsiveness in normal weight men and women. *Appetite* **19**, 175.

Drewnowski, A., Massien, C., Louis-Sylvestre, J., Fricker, J., Chapelot, D. and Apfelbaum, M. (1994a) Comparing the effects of aspartame and sucrose on motivational ratings, taste preferences and energy intakes in humans. *American Journal of Clinical Nutrition* **59**, 338–345.

Drewnowski, A., Massien, C., Louis-Sylvestre, J., Fricker, J., Chapelot, D. and Apfelbaum, M. (1994b) The effects of aspartame versus sucrose on motivational ratings, taste preferences, and energy intake in obese and lean women. *International Journal of Obesity* **18**, 570–578.

Evans, E. (1989) Effect of withdrawal of artificial sweeteners on energy intake of stabilised postobese women. *International Journal of Obesity* **13**, A111.

Härtel, B., Graubaum, H.-J. and Schneider, B. (1993) Einfluss von süsstoff-lösungen auf die insulinsekretion und den blutglucosepiegel. *Ernährungs-Umschau (Ernährungsphysiologie)* **40**, 152–155.

Horwitz, D.L., McLane, M. and Kobe, P. (1988) Response to single dose of aspartame or saccharin in NIDDM patients. *Diabetes Care* **11**, 230–234.

Ionescu, E., Rohner-Jeanrenaud, F., Proietto, J., Rivest, R.W. and Jeanrenaud, B. (1988) Taste-induced changes in plasma insulin and glucose turnover in lean and genetically obese rats. *Diabetes* **37**, 773–779.

Kanders, B.S., Lavin, P.T., Kowalchuk, M.B., Greenberg, I. and Blackburn, G.L. (1988) An evaluation of aspartame and weight loss. *Appetite* **11**, 73–84.

Kanders, B.S., Blackburn, G.L., Lavin, P.T., Whatley, J. and Pontes, M. (1990a) Does aspartame aid in weight loss? *International Journal of Obesity Supplement* **2**, 14:1F–16.

Kanders, B.S., Lavin, P.T., Kowalchuk, M. and Blackburn, G.L. (1990b) Do aspartame (APM)-sweetened foods and beverages aid in long-term control of body weight? *American Journal of Clinical Nutrition* **51**, 515.

Kanders, B.S., Blackburn, G.L., Lavin, P.T., Whatley, J., Keller, S. and Conaway, B. (1991) Aspartame facilitates long-term weight maintenance in a population of reduced obese women undergoing multidisciplinary treatment for obesity. *International Journal of Obesity Supplement* **3**, 15:P96, 61.

Karstaedt, P.J. and Pincus, J.H. (1993) Aspartame use in Parkinson's disease. *Neurology* **43**, 611–613.

Lavin, P.T., Sanders, P.G., Mackey, M.A. and Kotsonis, F.N. (1994) Intense sweeteners use and weight change among women: A critique of the Stellman and Garfinkel study. *Journal of the American College of Nutrition* **13**, 102–105.

Leon, A., Hunninghake, D., Bell, C., Rassin, D.K. and Tephley, T.R. (1989) Safety of long-term large doses of aspartame. *Archives of Internal Medicine* **149**, 2318–2324.

Mattes, R. (1990) Effects of aspartame and sucrose on hunger and energy intake in humans. *Physiology and Behavior* **47**, 1037–1044.

Monneuse, M.O., Bellisle, F. and Louise-Sylvestre, J. (1991) Responses to an intense sweetener in humans: Immediate preference and delayed effects on intake. *Physiology and Behavior* **49**, 325–330.

Morris, D.H., Cuneo, P., Stuart, M.J., Mance, M.J., Bell, K.J., Puleo, E., Ahmadi, S., Ward, A. and Rippe, J.M. (1993) High-intensity sweetener, energy and nutrient intakes of overweight women and men participating in a weight loss program. *Nutrition Research* **13**, 123–132.

Naismith, D.G. and Rhodes, C. (1995) Adjustment in energy intake following the covert removal of sugar from the diet. *Journal of Human Nutrition and Dietetics* **8**, 167–175.

Okuno G., Kawakami, F., Tako, H., Kashihara, T., Shibamoto, S., Yamazaki, T.,

Yamamoto, K. and Saeki, M. (1986) Glucose tolerance, blood lipid, insulin and glucagon concentration after single and continuous administration of aspartame in diabetics. *Diabetes Research and Clinical Practice* **2**, 23–27.

Parham, E.S. and Parham, A.R. (1980) Saccharin use and sugar intake by college students. *Journal of American Dietetic Association 86*, 560–563.

Porikos, K.P. and Pi-Sunyer, F.X. (1984) Regulation of food intake in human obesity: Studies with caloric dilution and exercise. *Clinical Endocrinology and Metabolism* **13**, 547–561.

Porikos, K.P., Booth, G. and Van Itallie, T.B. (1977) Effect of covert nutritive dilution on the spontaneous food intake of obese individuals. *American Journal of Clinical Nutrition* **30**, 1638–1644.

Porikos, K.P., Hesser, M.F. and Van Itallie, T.B. (1982) Caloric regulation in normal-weight men maintained on a palatable diet of conventional foods. *Physiology and Behavior* **29**, 293–300.

Renwick, A.G. (1994) Intense sweeteners, food intake, and the weight of a body of evidence. *Physiology and Behavior* **55**, 139–143.

Rodin, J. (1990) Comparative effects of fructose, aspartame, glucose and water preloads on calorie and macronutrient intake. *American Journal of Clinical Nutrition* **51**, 428–435.

Rogers, P.J. and Blundell, J.E. (1989) Separating the action of sweetness and calories: Effects of saccharin and carbohydrates on hunger and food intake in human beings. *Physiology and Behavior* **45**, 1093–1099.

Rogers, P.J. and Blundell, J.E. (1993) Intense sweeteners and appetite. *American Journal of Clinical Nutrition* **58**, 120–121.

Rogers, P.J. and Blundell, J.E. (1994) Reanalysis of the effects of phenylalanine, alanine and aspartame on food intake in human subjects. *Physiology and Behavior* **56**, 247–250.

Rogers, P.J., Carlyle, J., Hill, A.J. and Blundell, J.E. (1988) Uncoupling sweet taste and calories: Comparison of the effects of glucose and three intense sweeteners on hunger and food intake. *Physiology and Behavior* **43**, 547–552.

Rogers, P.J., Pleming, H.C. and Blundell, J.E. (1990) Aspartame ingestion without tasting inhibits hunger and food intake. *Physiology and Behavior* **47**, 1239–1243.

Rogers, P.J., Keedwell, P. and Blundell, J.E. (1991) Further analysis of the short-term inhibition of food intake by the dipeptide L-aspartyl-L-phenylalanine methyl ester (aspartame). *Physiology and Behavior* **49**, 739–743.

Rolls, B.J. (1991) Effects of intense sweeteners on hunger, food intake, and body weight: a review. *American Journal of Clinical Nutrition* **53**, 872–878.

Rolls, B.J., Laster, L.J. and Summerfelt, A. (1989) Hunger and food intake following consumption of low calorie foods. *Appetite* **13**, 115–127.

Rolls, B.J., Kim, S. and Fedoroff, I.C. (1990) Effects of drinks sweetened with sucrose or aspartame on hunger, thirst and food intake in men. *Physiology and Behavior* **48**, 19–26.

Ryan-Harshman, M., Leiter, L.A. and Anderson, G.H. (1987) Phenylalanine and aspartame fail to alter feeding behavior, mood and arousal in men. *Physiology and Behavior* **39**, 247–253.

Shigeta, H., Yoshida, T. and Nakai, M. (1985) Effects of aspartame on diabetic rats and diabetic patients. *Journal of Nutritional Science and Vitaminology* **31**, 533–540.

Smith, G.P., Young, R.C. and Gibbs, J. (1973) Cholocystokinin decreases food intake in rats. *Journal of Comparative Physiology and Psychology* **84**, 488–495.

Stellman, S.D. and Garfinkel, L. (1986) Artificial sweetener use and one-year weight change among women. *Preventive Medicine* **15**, 195–202.

Teff, K.L., Mattes, R.D. and Engelman, K. (1991) Cephalic phase insulin release in normal weight males: Verification and reliability. *American Journal of Physiology, Endocrinology and Metabolism* **26**, E430–E436.

Tordoff, M.G. and Elleva, A.M. (1990a) Oral stimulation with aspartame increases hunger. *Physiology and Behavior* **47**, 555–559.

Tordoff, M.G. and Alleva, A.M. (1990b) Effect of drinking soda sweetened with aspartame or high fructose corn-syrup on food intake and body weight. *American Journal of Clinical Nutrition* **51**, 963–969.

Tordoff, M.G. and Friedman, M.I. (1989) Drinking saccharin increases food intakes and preference – IV Cephalic phase and metabolic factors. *Appetite* **12**, 37–56.

11 Features of alitame™ as a new high intensity sweetener

M.E. HENDRICK, H.L. MITCHELL and P.R. MURRAY

11.1 Introduction

Sweetness is a subjective measurement. It is also dependent on several factors, including the concentration of the sweetener, temperature, pH, type of medium used, and sensitivity of the taster. Sucrose is the usual standard. The evaluation of the sweetness of a given substance in relation to sucrose is made on a weight basis (O'Brien Nabors and Gelardi, 1986). There are many alternative sweeteners to sucrose, but only a few have found wide application in foods and beverages to date. The chemical structures of alternative sweeteners are so diverse that it is difficult to predict which chemical structure will produce sweetness, although much work is in progress to improve our understanding of this subject. The known sweeteners do not belong to a single chemical class or have a common functional group.

Since the accidental discovery of aspartame in 1965 the potential of dipeptides as high intensity sweeteners has been an area of increasing interest. Based on a thorough study of the published dipeptide literature, it became apparent that an unmodified L-aspartic acid end group was required to achieve high sweetness potency.

Starting in the early 1970s an intensive and systematic programme to develop a superior dipeptide sweetener was carried out by Pfizer Central Research. Very large numbers of dipeptides with diverse chemical structures were synthesized, culminating in the development of alitame in 1979. Alitame and structurally related peptide sweeteners are the subject of a number of Pfizer patents (Brennan and Hendrick, 1983). Alitame has been shown to be sufficiently stable for use in many high temperature applications, and in this respect it is superior to other dipeptide sweeteners.

Sweeteners differ in taste characteristics and also in their stability to food processing, storage and handling conditions. Alitame has an excellent taste quality combined with good elevated temperature stability.

Alitame™ is a registered trademark of Pfizer Inc.

11.2 Development rationale

To increase the sweetness of dipeptides, it was proposed by Pfizer's research team that both sterically small and large groups with a specific spatial orientation are required on the second amino acid. The large group was designed as an amide, known to be more hydrolytically stable than the methyl ester group in the aspartame molecule. The small group was set as alkyl, in the D-amino acid series to maintain correct spatial configuration. With a constant amide structure, it was found that D-alanine had the highest sweetness potency of the various amino acids tested. Compounds derived from corresponding L-amino acids were found not to be sweet. Having established the L-aspartyl-D-alanine amide structure as a suitable base, a wide range of amines were added to it, with over 150 different amine end groups were evaluated for sweetness potency. Sweetness potencies were found to be highly dependent on the *N*-amide moiety. The structural possibilities for the amines were investigated systematically and were categorized according to structural features, such as ring size, branching and polarity.

Comparing the 5- and 6-membered cyclic amines with their alpha-methyl branched analogues revealed no strong dependence on ring size but showed that bulky α-branching in small rings can lead to high sweetness potencies. Carbocyclic rings larger than six or smaller than four resulted in reduced sweetness potencies.

An important discovery was the great sensitivity of sweetness potency to amine branching, as exemplified by a series of aliphatic amine end groups. Increasing steric bulk, either from trimethyl or cyclopropyl substitution at the α-carbon, was found to be required for high potency. Where only one branch contained a bulky substituent sweetness was greatly reduced. Finally an additional structural refinement was the inclusion of hetero-atoms, and thus increased polarity in a hindered amine series, which led to the identification of the thietane amine as an important element in achieving extremely high sweetness potency in dipeptides as shown in Figure 11.1. This knowledge led to the discovery of the unique chemical structure and sweetness intensity of alitame.

11.3 Physical and chemical characteristics

Alitame is a crystalline, non-hygroscopic powder, with a faint characteristic odour, formed from the amino acids L-aspartic acid and D-alanine, with a novel *C*-terminal amide moiety. It is this novel amide moiety (formed from 2,2,4,4-tetramethylthietanylamine) that is the key to the very high sweetness potency of alitame. Within the series of L-aspartyl-D-alanine amides, those structural features which were found to be most conducive to high sweetness potency include small to moderate ring size, presence of

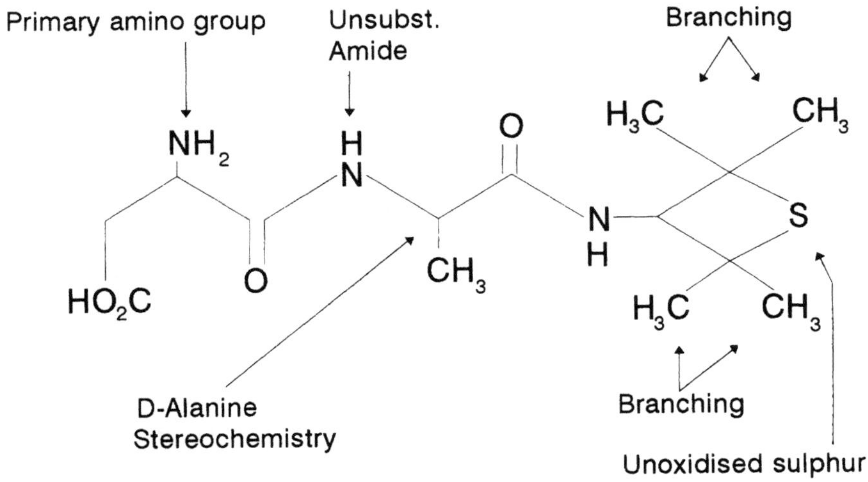

Figure 11.1 Structural items essential for sweetness in alitame.

small-chain branching alpha to the amine-bearing carbon, and the introduction of the sulphur atom into the carbocyclic ring. The chemical structure of alitame is shown in Figure 11.2.

The molecule is similar to aspartame in that it is an aspartic acid-containing dipeptide. However, there are important structural differences

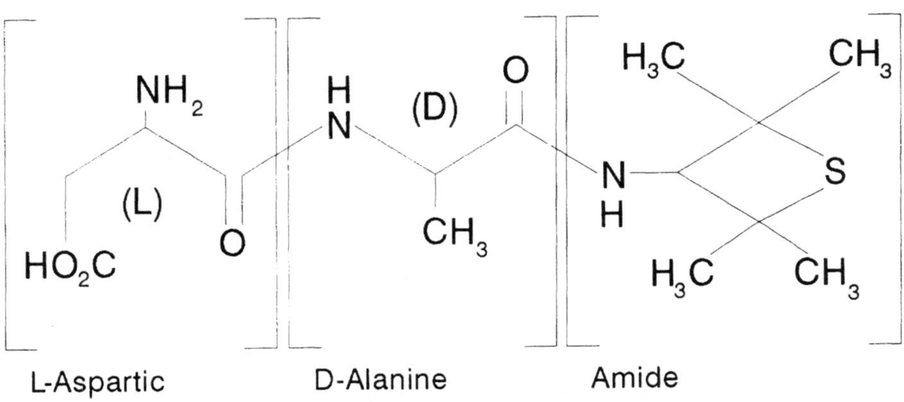

L-α-Aspartyl-N (2,2,4,4- tetramethyl-3-thietanyl)-D-alaninamide hydrate (2:5)

$C_{14}H_{25}N_3O_4S \cdot 2.5H_2O$

mw: 376.5

Figure 11.2 Alitame chemical structure.

which give alitame improved properties compared with aspartame. Most notable is the terminal amide of alitame, which provides better hydrolytic stability than the methyl ester of aspartame.

11.3.1 Solubility

At the isoelectric pH (generally the point of least solubility) alitame is very soluble in water. Solubility is also excellent in other polar solvents, as shown in Table 11.1.

Alitame is virtually insoluble in lipophilic solvents. In aqueous solutions the solubility rapidly increases with temperature and as the pH deviates from the isoelectric pH, as shown in Table 11.2.

11.3.2 Thermal stability

Aqueous stability measurements at 100°C in the typical pH range of 7–8 encountered in baked goods show a half-life of hours or days for alitame, compared with minutes or seconds for aspartame, as shown in Table 11.3. In practice, Maillard reactions between the aspartic unit and carbohydrates can occur with either alitame or aspartame.

Table 11.1 Solubility of alitame in various solvents at 25°C

Solvent	Solubility (% w/v)
Water	13.1[a]
Methanol	41.9
Ethanol	61.0
Propylene glycol	53.7
Chloroform	0.02
N-Heptane	0.001

[a] Isoelectric pH 5.6.

Table 11.2 Solubility[a] of alitame as a function of pH and temperature

pH	5°C	20°C	30°C	40°C	50°C
2.0	41.7	48.7	56.4	50.3	54.0
3.0	32.2	39.2	46.5	50.9	53.9
4.0	12.9	13.9	17.3	20.4	37.6
5.0	11.7	12.8	14.9	16.8	29.2
6.0	11.6	13.2	14.9	19.5	32.8
7.0	11.8	14.3	17.6	29.5	51.8
8.0	14.8	24.9	46.8	56.2	52.1

[a] Water solubility (%w/v).

Table 11.3 Stability of alitame and aspartame at elevated temperature in aqueous solution

Sweetener	Temp °C	Half-life (hours)		
		pH 6	pH 7	pH 8
Alitame	100	13.5	13.4	12.6
Alitame	115	2.1	2.1	2.1
Aspartame	100	0.4	0.1	0.04
Aspartame	115	0.1	0.03	0.02

11.3.3 Aqueous buffer stability

In the pH 2–4 range normally encountered in carbonated beverages, alitame is 2–3 times more stable than aspartame. At pH values over 5 the relative stability of alitame increases dramatically due to the absence of diketopiperazine formation and ester saponification decomposition pathways. In Figure 11.3 the half-lives of alitame in buffer solutions of various pH values are illustrated. As can be seen, the solution stability of alitame approaches the optimum for an aspartic acid dipeptide. At acid pH (2–4), alitame solution half-lives are more than twice those of aspartame. As the pH increases, this stability advantage increases sharply. In particular in the neutral pH range (5–8) alitame is completely stable for more than one year at room temperature.

A summary of the physical and chemical characteristics of alitame is given in Table 11.4.

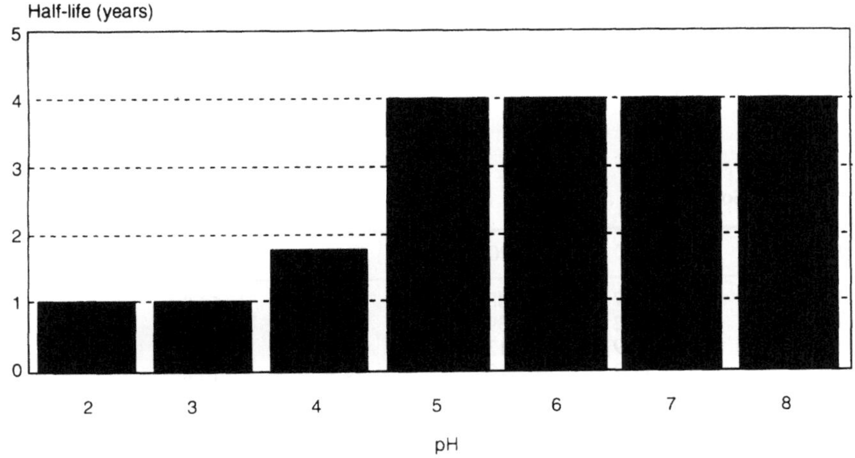

Figure 11.3 Stability of alitame in buffer solutions of pH 2–8 at 23°C.

Table 11.4 Summary of the physical and chemical characteristics of alitame

Chemical name:	L-alpha-aspartyl-*N*-(2,2,4,4-tetramethyl-3-thietanyl)-D-alanine amide hydrate (2.5)
Chemical family:	Dipeptide amide
Formula:	$C_{14}H_{25}N_3O_4S \cdot 2.5H_2O$
Molecular weight:	376.5
Melting point:	136–147°C
Solubility in water:	13.1% w/v (at 20°C)
Evaporation rate:	Not volatile
pH:	6.2 (0.1% w/v solution)
Isoelectric point:	pH 5.6
Appearance:	White, crystalline solid
Odour:	Faint, characteristic odour
Taste	Intensely sweet
Water (Karl–Fischer):	11–13%
Specific rotation $[\alpha]^{25}{}_D$ (1% in water):	(+40°)–(+50°)

11.4 Temporal profile – sweetness quality

All sweeteners have characteristic time/intensity profiles which are often used to help quantify the difference in the quality of the sweeteners. Table 11.5 shows how alitame compares with a range of other high intensity sweeteners.

Its sweetness potency, determined by comparison of the sweetness intensity of alitame solutions with concentrations in the range of 50 μg/ml to a 10% solution of sucrose, is approximately 2000 times that of sucrose. When compared with threshold concentrations of sucrose (typically 2–3%) the potency of alitame increases to about 2900 times that of sucrose. Table 11.6 shows the maximum potency of some high intensity sweeteners compared to sucrose.

Table 11.5 Sweetness and bitterness of selected sweeteners in water

	Sucrose	Alitame	Aspartame	Acesulfame-K
Sweetness				
Max. intensity (I_{max})	63.3	63.9	69.1	58.3
Time to I_{max} (sec)	9.7	9.3	9.6	7.9
Total duration (sec)	36.3	39.7	53.0	34.7
Rate max.	7.1	7.6	7.8	8.4
Aftertaste	1.6	1.7	2.8	1.4
Bitterness				
Max. intensity (I_{max})	11.9	16.9	16.4	41.8
Time to I_{max} (sec)	8.8	10.8	10.3	13.5
Total duration (sec)	24.1	29.4	29.4	46.9
Rate max.	1.4	1.9	1.9	4.3
Aftertaste	1.0	1.5	1.5	3.7

Source: Ott *et al*. (1991).

Table 11.6 Maximum sweetness potency of selected sweetners in water

Potency frequently reported		Sucrose potency at 2.0%	8%	10%
Sucrose	1	1	1	1
Alitame	2000	4500	2355	1640
Aspartame	200	250	143	107
Acesulfame-K	200	204	77	34
Na Cyclamate	30	26	27	18
Na Saccharin	300	510	188	0
Sucralose	600	614	520	385

Source: Dubois *et al*. (1991).

The sweetness of alitame is of a very high sucrose-like quality, without the accompanying bitter or metallic notes often experienced with other high-intensity sweeteners. The sweetness of alitame develops rapidly in the mouth and lingers slightly, in a manner very similar to that of aspartame.

11.5 Synergism – admixture potential

When two or more sweeteners are combined the effective sweetness of the mixture can be different from that found when they are tasted individually. In the case of alitame and acesulfame-K the sweetness intensity of the mixture is greater than predicted from intensity–concentration curves, and therefore there is real synergism between them. Alitame has been found to exhibit synergy when combined with both acesulfame-K and cyclamate. High quality blends may be obtained with these and other sweeteners. Where synergy is experienced, potential ingredient savings are possible. Synergistic effects in mixtures of alitame with acesulfame-K and cyclamate have demonstrated potential ingredient savings in the range of 25%. A three-way mixture of alitame, aspartame and acesulfame-K showed an even more pronounced synergy, with potential ingredient savings in the range of 40% (Lawless and Horne, 1994). Table 11.7 shows how the sweetness potency and taste quality of mixtures of alitame and acesulfame-K compare to a 10% w/v sucrose solution (aqueous).

11.6 Utility in food systems

Alitame is sufficiently stable for use in hard and soft candies, many heat-pasteurized foods, and in neutral pH foods processed at high temperatures, such as sweet baked goods. Table 11.8 shows how 95% of alitame is recovered after high temperature short time (HTST) processing of a low

Table 11.7 Sweetness potency and taste quality of mixtures of alitame and acesulfame-K when compared to 10% sucrose in water

% Composition (alitame: acesulfame-K)	Concentration of sweetener blend (g/100 ml) *vs.* 10% sucrose solution	Potency of mixture in relation to sucrose	Predicted potency in relation to sucrose	% Synergism	Taste quality
8:92	0.0235	425×	358×	19	Clean, sweet taste. No trace of bitterness.
10:90	0.0200	500×	436×	15	Clean, sweet taste. Slow onset of sweetness perception.
12:88	0.0174	574×	473×	21	Clean, sweet taste. Slow onset of sweetness perception.
14:86	0.152	654×	529×	24	Clean, sweet taste. Slow onset of sweetness perception.
16:84	0.0143	700×	573×	22	Clean, sweet taste. Slow onset of sweetness perception.
18:82	0.0137	723×	611×	19	Clean, sweet taste. Slow onset of sweetness perception.
28:80	0.0133	751×	643×	17	Clean, sweet taste. Slow onset of sweetness perception.

Table 11.8 Alitame recovery from certain prototype applications (no added sugar)

Application	Amount added (ppm)	Amount recovered (ppm)	% Recovery
Hard candy	162	137	85
Frozen yogurt (2% fat)	47	44	95
Yellow cake	100	75	75
Vanilla cookie	70	53	75
Oatmeal cookie	150	108	72

fat frozen yogurt (79–82°C for 25 sec), and 85% of alitame is recovered after sugar-free hard candy processing at 140°C.

11.6.1 Properties

Alitame has good stability and excellent taste quality in baked goods applications and in general exhibits excellent functionality because it is compatible with a wide variety of foods. Where high levels of reducing sugars are present, such as glucose and lactose, they may react with alitame to form Maillard reaction products. This reaction is also typical of other aspartyl dipeptides. The compatibility of alitame in a given recipe will depend on the other ingredients present and thermal and pH conditions in the manufacturing process.

As alitame is a dipeptide like aspartame, it is susceptible to degradation

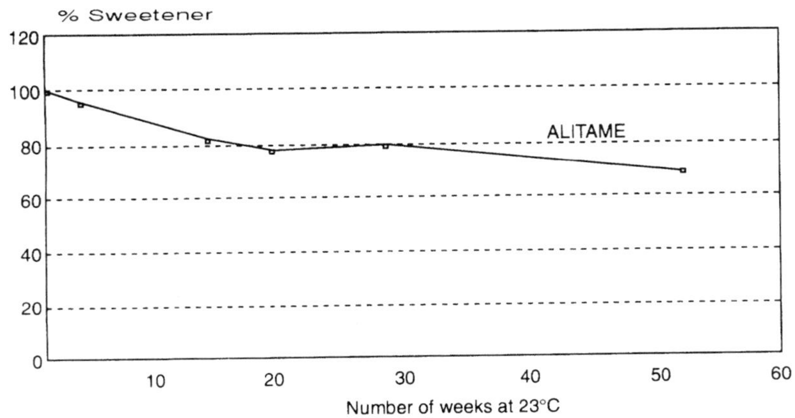

Week	0	4	13	19	26	52
Alitame	100	94	84	78	80	70

Figure 11.4 Alitame stability: carbonated lemon-lime soft drink at 23°C.

by Maillard reactions which may result in a loss of sweetness. Substances which may produce off-flavours on storage with alitame in liquid products are sodium bisulphite and ascorbic acid, and some types of caramel colour at pH's below 4.0. A research programme on this is in progress. The enhanced stability of alitame in carbonated beverages is clearly demonstrated in Figure 11.4. 70% of the sweetener was recovered after storage at 23°C for 52 weeks in a lemon-lime drink.

Table 11.9 shows some typical use levels in various food applications.

11.6.2 Handling suggestions

Since alitame has a sweetness intensity of approximately 2000× that of sucrose, it is recommended for use in solution form for most applications (1–3% w/v is optimum for convenience, but up to 10–12% w/v is possible).

Table 11.9 Alitame – suggested use levels in selected foods

Application	Use level (ppm)
Cola	30–45
Root beer	50–70
Lemon-lime drink	45–55
Orange drink	50–70
Dry mix breakfast drinks	50–85
Fruit juice drinks	40–50
Wine coolers	25–35
Iced tea (from mix)	20–30
Hot beverages	12–20
Flavoured milk	25–40
Hot cocoa mix	45–50
Milk shakes	75–85
Ice cream	35–100
Whipped topping	75–85
Gelatin desserts	80–100
Puddings, pie fillings, custards	50–75
Hard candy	140–160
Soft candy	200–300
Chocolate	225–400
Chewing gum	300–400
Fruit spreads	35–75
Fruit fillings	35–50
RTE cereal	100–200
Doughnuts	60–100
Fruit pies	40–50
Cookies	75–200
Muffins	150–225
Cake	80–200
Dessert topping	200–220
Maple syrup	150–400
Chocolate syrup	100–165
Flavoured yogurt	35–50

Table 11.10 Energy-reduced[a]/reduced-fat[b]/no sugar added[c] chocolate cake

	Ingredients	Per cent
Part 1	Litesse™ [d]	7.04
	Sorbitol (crystalline)	8.74
	Maltodextrin[e]	8.74
	Skimmed milk mix[f]	4.11
	Chocolate flavour	0.30
	Vanilla flavour	0.05
Part 2	Cake flour	19.49
	Cocoa, dutched (10–12%)	3.51
	Sodium bicarbonate	0.53
	Dicalcium phosphate	0.30
	Sodium aluminium phosphate	0.30
	Salt	0.30
	Xanthan gum	0.20
Part 3	Whole eggs	21.80
	Water	14.47
Part 4	Water	10.12
	Alitame	(100–200 ppm)

[a] Approx 983 kJ/100 g. [b] Approx 3.91 g fat/100 g. [c] No added sugar. [d] Improved Polydextrose (Pfizer Food Science). [e] Maltrin 100. [f] N-Flate (National Starch and Chemical Co.).

A stock solution minimizes repeated handling of small quantities, reduces exposure to dusting and ensures an easy and uniform delivery of the desired amount of alitame. Solutions should be stored in amber glass bottles to limit exposure to light, and may remain refrigerated or at ambient temperatures when in light-protecting bottles. The use of distilled water is recommended, as water containing more than 1 ppm of iron could result in off-odour development. Alitame is very soluble in water, and at its isoelectric point (pH 5.6), its solubility is 12% w/v at 5°C and 13% w/v at 25°C.

When a dry powder system is required, then alitame can be blended with other dry ingredients, using them as carriers or as dispersants. Alitame may also be combined with liquid flavour systems if preferred.

It is important to realize that other ingredients within a formulation may contribute to sweetness, and that this may have an impact on the use level of alitame. For instance, polyols and flavour enhancers may contribute to or enhance the overall sweetness.

Table 11.10 shows a typical formulation for a reduced-calorie, reduced-fat, no sugar added chocolate cake using Litesse™ as a bulking agent and alimate as sweetener.

Litesse™ is a registered trademark of Pfizer Inc.

11.7 Decomposition pathways

The principal pathways in water involve the hydrolysis of the aspartyl-alanine dipeptide bond to give aspartic acid and analyl-2,2,4,4-tetramethyl-thietane amide. The α,β-aspartic rearrangement, common to all peptides bearing *N*-terminal aspartic acid, also takes place, giving the β-aspartic isomer of alitame. This rearranged dipeptide hydrolyses at a slower rate than alitame to give the same products as those arising from alitame, namely aspartic acid and alanine amide. No cyclization to diketopiperazine or hydrolysis of the alanine amide bond is detectable in solutions of alitame that have undergone up to 90% decomposition. All three major decomposition products are completely tasteless at typical use levels in foods. Figure 11.5 shows the main degradation pathways of alitame.

11.8 Metabolism

The metabolic fate of alitame has been studied by means of ^{14}C labelled material in three animal species and in man. In all species, the principal

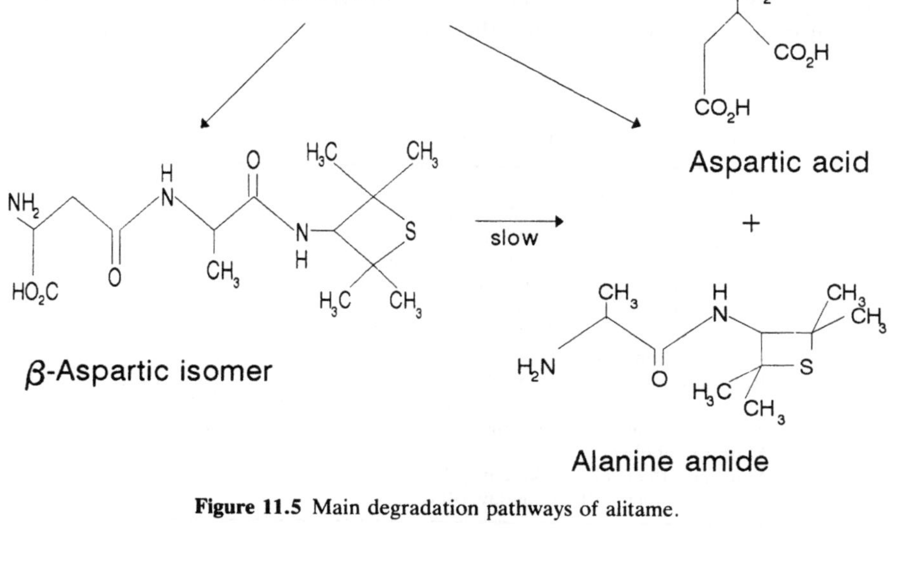

Figure 11.5 Main degradation pathways of alitame.

pathway involves cleavage of aspartic acid, which enters the amino acid pool. The remaining alanine amide moiety is excreted directly or as a glucuronide, and a minor fraction is oxidized to the corresponding sulphoxides and sulphone. About 5–20% of an administered alitame dose is excreted in the faeces as unchanged alitame and alanine amide. No cleavage of the alanine amide or rupture of the thietane ring is observed. Since the aspartic acid portion of the molecule is available for normal amino acid metabolism, alitame is partially caloric. The maxumum caloric contribution of alitame is 1.4 Calories per g (6 kJ per g). Alitame is used typically at between 20 and 200 ppm; therefore its energy contribution is clearly insignificant at these levels of use.

Alitame does not promote tooth decay and is therefore not cariogenic.

11.9 Safety

Extensive testing in Pfizer's Drug Safety Evaluation Department has demonstrated the safety of alitame at levels of more than 300× the estimated mean chronic daily intake of 0.34 mg/kg body weight (MRCA estimate). This intake estimation assumes that alitame is the only sweetener used in all 16 food categories requested in the US Food Additive Petition. A list of major studies already completed is shown in Table 11.11. In all of them the no-effect level (NOEL) was consistently above 100 mg/kg b.w./day, or greater than 300× the mean chronic human exposure. There was no evidence of carcinogenic potential in rats and mice treated for 2 years with doses as high as 564 and 1055 mg/kg/day, respectively. In doses as high as 1000 mg/kg/day during organogenesis, alitame was devoid of embryotoxic or teratogenic potential in both rats and rabbits. In two-generation reproduction studies there were no compound effects on mating

Table 11.11 Major safety studies completed

Genetic Toxicology
Three-month dog (0.5, 1, 2% of the diet)
Three-month rat (0.5, 1, 2% of the diet)
Rat teratology (100, 300, 1000 mg/kg/day)
Rabbit teratology (100, 300, 1000 mg/kg/day)
One-month mouse (1, 2, 5% of the diet)
Twelve-month rat (0.01, 0.03, 0.1, 0.3, 1% of the diet)
Eighteen-month dog (10, 30, 100, 500 mg/kg/day)
Rat reproduction (0.1, 0.3, 1% of the diet)
Twenty-four-month mouse oncogenicity (0.1, 0.3, 0.7% of the diet)
Twenty-four-month rat oncogenicity (0.1, 0.3, 1% of the diet)
Man, metabolism
Man, no effects (15 mg/kg/day, 14 days)
Man, no effects (10 mg/kg/day, 90 days)

behaviour, pregnancy rate, the course of gestation, litter size, or upon maternal or offspring survival.

Mutagenicity assays, both *in vitro* and *in vivo*, revealed no genotoxicity at gene or chromosomal levels in microbial and mammalian test systems. Alitame had no detectable activity in a battery of pharmacological tests employed to assess autonomic, gastrointestinal, renal and central nervous system functions. There were no effects on fasting blood glucose or upon the disposition of an oral glucose load. No need for warning statements on food labels is anticipated.

11.10 Regulatory status

In 1986 a Food Additive Petition was submitted to the US Food and Drug Administration, requesting broad clearance for alitame. In addition to the USA, permission for use of alitame in food has been requested from a number of other countries, including the European Union (EU). Alitame was approved in Australia, December 1993; Mexico, May 1994; New Zealand, October 1994; and the People's Republic of China, October 1994.

References

Brennan, T.M. and Hendrick, M.E., US Patent No. 4 411 925 (October 25, 1983).

Dubois, G.E. (1991) Sweeteners, Nonnutritive, in *Encyclopedia of Food Science and Technology* (ed. Y.H. Hui), John Wiley & Sons, New York, pp. 2470–2487.

Lawless, H. and Horne, J. (1994) *Sweetener Mixture Research (First Report) for Pfizer Food Science*. Sensory Research Laboratory, Food Science Department, Cornell University, New York.

O'Brien Nabors, L. and Gelardi, R. (eds) (1986) *Alternative Sweeteners*, Marcel Dekker Inc., New York and Basel.

Ott, D.B., Edwards, C.L. and Palmer, S.J. (1991) Perceived intensity and duration of nutritive and non-nutritive sweeteners in water using time intensity (T–I) evaluations. *Journal of Food Science* **56**, 535–542.

12 Neohesperidine dihydrochalcone: recent findings and technical advances

M.G. LINDLEY

12.1 Introduction

In the late 1950s, Gentili, working with Horowitz in the Western Regional Research Laboratory of the US Department of Agriculture, was preparing derivatives of phenolic glycosides as part of a programme to help in understanding the relationships between chemical structure and bitter taste. As part of this research project, neohesperidine dihydrochalcone was prepared from the flavanone neohesperidin. On tasting, neohesperidine DC was discovered to be sweet rather than bitter, and this led to a patent being issued to the USDA in 1963 (Horowitz and Gentili, 1963).

While this discovery attracted much academic and commercial interest it was quickly realised that the taste characteristics of neohesperidine DC were substantially different from those delivered by the main sweeteners of that time, namely sucrose, saccharin and cyclamate. The taste profile is characterised by a distinct delay in perceiving sweetness, followed by a lingering sweet, cooling aftertaste. Comprehensive attempts to try to alter these intrinsic taste characteristics through structural modification of the neohesperidine DC molecule were not sufficiently successful for this sweetener to be considered suitable for broad application within the food industry. Neohesperidine DC has recently been allotted an ADI of 0–5 mg/kg body weight by the Scientific Committee for Food of the European Union, and the sweetener is included in the European Parliament and Council Directive on sweeteners for use in foodstuffs.

The impetus for this regulatory re-evaluation of neohesperidine DC followed after the generation of additional safety data and expiry of the original USDA patents. These factors, together with an increase in the practice of blending sweeteners in Europe, meant that industry could take advantage of this sweetener's high potency, without suffering its disadvantages of lingering sweetness and cooling aftertaste. For these reasons, industry successfully supported the retention of neohesperidine DC on draft versions of the Sweeteners Directive.

With the increasing interest in neohesperidine DC came a growing awareness of its potential to modify and improve the flavour of food products when used at very low sub-sweetness addition levels. Detailed examination of this characteristic led ultimately to the inclusion of neohesperidine DC in the Miscellaneous Additives Directive of the European Union for use as a flavour modifier.

The market development of this interesting sweetener and flavour modifier has now begun and its full potential is being examined closely, so that a review of current knowledge of the characteristics and technical applications of neohesperidine DC is entirely appropriate.

12.2 Origin, production and related substances

Neohesperidine DC may be prepared on a commercial scale using either of the flavanones, neohesperidin or naringin, as starting material. Neohesperidin occurs naturally in bitter (Seville) oranges (*Citrus aurantium*), particularly in the peel of the immature fruit, and naringin in the peels of both grapefruit (*Citrus paradisii*) and bitter orange.

The preparation of neohesperidine DC on a commercial scale is relatively straightforward, particularly when neohesperidin is used as starting material. In this case, the flavanone is converted to the corresponding chalcone by addition of alkali, and the chalcone double bond is subsequently hydrogenated to form the dihydrochalcone.

Neohesperidine DC has not yet been shown to occur naturally, although several structurally related compounds have been identified in approximately 20 families of plants (Borrego *et al.*, 1995). For example, the compounds phloridzin and triloblatin have been found in the leaves of *Lithocarpus litseifolius*, a variety of sweet tea. Other sweet dihydrochalcones are reported in the fruit of *Iryanthera laevis*, used in the preparation of a sweet food from Colombia.

12.3 Regulatory developments

The safety assessment of neohesperidine DC has been based on results from a number of toxicological studies, including acute, sub-chronic and chronic feeding studies in rats and dogs, reproductive and teratogenic studies in rats, mutagenicity studies *in vitro* and *in vivo*, and studies on chromosome aberration (Horowitz and Gentili, 1991).

In 1987 the Scientific Committee for Food of the European Community assessed the safety of neohesperidine DC and concluded that the lowest 'no observable effect level' (NOEL) from all studies available was 500 mg/kg body weight in the rat. Hence the committee assigned a 5 mg/kg body weight ADI (Commission of the EC, 1987).

With recent publication and adoption of the EU Sweeteners Directive

(*Official J. European Communities*, 1994a), neohesperidine DC has broad sanction for use within the European Union, with those products in which it may be used defined, along with the maximum usable dose in each product (Table 12.1). These uses are in addition to those applications for which neohesperidine DC has already been sanctioned, including its use in chewing gum and lemonades in Belgium (Belgisch Staatsblad, 1982), in soft drinks in Spain (BOE, 1992) and in a broad array of foods in Switzerland.

In addition to its sanction as a sweetener in Europe, neohesperidine DC has also received approval for use as a flavour modifier/enhancer at concentrations below the sweet taste threshold in a variety of sweet and savoury food products. It is included in the Directive on Food Additives other than Colours and Sweeteners (*Official J. European Communities*, 1994b). These approvals were based on demonstration of technical value and industrial support, and are summarised in Table 12.2.

Table 12.1 Food products in which neohesperidine DC is permitted for use as a sweetener in Europe

Foodstuff	Maximum usable dose
Non-alcoholic drinks	50 mg/kg
Desserts	50 mg/kg
Confectionery	150 mg/kg
Chewing gum	400 mg/kg
Cider/perry	20 mg/kg
Other alcoholic drinks	10 mg/kg
Edible ices	50 mg/kg
Canned/bottled fruit	50 mg/kg
Jams, jellies, marmalades	50 mg/kg
Fruit/vegetable preserves	100 mg/kg
Fruit/vegetable preparations	50 mg/kg
Fish preparations	30 mg/kg
Sauces	50 mg/kg
Mustard	50 mg/kg
Fine bakery products	150 mg/kg
Meal replacers	100 mg/kg
Food supplements	100 mg/kg

Table 12.2 Food products in which neohesperidine DC is permitted for use as a flavour modifier in Europe

Foodstuff	Maximum usable dose
Margarines	5 mg/kg
Minarines	5 mg/kg
Meat products	5 mg/kg
Fruit jellies	5 mg/kg
Vegetable proteins	5 mg/kg

12.4 Stability

The stability of food ingredients under normal conditions of use and storage is a critical determinant of suitability for any particular product application. Obviously, functional properties must not change at rates which compromise the required shelf-life of products. This is particularly important for sweeteners, since absence or significant reduction in sweetness is detrimental to product quality and readily apparent to consumers.

Published data on the stability of neohesperidine DC is sparse, but recent studies have begun to fill this gap. This lack of detailed basic information has been addressed in an intial study in which the stability of neohesperidine DC has been evaluated in model systems. Solutions of neohesperidine DC in aqueous buffers ranging from pH 1.0–7.0 were stored at temperatures from 30°C to 60°C over periods up to 20 weeks and the residual neohesperidine DC analysed by HPLC (Canales *et al.*, 1993). Examination of the data shows that the breakdown of neohesperidine DC can be represented as a pseudo-first order reaction at all temperatures and pHs studied. Maximum stability was observed at approximately pH 4.0, although stability at the other pHs was excellent suggesting that neohesperidine DC is unlikely to pose stability problems.

The very stable nature of the neohesperidine DC molecule requires that an Arrhenius plot be used to predict rate constants of breakdown under what might be considered 'normal' storage conditions. Using this Arrhenius calculation, it has been estimated that following 12 months' storage at pH 3.0 and 20°C, only 3% of the neohesperidine DC originally present in solution would have degraded. At pH 4.0 under the same conditions, 97.2% of the neohesperidine DC originally present will remain in solution unchanged. Admittedly, this study was performed in a simplified aqueous system which will eliminate the potential for interactions to occur between neohesperidine DC and any food components. However, even allowing for such a study to represent 'best case', these results suggest that neohesperidine DC would be sufficiently stable to withstand most reasonable requirements of stability to food processing and subsequent storage. With the principal markets for low-calorie sweeteners being beverages, most of which are formulated within the pH range of 2.5–4.0, it is obvious from this study that neohesperidine DC would continue to deliver both sweetening and flavouring characteristics throughout even the most stringent product shelf-life requirements.

Further confirmation of these results has been obtained from a study of the stability of neohesperidine DC during the manufacturing process of a blackcurrent fruit jam, and following 18 months' storage of the prepared product (Tomas-Barberan *et al.*, 1995). During the preparation of this jam, to which 10 ppm neohesperidine DC had been added, boiling temperature

Table 12.3 Residual neohesperidine DC in blackcurrant jam following preparation and storage at room temperature

Storage time	Residual neohesperidine DC (ppm)
Zero time	10.0 ± 50.42
3 months	9.8 ± 0.47
6 months	9.2 ± 0.77
18 months	8.9 ± 0.55

was maintained for 40 min until the jam had reached 65°Brix, at which point the temperature had reached 106°C. The final pH was 3.1. Samples were packed in glass and stored at 25°C for 18 months, with HPLC analyses for residual neohesperidine DC conducted at intervals. The results are summarised in Table 12.3.

These data were found to be in good agreement with those calculated from the earlier model system study, from which residual neohesperidine DC following storage of a product at pH 3.1 for 18 months and at 25°C would be expected to be *c*. 92%. That 89% of the neohesperidine DC originally present remained demonstrates the value of this model system as a useful predictor of residual neohesperidine DC under conditions of food manufacture and storage.

A follow-on study has examined the stability of neohesperidine DC during yogurt manufacture and its subsequent storage. Again, no loss of neohesperidine DC was found during milk pasteurisation, throughout fermentation or following storage over 6 weeks at 3°C. Additionally, it was shown that the maximum legally permitted concentration of neohesperidine DC did not affect the rate of acidification by lactic acid bacteria during fermentation.

Taken together, all these data suggest that neohesperidine DC is unlikely to pose any significant stability problems for food manufacturers, even though there is a theoretical possibility of hydrolytic cleavage of the glycosidic linkage. They should therefore provide a starting point for manufacturers to make evaluations in their own products.

There is less certainty regarding its suitability for use in some non-food applications: in common with many other food components, neohesperidine DC is more stable under acid conditions than under the neutral or even basic pH requirements of toothpastes, mouthwashes and some other OTC products. However, to date no studies have been reported which examine stability in systems relevant to these types of products, so it is not yet possible to predict the potential of neohesperidine DC as a sweetener/ flavour modifier for these markets.

12.5 Sensory properties

Neohesperidine DC is able to supply sweetness at very low levels, and is generally described as several hundred times as sweet as sucrose. The threshold concentration is variously described as ranging from 0.7–7.0 ppm, with the lower value being that most widely accepted. At this concentration it is judged to be 1800 times as sweet as sucrose. In common with all low-calorie sweeteners, sweetness intensity relative to sucrose falls as concentration increases. Thus at practical use concentrations, neohesperidine DC is quoted as being around 400–600 times as sweet as sucrose.

Measurement of the relative sweetness of neohesperidine DC is complicated by its taste profile, and the probably explains some of the inconsistencies in the literature. For example, DuBois *et al.* (1977) reported the sweetness of neohesperidine DC to be over 600 times that of sucrose at 8.5% sucrose equivalence, but this was later revised. From a practical standpoint, however, maximum sweetness contribution to most products from neohesperidine DC will rarely exceed that delivered by around 5% of sucrose. This is due to the 50 ppm maximum permitted levels, as laid down in the EU Directive, in most foods and beverages which traditionally use low-calorie sweeteners. At these and lower sucrose equivalences, neohesperidine DC is described as being approximately 400–900 times as sweet as sucrose (Horowitz and Gentili, 1991).

The sweet taste delivered by neohesperidine DC is perceived more slowly than that from most other sweeteners including the carbohydrates, aspartame, saccharin, acesulfame-K and cyclamate. Various attempts have been described to overcome this initial lag in perception, but the most successful approach, and one which must be employed when using neohesperidine DC for its sweetening properties, is undoubtedly to blend it with other low-calorie sweeteners. With the appropriate choice of blend concentrations, this approach proves to be a fully successful way of overcoming the negative, slow onset of sweetness.

In addition to a perceptible delay in sweetness, neohesperidine DC also elicits a lingering sweet, cooling aftertaste. DuBois and co-workers have attempted to quantify the temporal profile of sweetness delivery by neohesperidine DC, and they compared their findings with similar data for sucrose, saccharin and mono-ammonium glycyrrhizinate. These data have been adapted and are reported by Horowitz and Gentili (1991) (see Table 12.4).

It is obvious that the temporal characteristics of neohesperidine DC differ substantially from those of sucrose and saccharin. Clearly, the timings reported can only be considered as approximations, and their absolute values are heavily dependent on the concentrations at which tasting is carried out. However, they do indicate the extent of the differences in temporal properties between neohesperidine DC and some

Table 12.4 Time course of sweetness induced by selected sweeteners

Compound	Time (seconds) for taste to: reach a maximum	decay
Sucrose	4	13
Saccharin	4	14
Neohesperidine DC	9	41
Ammonium glycyrrhizinate	16	69

other sweeteners, and it is these temporal differences which control and limit its method of use and its potential applications.

Additional studies on the sensory characteristics of many sweeteners, including neohesperidine DC, carried out by workers at Duke University, North Carolina, in co-operation with the NutraSweet Company, have been reported (Kim and DuBois, 1991). Attempts were made to break down the taste of neohesperidine DC into its component attributes. The findings showed clearly that neohesperidine DC delivers a taste described as predominantly sweet, but it also delivers a relatively strong liquorice taste. Also, following exposure to concentrations in water of >20 ppm, a quite long-lasting cooling sensation is apparent, perceived on breathing in through the mouth. While it would be interesting to understand the physiological mechanism involved in this sensation, since it is generally undesirable, the practical consequences are limited concentration levels in food and drinks, which are relatively easy to identify.

It has frequently been suggested that the presence of these non-sweet components in the overall profile of neohesperidine DC will constrain its market potential severely. However, recent studies indicate that when evaluated in foods and drinks, neohesperidine DC does deliver important beneficial sensory characteristics to many products. Although there are few published data on sweetness synergy between neohesperidine DC and other low-calorie sweeteners, there are some clear indications that it can make an important contribution through this. For example, von Rymon-Lipinski and Luck (1977) describe the synergistic effects of blending acesulfame-K and neohesperidine DC. Additionally, Jenner (1992) describes the synergy between low concentrations of neohesperidine DC and the sweetener sucralose. Although a simple sensory methodology was employed, some quite substantial enhancements of sweetness were observed in blends of these two sweeteners. Given that sucralose is also synergistic with other sweeteners, including aspartame, cyclamate and saccharin (Lindley *et al.*, 1991), it seems probable that sweetness synergy, and hence economic benefits, are also likely between blends of neohesperidine DC and these other low-calorie sweeteners.

In addition to delivering some sweetness and providing economic benefits through sweetness synergy phenomena, NHDC also:

- modifies and enhances flavours, particularly fruit flavours, frequently with significant beneficial consequences
- enhances the mouthfeel of many 'diet' or sugar-reduced products and even of fat-reduced foods
- modifies the sweetness profile of other low-calorie sweeteners to help them deliver a more sucrose-like sensory profile
- reduces the perception of bitterness

12.5.1 Flavour modification

It is, of course, quite common for sweeteners simultaneously to modify or enhance flavour while also eliciting sweetness. Sucrose and fructose have been shown to enhance fruit flavours, particularly citrus varieties (Hyvonen, 1982; Wiseman and McDaniel, 1991). There are also frequent references to flavour enhancement by aspartame, again particularly with fruit varieties (Beck, 1974; Homler, 1984; Homler *et al.*, 1991). In all these reports, however, each sweetener was present at supra-threshold concentration, so their flavour modifying characteristics could be considered to be a supplementary consequence of eliciting sweetness, rather than due to any structure-induced effects. In contrast, neohesperidine DC has been found to modify flavours both at supra- and sub-threshold addition concentrations, a feature which makes it potentially a more versatile and useful ingredient.

The first report of flavour modifying effects induced by neohesperidine DC (Inglett *et al.*, 1969) described its effects on samples of powdered beverages flavoured with grape, root beer and cherry. In all cases, neohesperidine DC was being evaluated as part of the sweetening system, and its effects were merely described as generally beneficial. Recent work (Lindley *et al.*, 1993) has employed more rigorous descriptive sensory analysis procedures to explore the flavour-modifying effects of sub-sweetness threshold concentrations in detail. A broad range of both sweet and non-sweet foods and beverages was prepared with added neohesperidine DC. The products evaluated were a non-carbonated blackcurrent beverage, lemon flavoured water-ice, soya milk, tomato ketchup and low-fat margarine spread. Standard descriptive sensory techniques were used in which a panel of highly trained tasters was employed and the sensory attributes of each product were described, with reference products identified and standardised. Panellists then carried out sensory evaluation of each product in triplicate, and their quantitative results were analysed by standard ANOVA techniques.

Significant differences were found between each test and control pair evaluated. Neohesperidine DC added at levels up to 4 ppm induced significant changes to some of the flavour attributes of all these diverse products. Those attributes modified by neohesperidine DC, and the direction of modification, are listed in Table 12.5.

Table 12.5 Flavour attributes modified by neohesperidine DC

Product	NHDC (ppm)	Attribute	Enhanced/reduced
1. Blackcurrant drink	2.0	Blackcurrant odour	Enhanced
		Juicy taste	Enhanced
		Syrupy mouthfeel	Enhanced
		Juicy aftertaste	Enhanced
		Bitter taste	Reduced
		Drying taste	Reduced
		Drying aftertaste	Reduced
		Cooling aftertaste	Reduced
2. Lemon water ice	1.0	Total flavour (bite)	Enhanced
		Length of aftertaste	Enhanced
		Acidity (suck)	Reduced
		Bitterness (suck)	Reduced
3. Tomato ketchup	4.0	Sweet odour	Enhanced
		Tomato odour	Enhanced
		Tomato flavour	Enhanced
		Vinegar odour	Reduced
		Spicy odour	Reduced
		Spicy flavour	Reduced
		Throat catching aftertaste	Reduced
4. Soya milk	4.0	Thickness/mouthfeel	Enhanced
		Walnut aftertaste	Reduced
5. Low-fat margarine	3.0	Buttery odour	Enhanced
		Butter flavour	Enhanced
		Floury odour	Reduced
		Salty flavour	Reduced

While many of these flavour changes were reported to be slight, and most were considered beneficial to overall product quality, it was acknowledged that some modifications were probably detrimental to product impact. In particular, in tomato ketchup, reductions in the perceived intensity of what can be considered to be key distinguishing attributes, namely, vinegar and spicy odours, spicy flavour and the throat-catching characteristic, may all detract from the expected impact of this product. Thus it must be concluded that the use of neohesperidine DC for its flavour-modifying effects should be examined in each individual product, and that it cannot be assumed that its inclusion will automatically be beneficial to product quality.

The inclusion of both sweet and non-sweet foods in this study, with significant flavour magnitude changes demonstrated in all products, led the authors to suggest that these flavour modifications were probably independent of sweetness induction. Although the sweet taste threshold of neohesperidine DC in water has been reported to be as low as 0.7 ppm (Guadagni *et al.*, 1974), in no product was a statistically significant increase in sweetness recorded, and the sweetness threshold concentration of neohesperidine DC in flavoured food and drink products is highly likely to be much greater than that detected in water.

12.5.2 Mouthfeel enhancement

It will be noted that in some products evaluated as part of the flavour modification study (Lindley *et al.*, 1993), mouthfeel or perceived thickness of products was identified as a consequence of including small quantities of neohesperidine DC. In particular, the perceived mouthfeel of both the blackcurrent beverage and the soya milk were judged to benefit from its addition. These observations are consistent with those claimed by Foguet *et al.* (1994), in which improvements to perceived mouthfeel and body of a range of products are described. For example, addition of 1 ppm neohesperidine DC to an aspartame-sweetened cola beverage was described as inducing improved mouthfeel and body, with a more sucrose-like sweetness quality. Similar benefits were described in a tonic water drink. Benefits in sugar-free dessert formulations were described as improved mouthfeel, enhanced creaminess and improved sweetness quality. Identical observations were also made with sugar-free confectionery products and tomato ketchup.

In all these examples, neohesperidine DC was incorporated in the various products at levels between 0.1 and 4.0 ppm. These concentrations are clearly unable to affect measurable phyical viscosity, leading to the conclusion that the improvements to perceived mouthfeel must be a consequence of improvement in sweet taste quality. The corollary to this is that, at least in part, mouthfeel must involve a chemical perception mechanism akin to the ligand–receptor mechanism believed to be responsible for initial sweet taste perception.

12.5.3 Sweet taste quality enhancement

As described in the previous section, low addition levels of neohesperidine DC have been noted frequently to induce improvements in sweet taste quality. These improvements are seen as a measurable conversion of the sweet taste quality delivered by low calorie sweeteners such as aspartame, acesulfame-K and saccharin to a fuller, more rounded and more sucrose-like taste. Independent confirmation of these observations has been reported by von Rymon-Lipinski and Luck (1979) who describe blends of acesulfame-K and neohesperidine DC as being more sucrose-like than acesulfame-K alone. Similarly, Eisenstadt (1978) describes combinations of neohesperidine DC and saccharin as tasting more like sucrose than does saccharin alone.

12.5.4 Bitterness reduction

It is possible that these 'non-sweetness' sensory effects of adding neohesperidine DC to foods and beverages follow directly from its ability to reduce the perception of bitterness. Guadagni *et al.* (1974) describe the bitterness-reducing effects of neohesperidine DC on limonine and naringin. The properties to alter the taste qualities of both acesulfame-K and saccharin to more closely mimic sucrose are also probably a direct consequence of bitterness amelioration. Thus this well-documented phenomenon may suggest potential applications for neohesperidine DC, not only in food products, but also as a component in the flavouring of pharmaceuticals and OTC medicines, which may frequently be perceived as bitter.

It has also been speculated (Lindley *et al.*, 1993) that many of the flavour modification effects of neohesperidine DC may be a direct consequence of bitterness reduction. Where bitterness is reduced, reversal of the phenomenon of 'mixture suppression' originally caused by threshold bitterness, may explain why many of the flavour modifications which have been reported are perceived as an enhancement. In addition, this explanation is consistent with the observed reduction in impact of some other flavour attributes.

Clearly, there is real potential to use neohesperidine DC to induce a variety of sensory effects, many of which should prove to be beneficial to sweet taste and flavour qualities. However, it is important to realise that there are no hard and fast rules which can guide product formulators; each potential application of neohesperidine DC must be examined individually.

12.6 Product applications

Many potential applications of neohesperidine DC are described in the scientific and patent literature. Included are applications in foods and beverages, pharmaceuticals and animal feeds. Actual uses at the moment are not as broad as potential uses, but neohesperidine DC is being used quite widely to improve the flavour and palatability of animal feed preparations. A number of pharmaceutical products containing neohesperidine DC are commercial realities across Europe. In this application its bitterness-ameliorating character is used to advantage, and many consumer trials for its use in foods and beverages are in progress. Neohesperidine DC is used commercially in some foods on sale in Italy, Spain and Portugal. Product applications include bakery products, beverages and chewing gum. Market development in early 1995 is still hindered by the time taken by national authorities to convert the

Sweetener Directive provisions into national laws throughout the European Union. However, the current successes in Italy, Spain and Portugal bode well for this sweetener as national laws are adjusted.

There is a satisfying consistency in the current and potential applications of neohesperidine DC which demonstrates its beneficial properties. Use in animal feeds depends on its flavour-modification characteristic having a clearly positive impact on feed palatability. Pharmaceutical applications rely largely on the improvement of palatability through an amelioration of bitterness. Food and beverage potential should be fulfilled by harnessing its flavour-modifying and sweet taste quality characteristics effectively.

References

Beck, C.I. (1974) Sweetness, character and applications of aspartic acid based sweeteners, in *Symposium: Sweeteners* (ed. G.E. Inglett), AVI, Westport, p. 164.

Belgisch Staatsblad (1982) July 30, p. 8737.

BOE (1992) January 28, p. 2517.

Borrego, F., Canales, I. and Lindley, M.G. (1995) Neohesperidin dihydrochalcone: state of knowledge review. *Lebensm. Unters. u.-Foresch.* **200**, 32–37.

Commission of the EC (1987) SCF report of the Scientific Committee for food on sweeteners (CS/EDUL/56-final).

Canales, I., Borrego, F. and Lindley, M.G. (1993) Neohesperidin dihydrochalcone stability in aqueous buffer solutions. *J. Food Sci.* **58**, 589–591, 643.

DuBois, G.E., Crosby, G.A., Stephenson, R.A. and Wingard, R.E., Jr. (1977) Dihydrochalcone sweeteners: Synthesis and sensory evaluation of sulfonate derivatives. *J. Agric. Food Chem.* **25**, 763–771.

Eisenstadt, M.E. (1978) Neohesperidin dihydrochalocone sweetening compositions. *US Patent 4 085 232.*

Foguet, R., Cistero, A. and Borrego, F. (1992) Body and mouthfeel potentiated foods and beverages containing neohesperidin dihydrochalcone. *European Patent Application 0 500 977 A1.*

Guadagni, D.G., Maier, V.P. and Turnbaugh, J.H. (1974) Some factors affecting sensory threshold and relative bitterness of limonin and naringin. *J. Sci. Food Agric.* **23**, 1199–1205.

Homler, B.E. (1984) Aspartame: Implications for the food scientist, in *Aspartame: Physiology and Biochemistry*, Marcel Dekker, New York. p. 247.

Homler, B.E., Deis, R.C. and Shazer, W.H. (1991) Aspartame, in *Alternative Sweeteners*, 2nd Edition (eds. L. O'Brien-Nabors and R.C. Gelardi), Marcel Dekker, New York, pp. 39–70.

Horowitz, R.M. and Gentili, B. (1963) Dihydrochalcone derivatives and their use as sweetening agents. *US Patent 3 087 821.*

Horowitz, R.M. and Gentili, B. (1991) Dihydrochalcone sweeteners from citrus flavanones, in *Alternative Sweeteners*, 2nd Edition (eds. L. O'Brien-Nabors and R.C. Gelardi), Marcel Dekker, New York. pp. 97–116.

Hyvonen, L. (1982) Fructose in food systems, in *Nutritive Sweeteners* (eds. G.G. Birch and K.J. Parker), Applied Science, London, pp. 133–144.

Inglett, G.E., Krbechek, L. Dowling, B. and Wagner, R. (1969) Dihydrochalcone sweeteners. Sensory and stability evaluation. *J. Food Sci.* **34**, 101–103.

Jenner, M.R. (1992) Sweetening agents, *European Patent Application 0 507 598 A1.*

Kim, S.-H. and DuBois, G.E. (1991). Natural high potency sweeteners, In *Handbook of Sweeteners* (eds. S. Marie and J.R. Piggott), Blackie A&P, Glasgow, pp. 116–185.

Lindley, M.G., Beyts, P.K., Canales, I. and Borrego, F. (1993) Flavor modifying characteristics of the intense sweetener neohesperidin dihydrochalcone. *J. Food Sci.* **58**, 592–594, 666.

Official J. European Communities (1994a) September 10, No. L237/3.
Official J. European Communities (1994b) June 24, No. C172/4.
Tomas-Barberan, F.A., Borrego, F., Ferreres, F. and Lindley, M.G. (1995) Stability of the intense sweetener neohesperidin dihydrochalcone in blackcurrant jams. *Food Chem.*, in press.
von Rymon-Lipinski, G.-W. and Luck, E. (1977) Sweetener mixture, *US Patent 4 158 068*.
Wiseman, J.J. and McDaniel, M.R. (1991) Modification of fruit flavors by aspartame and sucrose, *J. Food Sci.* **56**, 1668–1670.

13 The uses and commercial development of sucralose

M.R. JENNER

13.1 Introduction

In the last three decades many thousands of intensely sweet compounds have been discovered. Some have been found by accident while others have emerged as the result of intensive research campaigns by academia and industry. Of course the vast majority of these sweeteners will always remain as academic curiosities providing opportunities for structure–activity relationship studies involving advanced computer-driven molecular mechanics leading to ever more detailed speculation on the nature, function and, indeed, origin, of the sweet taste receptor.

One chemical that had long intrigued scientists in this field was sucrose itself. The factors responsible for its exceptionally pleasant sweet taste had been the subject of several research studies, and the fact that sucrose contains a multitude of potentially reactive sites presented chemists with a wonderfully versatile building block. The stage was set for an exploration of the chemical, physical and structural properties of sucrose (Figure 13.1).

In 1976 the possibility of enhancing the sweetness of sucrose by chemical manipulation was discovered during a collaborative research programme between Prof L. Hough at Queen Elizabeth College in the University of London, and scientists at Tate & Lyle (Hough, 1976). Prior to this time, all derivatives of sucrose that had been tasted had been found to be tasteless or bitter. For example, work reported by Lindley (1976) showed that substitution of the hydroxyl groups at the 4- and 1′- positions by methyl groups had a significant effect in reducing the perception of sweetness. Somewhat surprisingly, Hough and his colleagues showed that selective chlorination of certain hydroxyl groups in sucrose resulted in a significant increase in the perception of sweetness. This discovery stimulated an extensive research programme (Jenner, 1991). Many hundreds of compounds were prepared in a classic structure–activity relationship study. The factors important for intensification of the sweetness of sucrose were found to be the presence of hydroxyl groups at the 2, 3, 6 and 3′ positions, with large, highly electronegative groups, at the remaining positions (Figure 13.2). The sweetest compound discovered in this programme was the 4,1′,4′,6′-tetrabromo compound, which is approximately 7500 times sweeter than sugar itself.

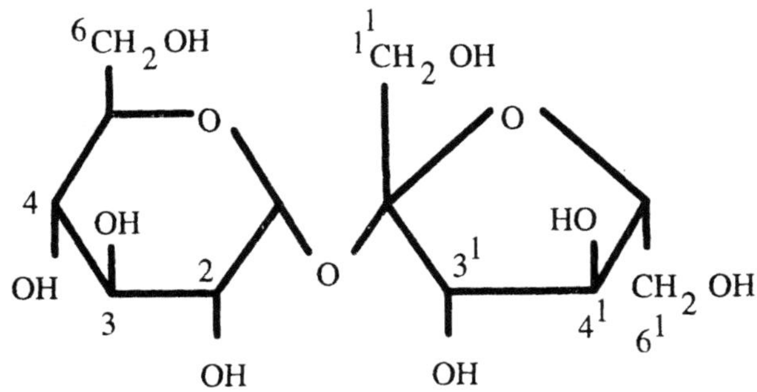

Figure 13.1 Sucrose.

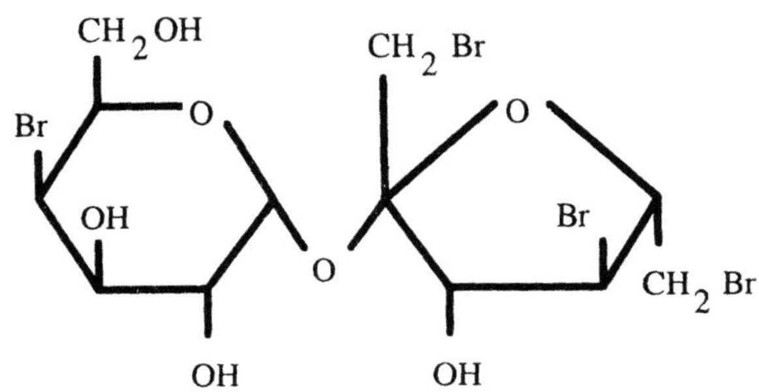

Figure 13.2 4,1′,4′,6′-Tetrabromogalactosucrose.

Derivatives of sucrose	Sweetening power (× sucrose)
4-chlorogalactosucrose	5
1^1-chlorosucrose	20
6^1-chlorosucrose	20
6,1^1,6^1-trichlorosucrose	25
4,1^1,6^1-trifluorogalactosucrose	40
1^1,6^1-dichlorosucrose	80
4,1^1-dichlorogalactosucrose	120
4,1^1,6^1-triiodogalactosucrose	120
4,1^1,6^1-trichlorogalactosucrose	600
4,1^1,6^1-tribromogalactosucrose	800
4,1^1,4^1,6^1-tetrachlorogalactosucrose	2200
4,1^1,4^1,6^1-tetrabromogalactosucrose	7500

Figure 13.3 Sweetening power of some sucrose derivatives.

Figure 13.3 shows a selection of some of the sweet compounds produced and their relative sweetening power. As can be seen, the range is very large, with very small changes in the configuration and conformation having a profound effect on the sweetness level. Thus a single chlorine substitution at the 4-, 1′- or 6′- position gives a modest enhancement of sweetness, which di-substiution at the 1′,6′- or 4,1′- positions increases the

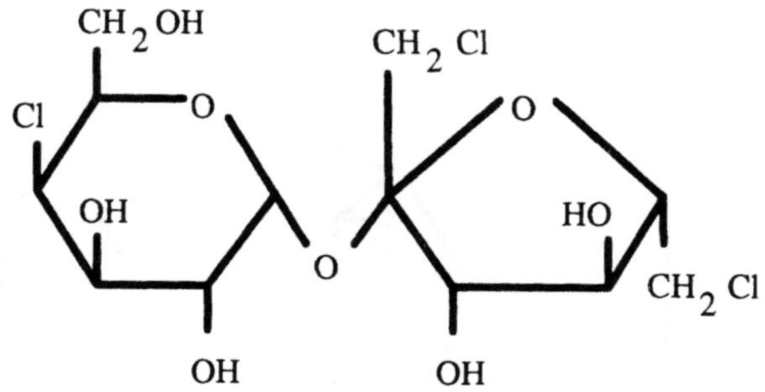

Sucralose

- 600× sweeter than sugar
- Excellent sugar-like taste
- Stable in foods and beverages
- Low acute toxicity
- Water soluble
- Biodegradable

Figure 13.4 Sucralose.

sweetness to about 100 times that of sucrose. Additional chlorine substitution to give the tri-substituted product raises the sweetness to about 600 times and tetra-chlorine substitution gives further enhancement to over 2000 times that of sucrose.

Many compounds containing mixed halogens were produced, some by long and exotic multistage synthetic processes with very low final yields. For example, the 4′-iodo derivative of 4,1′,6′-trichlorogalactosucrose was found to be about 7000 times sweeter than sucrose. Eventually, one compound emerged from this programme with clear, demonstrable advantages over other analogues.

Sucralose (Figure 13.4) was chosen for further development on the basis that it has a high quality of sweet taste, is stable to the processing and storage conditions used for foods and beverages, has a very low acute toxicity and is highly water soluble (Jenner, 1989). It is roughly 600 times sweeter than sugar with an approximate range of 400–800 times, depending on the level of sweetness desired and the individual product application. Furthermore it is biodegradable in the environment.

13.2 Production

Sucralose is made from sugar by a process involving selective chlorination (Figure 13.5). There are several chemical strategies that can be employed for this transformation, but they mainly involve variations on the theme of

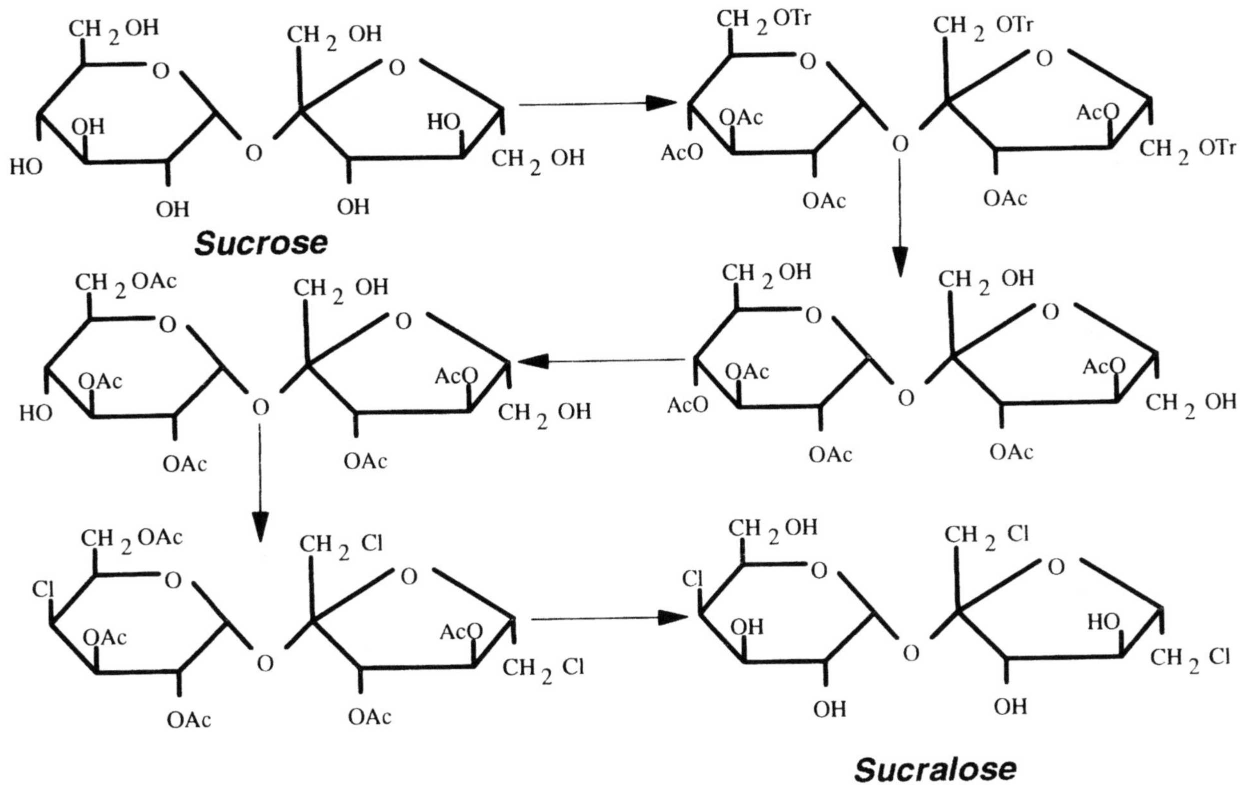

Figure 13.5 Sucralose manufacture.

Form	White crystalline powder
Odour	Odourless
Taste	Intensely sweet
Taste intensity	600 times sucrose (range 400–800)
Calorie content	Zero
Cariogenicity	Non-cariogenic
Solubility	Freely soluble in water and ethanol Insoluble in corn oil
Molecular weight	397.64
pH of 10% aqueous solution	5–8

Figure 13.6 Physico-chemical properties of sucralose.

selective protection of the relatively reactive 6-position of sucrose, followed by halogenation and subsequent removal of protection. This example shows the sequence which involves protection of the primary hydroxyls by trityl ethers and the secondary hydroxyls by acetates. Detritylation followed by acetyl migration gives the sucrose penta-acetate with the desired reaction sites unprotected. Chlorination using a system which inverts the configuration at carbon 4 from the gluco- to the galacto-analogue followed by de-acetylation leads to the required product, sucralose.

13.2.1 Properties

Sucralose is a white, crystalline, free-flowing powder (Figure 13.6). It is intensely sweet, odourless, freely soluble in water and has a negligible effect on the pH of solutions (Jenner, 1989). Sucralose contributes no calories to the diet and as it cannot support the growth of *Streptococcus mutans* in the oral cavity, it is also non-cariogenic (Bowen, 1990).

One particularly attractive feature of sucralose concerns its stability profile (Miller 1991) (Figure 13.7). Under extreme conditions of acid and heat, sucralose will slowly hydrolyse to its component monosaccharide derivatives. The mechanism is essentially the same as that for the conversion of sugar to glucose and fructose. However in sucralose the chlorine atoms have the effect of stabilising the glycosidic linkage towards protonation and thereby reducing the rate of this reaction by approximately

Cola	Storage	pH 3	30°C	1 year
Tomato ketchup	Pasteurisation	pH 3.8	93°C	51 min
Baked beans	Sterilisation	pH 5.6	121°C	80 min
Dairy product	UHT	pH 6.7	140°C	15 sec
Sponge cake	Baking		180°C	25 min
Biscuits	Baking		210°C	8 min
Crackers	Baking		230°C	4 min

Figure 13.7 Sucralose stability proven in these systems.

two orders of magnitude. For all practical purposes, therefore, the loss of sucralose by acidic hydrolysis in food systems is negligible. Tests using both model systems and product prototypes have clearly demonstrated that sucralose withstands the processing and storage conditions appropriate for a wide range of foods and beverages. For example, sucralose-sweetened soft drinks show no loss of sweetness after storage for 1 year at 20°C (Quinlan, 1990). This provides a clear benefit to the food manufacturer as well as to the consumer. Consistent products can be marketed, with long shelf lives where there is no reduction of sweetness as a result of degradation. Furthermore, sucralose can be successfully used in higher-temperature applications. Sucralose withstands UHT and pasteurisation processes, and can even be used in cooking, baking and extrusion.

Cakes and biscuits can be formulated with sucralose provided that the bulk of the displaced sugar is substituted, and extruded products such as cornflakes can be successfully produced by incorporating sucralose into the mix prior to extrusion. As well as being highly stable, sucralose is compatible with all common food ingredients, as it does not undergo any interactions with any colours, flavours or other components.

Obviously the quality of the sweet taste of a product such as sucralose will influence how widely the product is likely to be used by food manufacturers and how well it will be accepted by consumers. Fortunately sucralose has a remarkably similar taste to that of sugar. Some sweeteners have a delayed onset of sweet taste and others are characterised by a long, lingering sweet taste. Sucralose is close to ideal, with a rapid onset of sweetness perception and a persistence similar to that of sugar. Furthermore, the flavour profile of sucralose is very similar to that of sugar as it does not give rise to the pronounced bitter, metallic or astringent sensations sometimes associated with other low calorie sweeteners.

13.3 Regulatory situation

The data requirements for a food additive regulatory petition are necessarily extensive. Typically a food additive petition covers three major areas: (i) additive identification and characterisation; (ii) need, applications and projected intake; and (iii) toxicological test data. Figure 13.8 shows in a little more detail an outline of the typical product data requirements for a major new food additive.

Analytical chemistry plays a major part in the identification and characterisation section. It is critical that the product in question be properly characterised so that all the physico-chemical and safety studies are conducted on material of known composition and quality. Analytical methods have to be developed to set up a specification and search for potential impurities. Chemical and physical properties must be measured

Additive Identification and Characterisation
- Name, formula, structure
- Specification, potential impurities, analytical methods
- Chemical and physical properties
- Stability characteristics
- Manufacturing methods and quality control checks

Likely Applications and Potential Intakes
- Functionality in petitioned applications
- Advantages to consumers and manufacturers
- Use levels, stability, interactions, analysis
- Possible abuse situations, nutritional considerations
- Per capita intake, mean and extreme values
- Intake in special subgroups, e.g. children, diabetics

Toxicological Tests
- Acute
- Sub-acute
- Chronic
- Metabolism and pharmacokinetics
- Reproductive toxicology
- Carcinogenicity
- Special studies
- Ecotoxicology, biodegradability, environmental impact

Figure 13.8 Typical data requirements for a major new food additive.

and shown to be consistent. Stability characteristics may require extensive analytical capability, and of course quality control checks on all material produced must be rigorously applied.

In the food science and food technology areas, functionality in all likely applications must be understood, including use levels, stability profiles in use and potential interactions, and of course this also requires extensive analytical research. It is helpful to be able to demonstrate clear advantages over existing products and to show the relevance of these advantages to both the manufacturers of processed foods and to the consumer. Potential intakes must be calculated to include extreme values and consumption levels for special subgroups such as diabetics, in the case of sweeteners. This may also necessitate an evaluation of possible abuse situations where certain sections of the population may be consuming extreme diets. In addition there may be nutritional considerations when the reduction in calories from one component of the diet may lead to a relative increase in the calories derived from the other dietary elements.

As for toxicology testing, extensive studies must be conducted in several species to search for potential adverse effects. Many tests are standard, but inevitably the needs for evaluation of each new product are set individually through discussions between the petitioner and the regulatory agencies. The basic list requires acute, subacute and chronic toxicity studies, metabolism and pharmacokinetics, reproductive studies, carcinogenicity, special biochemical studies and ecotoxicology including biodegradability and environmental impact.

In the case of sucralose, in excess of 140 separate studies were conducted to evaluate these aspects.

This represents a formidable expenditure of money, manpower and time. An obvious consquence of the huge volume of data is that it takes regulatory agencies a lot of time to digest, question and evaluate the myriad of facts and figures. Bearing in mind that each regulatory agency has different criteria, relying on their own systems to evaluate new products, it is not surprising that some agencies are able to complete the process quicker than others. It follows that during the transition from a research and development project into a full commercial proposition, a new product such as sucralose will experience a progression of approvals in different territories (Figure 13.9).

Sucralose has so far been approved for food use in Canada, Australia, Russia, Romania, Mexico and Qatar. It has also been evaluated by JECFA, the Joint Expert Committee on Food Additives of the Food and Agriculture Organisation of the UN and the World Health Organisation, and was given an acceptable daily intake (ADI) of 15 mg/kg bodyweight/day.

Sucralose is gaining rapid acceptance in Canada where it is available in over 150 different products. It is sold to food manufacturers under the Splenda brand name, and the granular table-top Splenda low-calorie sweetener, has already captured 13.5% of the table-top market. The spoon-for-spoon version share of the market is already larger than that of all the other high-quality granular products combined, and it is accelerating the decline of the more traditional tablet and liquid forms. Repeat purchasing amongst customers is high, largely due to the ease of use and versatility of the product. The brand is supported by media activities and a consumer information centre with a panel of experts including nutritional and dietetic specialists. To date over 83 000 calls have been logged, the majority being from diabetics asking if the product is suitable for their particular needs. The remaining enquiries have mostly been from customers asking about recipes.

Canada	1991
Australia	1993
Russia	1993
Romania	1994
Mexico	1994
Qatar	1994
JECFA	1990

Figure13.9 Sucralose approvals as of February 1995.

Figure 13.10 shows the food categories in which sucralose can be used in Canada, together with the permitted levels (Knight, 1994). Not only can sucralose be used in traditional products such as beverages and chewing gum, but because of its heat stability it is suitable for pasteurised products such as sauces and canned fruit, UHT products such as dairy desserts and flavoured milks, and even extruded products such as breakfast cereals. Sucralose can be used in baked products, condiments, salad dressings, confectionery and table syrups.

Figure 13.11 presents a summary of the main physico-chemical properties of sucralose. It is produced as a white, odourless crystalline powder with an

Category	**Maximum level**
Tabletop sweeteners	GMP
Breakfast cereals	0.1%
Beverages	0.025%
Desserts, toppings, fillings	0.025%
Chewing gum, breath mints	0.15%
Fruit spreads	0.045%
Salad dressings	0.04%
Confectionary	0.07%
Bakery products	0.065%
Processed fruits and vegetables	0.015%
Alcoholic beverages	0.07%
Puddings	0.04%
Table syrups	0.15%

(GMP: Good Manufacturing Practice)

Figure 13.10 Approval categories of sucralose in Canada and permitted levels.

Physico-chemical properties of sucralose

Physical form	White crystalline powder
Odour	Odourless
Taste	Intensely sweet
Taste intensity	400–800 times sweetener than sugar
Calorie content	Zero
Cariogenicity	Non-cariogenic
Solubility	Freely soluble in water and ethanol Insoluble in corn oil (<0.1 g/100 g at 20°C)
Specific optical rotation	$(a)^{20}_{D}$+85.8° (C10, aqueous)
Octanol/water partition coefficient	0.32 (20°C)
Surface tension of aqueous solutions	71.8 mN/m (20°C, 0.1 g/100 ml)
Melting (decomposition) point	125°C (when heated from 115°C at 5°/min)
Specific gravity (10% aqueous solution)	1.04 (20°C)
Specific gravity (crystals)	1.66 (20°C)
Refractive index	Linear correlation with concentration
Molecular weight	397.64
Viscosity of aqueous solutions	Newtonian behaviour
pH of 10% aqueous solution	5–8

Figure 13.11 Summary of physico-chemical properties of sucralose.

intensely sweet taste. It is freely soluble in water, for example to the extent of 28.2 g/100 ml at 20°C, which indicates that it can be easily incorporated into foods and beverages using normal mixing procedures. Furthermore, a low octanol/water partition coefficient and very low solubility in corn oil demonstrate that sucralose will exhibit physical behavour similar to sugar in multiphase food systems, with a propensity to partition into the aqueous phase. Lowering of the surface tension of aqueous solutions by sucralose is negligible, so it will not promote foaming which might adversely effect high speed filling lines. In addition it has little or no effect on pH of aqueous solutions.

13.4 Conclusions

Sucralose is a versatile, high quality, low-calorie sweetener made from sugar and with a sweet taste similar to that of sugar. It is about 600 times sweeter than sugar and is stable to the conditions under which most foods are processed and stored. It is highly water soluble and is non-caloric as well as non-cariogenic. This combination should ensure that sucralose is rapidly accepted by both food producers and consumers.

References

Bowen, W.H., Young, D.A. and Pearson, S.K. (1990) The effects of sucralose on coronal and root-surface caries. *J. Dent. Res.* **69**(8), 1485–1487.

Hough, L. and Phadnis, S.P. (1976) Enhancement in the sweetness of sucrose. *Nature* **263** 800.

Jenner, M.R. (1989) Sucralose: Unveiling its properties and applications, in *Progress in Sweeteners* (ed. T.H. Grenby), Elsevier, London, pp. 121–141.

Jenner, M.R. (1991) Sucralose – How to make sugar sweeter, in *Sweeteners – Discovery, Molecular Design and Chemoreception* (eds. D.E. Walters, F.T. Orthoefer and G.E. DuBois), American Chemical Society, Washington, pp. 68–87.

Jenner, M.R., Bagley, L. and Heath, C.R. (1989) Sucralose – a unique high intensity sweetener. *Food Tech. Int. Europe*, 273–275.

Knight, I. (1994) The development and applications of sucralose, a new high intensity sweetener. *Can. J. Physiol. Pharmacol.* **72**, 435–439.

Lindley, M.G., Birch, G.G. and Khan, R.A. (1976) Sweetness of sugar and xylitol. Structural considerations. *J. Sci. Food Agric.* **27**, 140–144.

Miller, G.A. (1991) Sucralose, in *Alternative Sweeteners*, 2nd edn (eds L. O'Brien Nabors and R.C. Gelardi), Marcel Dekker, New York, pp. 173–195.

Quinlan, M.E. and Jenner, M.R. (1990) Analysis and stability of the sweetener sucralose in beverages. *J. Food Sci.* **55**(1), 244–246.

14 The blending of sweeteners – Applications and safety issues

G.-W. VON RYMON LIPINSKI

14.1 Introduction

Intense sweeteners have become more popular and their use has been expanding for many years. Whereas in the past they were chiefly used in diabetic products, with saccharin regarded as a cheap sugar substitute, intense sweeteners are now common ingredients in a range of low-calorie and calorie-reduced foods and beverages. In the course of the last 15 years the new sweeteners acesulfame-K (Sunett®, Hoechst AG) and aspartame (e.g. NutraSweet®, NutraSweet Co., Sanecta®, Holland Sweetener Co. or Aspartil®, Enzymologa S.A.) have been approved in many countries and are being increasingly used in a variety of applications. With the availability of improved sweeteners with a good taste profile, perceived by consumers as safe food ingredients, new and extended fields of application are being developed or will be possible in countries progressing towards approval for additional applications e.g. by implementing the EUC Sweetener Directive.

Approvals and newly devised applications for the new sweeteners have extended the market share of low-calorie and calorie-reduced sweet products in general, enabling the food and beverage industry to manufacture good tasting products with acceptable shelf-life at reasonable cost.

Sweetener blends have become important in the production of foods and beverages, making use of the benefits of multiple sweeteners, as synergistic taste enhancement and sweetness profile modifications offer advantages over the use of single sweeteners. Further sweetener blends will improve versatility and therefore play a key role in new sweetener applications.

14.2 Synergism

The most important reason for the use of sweetener blends is synergism. Synergism is a general description of positive sensory interactions resulting from some blends. Synergistic effects may be more or less pronounced, depending on sweetener type, contributing to a greater or lesser degree to the general perception of sweetness. Quantitative and qualitative synergistic effects are both important.

14.2.1 Synergistic taste enhancement

Synergistic taste enhancement is the term used for an intensification of sweetness. Quantitative synergism is particularly pronounced in certain blends of intense sweeteners. It can easily be demonstrated by preparing solutions of two intense sweeteners in concentrations providing the same sweetness. Blends of equal volumes of such solutions would be expected to show the same sweetness level, but instead the sweetness is distinctly enhanced.

A long established example for the use of such blends is a mixture of cyclamate and saccharin in a ratio of approximately 10:1. This has been generally used in countries approving both sweeteners as, due to the synergistic effects, the blend is approximately 100 times sweeter than sucrose, and therefore sweeter than the calculated sweetness intensity value of approximately 80 times. Thus concentrations are often required not to exceed the quantity restrictions imposed on these two sweeteners in countries approving their use, whereas a single sweetener in approved quantity would provide insufficient sweetness.

Synergistic taste enhancement is observed with a variety of sweetener blends (Frank *et al.*, 1989; Helgren and Kirchmeyer, 1955; Scott, 1971; von Rymon Lipinski and Lück, 1978; Beyts and Latymar, 1982; Stephens and Torres, 1984). The published information, however, is sometimes based on a limited number of blends and blend ratios only. More systematic investigations showed substantial differences in the degree of sweetness enhancement in different blends. Frank *et al.* (1989) studied eight blends of five different intense sweeteners. Whereas virtually no synergism was found in blends of acesulfame and saccharin or cyclamate and stevioside, taste enhancements of between 17 and 38% were reported for the other blends. One of the two potential blends for which no data were reported is the well known synergistic blend of saccharin and cyclamate. At least 7 of the 10 possible binary combinations showed synergistic taste enhancement, and no sweetness suppression was found in any blend of these sweeteners.

Calculations of the synergistic effects and quantity savings in synergistic

Table 14.1 Synergistic sweetener blends

Acesulfame–aspartame
Acesulfame–cyclamate
Acesulfame–stevioside
Aspartame–cyclamate
Aspartame–stevioside
Aspartame–saccharin

Source: Frank *et al.* (1989).

sweetener blends are sometimes possible. Tables published by Hoppe and Gaßmann (1991) give examples for sweetener concentrations having the same sweetness levels. In addition, instructions to calculate equisweet concentrations for bulk and intense sweeteners as well as blends for sweeteners in general are given. Whereas for most substances calculated values coincide quite well with the concentrations determined in sensory studies, some reported data from sensory studies indicate higher synergistic effects than calculated from the published tables.

The most important example for the use of synergistic sweeteners is now the blend of acesulfame and aspartame. As these sweeteners show a particularly good complementary behaviour, synergistic effects between them have probably been studied in more detail than between any two other sweeteners. This refers to different concentrations as well as blend ratios. Results of sensory studies on blends of acesulfame and aspartame were published in 1990 (von Rymon Lipinski, 1990). Curves representing the equi-sweet concentrations for acesulfame and sucrose, aspartame, a 1:1 blend and a 2:1 blend of the two intense sweeteners show the obvious synergism from the blends (Figure 14.1). A given quantity of the sweetener blend matches the sweet taste of a much higher sucrose concentration than the same quantity of a single sweetener. The gain in sweetness exceeds 30% for medium- and 40% for high-sweetness levels.

Another advantage of sweetener blends can be seen clearly from the curves. As described by Hoppe and Gaßmann (1991), for single sweeteners maximum sweetness intensities seem to exist, and these cannot be exceeded by increasing sweetener concentrations. With blends, however,

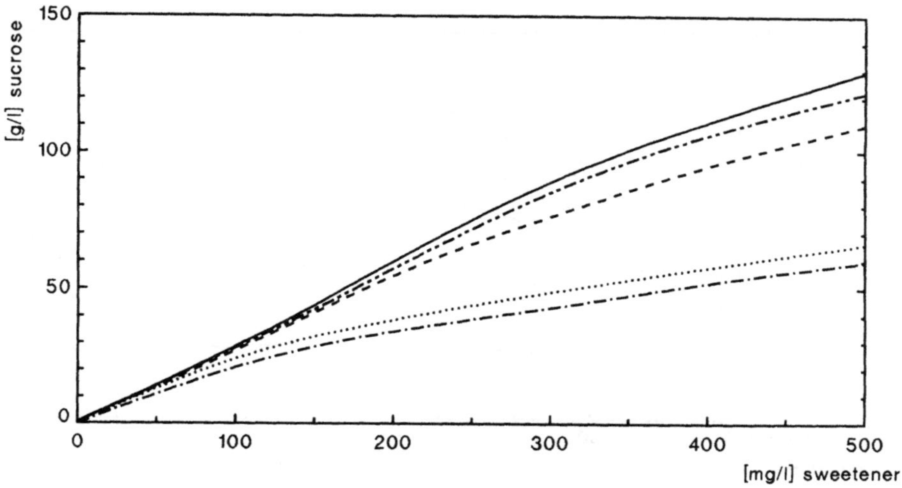

Figure 14.1 Equi-sweet concentrations of acesulfame (— · —), aspartame (· · ·), sweetener blends and sucrose. Blend ratio, – –, 2:1; — - -, 1:2; ——, 1:1.

much higher sweetness levels can be obtained than with the single sweeteners. In addition, the dose–response ratio is almost linear, so that an approximate calculation of quantities required for blends should be possible. Although slight adjustment of concentrations will be necessary in practice, such calculations will at least come close to the intended sweetness level and therefore facilitate product development.

Whereas according to literature data and our observations, the highest synergistic taste enhancement is generally found in blends to which both sweeteners contribute equal levels of sweetness, some synergistic taste enhancement is found for other blend ratios (Figure 14.2). Although two important advantages of blends, i.e. sweetness enhancement and simple calculation of concentrations required, are not fulfilled by other blend ratios, there may be reasons to use another than the optimal ratio. As intensification of the sweetness can still be demonstrated for other ratios, only a small proportion of the quantity and cost reduction has to be sacrificed should reasons exist to use other than the optimal blend ratio.

Quantitative data have been published in a similar way for blends of acesulfame and neohesperidin dihydrochalcone and acesulfame and cyclamate (Lotz and Meyer, 1995). In the case of the blend of acesulfame and cyclamate the limited approval of cyclamate will not allow general use. For the combination of acesulfame and neohesperidin dihydrochalcone, only limited quantities of NHDC show synergistic sweetness enhancement, whereas higher concentrations do not bring about any additional effect. Due to side-tastes neohesperidine dihydrochalcone has at higher concentrations, practical applications would be limited to levels of use specified in

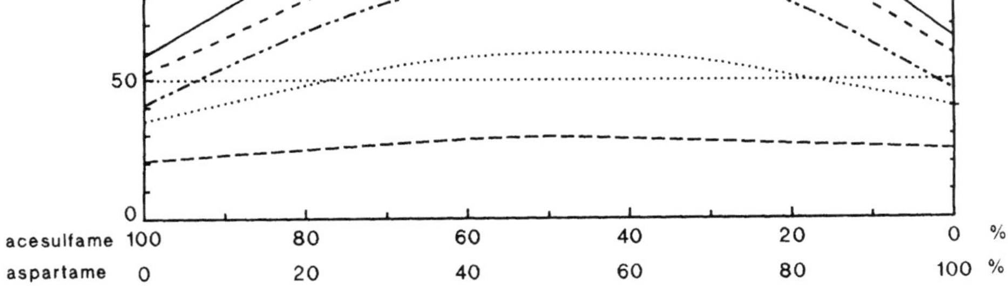

Figure 14.2 Equi-sweet concentrations (mg/l) of sucrose and of acesulfame–aspartame blends. – –, 100; – – – ·, 200; – - -, 300; – –, 400; ——, 500.

this study. With the low equisweet sucrose levels from the blends, use of these sweeteners together for synergy would not be possible in certain important fields of application. Due to the flavour modifying properties of neohesperidin dihydrochalcone, however, it may be added in concentrations not showing a pronounced sweetness enhancement but in order to modify and suppress side-tastes of other sweeteners.

Studies carried out on tertiary and quaternary blends often indicate that there is no additional taste enhancement beyond that from binary systems. No systematic data have been made available for such blends, however.

Systematic data published on synergistic sweetness enhancement are generally based on sensory studies performed with solutions of sweeteners in water. Although slight variations in synergism may be found between different foods and beverages, as a rule the effects in many food systems are similar to those in water. For beverages results obtained in water often require only minute adjustment.

Although synergistic sweetness enhancement is particularly pronounced with blends of intense sweeteners, blends of intense sweeteners and other sweet-tasting compounds often show less pronounced sweetness enhancement. Some examples are given in Table 14.2, in which calculation of synergistic enhancement is based on levels for equi-sweet concentrations taken from the literature and the levels observed in sensory studies. Sometimes the presence of small quantities of carbohydrates, in particular fructose, will contribute substantially to the total sweetness, indicating a high synergistic effect.

Table 14.2 Use of synergistic blends

Acesulfame-K (mg/l)	Aspartame (mg/l)	Sugar (g/l)	Fructose (g/l)	Equisweet sugar level (g/l)
Beverages				
140	140	–	–	85
200	100	–	–	85
160	160	–	–	100
240	120	–	–	100
200	–	50	–	100
70	70	50	–	100
130	130	–	10	100
Yoghurt				
200	200	–	–	100
160	160	10	–	100
Milk and cocoa drinks				
160	170	–	–	100
240	130	–	–	100
220	–	50	–	100
75	75	50	–	100

14.2.2 Synergistic taste improvement

In addition to the synergistic taste enhancement other taste effects may result from blending.

Qualitative improvement may be obtained by masking side-tastes and after-tastes. This well known effect has been used for decades in the common saccharin–cyclamate blends. Data published by Paulus and Braun (1988) indicate a substantial reduction of negative taste characteristics from this blend. As threshold values for undesired taste characteristics are normally high and the build-up of such tastes is evident at higher concentrations only, simple reduction in quantity may therefore be the reason for this type of taste improvement.

Time–intensity modification of sweetness perception is an important factor for quality improvement of sweeteners. Flavours of foods show specific time–intensity profiles. Normally sweetness and flavour should be matched in order to get optimal sensory quality. Should a sweetener have a delayed onset of sweetness and a delay be perceptible between flavour and sweetness detection it will impair product quality. Too fast a fading or a lasting sweetness which will be tasted after disappearance of other flavours may often not be regarded favourable either. Different time–intensity profiles of diverse sweeteners may therefore be used to modify the time–intensity profile of the sweetness. This property is particularly pronounced with blends of acesulfame and aspartame. When the sweeteners are used in the optimal ratios particularly high quality ratings are often obtained. But examples of products on the market demonstrate that not only blend ratios giving highest taste enhancement, but also other ratios providing an initial or a more lasting sweetness, are used in order to get the best adaptation of the sweetness to the product flavour. Synergistic sweetness intensification is still observed even when the blend ratio is not optimal, so that additional product improvement may be obtained without substantial additional cost. This important aspect of sweetener blends should receive greater attention, as with the availability of several sweeteners better adaptation of sweetness to various flavours has become possible.

14.3 Sweetness ability

Taste stability is an important quality criterion for many products. Taste changes during the normal shelf-life periods will commonly be regarded as undesirable and seen as a loss of quality. Sweetness stability is an important part of taste stability. Too fast a loss in sweetness will therefore adversely affect product quality after prolonged storage, or may necessitate shortening the expiry date.

Although some of the sweeteners available provide satisfactory sweetness stability, the use of blends of less stable and stable sweeteners will improve the shelf-life substantially compared to the single use of sweeteners that show some degradation during product storage. Shelf-life will be extended because part of the sweetness will be provided by the stable sweetener and will therefore not be subject to degradation, and only the sweetness of the less stable sweetener will fade. Theoretically, for blends without any synergistic taste enhancement, the extension in shelf-life should be mostly determined by the blend ratio, and perceptible taste changes would arise after twice as long a storage period, for a 1:1 blend on a sweetness basis in a hydrolysing product. In synergistic blends this pattern is changed only slightly. The example of acesulfame and aspartame blends demonstrates that changes of the blend ratio around the optimal ratio do have a less pronounced influence on the extent of synergism than changes of ratios in which one of the sweeteners dominates. Losses in sweetness will therefore be mostly determined by loss of one of the sweeteners and the absence of synergism will contribute only slightly to the total loss of sweetness. Shelf-life extension in synergistic blends will not be exactly as calculated theoretically, but not too far away. It should be kept in mind that a change of blend ratio from 1:1 to 2:1 on a weight basis would take place after 50% decomposition of the less stable product, which for single sweetener use would mean a substantial loss of sweetness.

Sweetness profiles change with a decline in one component in a blend, but sweetness loss would be much more evident and therefore change the taste impression much more prominently than alterations to the sweetness profile.

14.4 Selection criteria

Intense sweeteners have low functionality apart from their sweet taste, and therefore cannot be used as the only sweetening agents when other technological functions are important. Combinations of intense and bulk sweeteners will come close to sucrose and sweet carbohydrates in functionality and taste, and can therefore be considered as an alternative to sugar worth considering in applications requiring functional properties. However many sweet-tasting foods and beverages do not require the functionality of sucrose and sweet carbohydrates, and these products are the most important fields of applications of intense sweeteners.

The typical fields of application of intense sweeteners may be divided into several groups according to the main requirements (von Rymon Lipinski, 1991):

- products having a high sweetness level but not requiring sweetener functionality
- products undergoing heat treatment during production

- products with a long shelf-life
- products requiring functionality and therefore the use of bulk sweeteners in addition to intense sweeteners

In products having a high sweetness level intense sweeteners are among the most important ingredients. They have a strong influence on the taste characteristics of the product and may be one of the most important cost factors, too.

The sensory characteristics of the intense sweetener(s) to be used are of high importance for the products belonging to this group. A good, clean sweetness which does not linger is generally beneficial. As the different intense sweeteners have different sensory characteristics, and may have a side-taste at elevated concentrations and time-intensity profiles different from sucrose, high-quality sweetness is not easily achieved. Single intense sweeteners will often not provide characteristics satisfactory in all respects for products with a high sweetness level. Too fast a fading of the sweetness, too long a sweetness duration, and side-tastes will provide a kind of sweetness substantially different from the taste required.

Taste profiles of intense sweeteners vary, but they can be combined into blends to overcome disadvantages by choosing mixtures of appropriate proportions. Blends allow adaptation of time-intensity profiles and flavours and will in addition be highly economical as a result of synergism.

Stability against thermal degradation or hydrolytic decomposition at elevated temperatures is very important for sweeteners to be used in products that have to be heat-treated. For canned fruit, jams, marmalades and preserves intense sweeteners with good stability properties should be used. During pasteurisation or hot bottling of fruit juice beverages some loss of less stable sweeteners can take place, although such losses are lower than those during sterilisation and cooking. Should less stable sweeteners have to be used, blending with more stable sweeteners is appropriate to eliminate or minimise losses during heat processing.

Pasteurisation and UHT-treatment of dairy products do normally not cause too great a loss of sweeteners, even the less stable ones. It should be kept in mind, however, that at least for UHT-treated products storage for some time at room temperature is common, so that stability during processing as well as during storage is necessary.

Some products containing intense sweeteners need to have a long shelf-life. Therefore shelf stability of intense sweeteners is important for these applications, in particular for beverages. Sweetness is an important taste characteristic of soft drinks, and changes in this will impair their quality. Some of the intense sweeteners provide adequate sweetness for shelf-life and labelling requirements, and blending of less stable with stable sweeteners will reduce sweetness losses to acceptable levels within normal shelf-life periods. Unless a substantial proportion of one of the sweeteners is degraded, a substantial gain in sweetness stability is possible, compared with single sweetener use.

In products requiring functionality as well as sweetness choice of an appropriate bulk sweetener or bulk sweetener and bulking ingredient system is crucial for correct product characteristics. When bulk sugar substitutes are used to replace sugar they provide some sweetness. Intense sweeteners round and balance the sweetness profile and bring the sweetness to a higher level. Balancing flavour and sweetener profiles and sweetener stability are the key selection factors for these products, as they, like bakery products and confections, are often exposed to high temperatures during processing.

14.5 Safety

Intense sweeteners require food additive approvals. Extensive safety tests are necessary before food additive petitions can be filed. Such tests have to demonstrate the absence of chronic toxicity, carcinogenicity, mutagenicity, reproductive toxicity and unacceptable pharmacological effects. Substances for which the absence of such effects cannot clearly be demonstrated will not obtain food additive approvals. This applies even for tests at grossly exaggerated concentrations far exceeding normal human consumption levels, as for the calculation of the acceptable daily intake for humans the level of 'no observed effect' in safety studies is normally divided by a safety factor of 100.

The lack of indications of any substantial toxicity at normal use levels for the approved intense sweeteners, and the high safety margin between life-time 'no observed effect' levels in animals and acceptable daily intake, exclude any appreciable risk for human consumption of single sweeteners as well as of blends. Accordingly, the High Commission of the German Research Council (1989) concluded on evaluating the risks of sweetener blend use:

> Synergistic effects between some of the sweeteners allow use of a lower quantity of sweeteners in sweetener blends than for single sweetener use. Thus the total consumption of sweeteners can be reduced.
>
> From a toxicological point of view no objections against the use of sweetener blends exist. The High Commission for the Evaluation of Food Additives and Contaminants of the German Research Council has concluded that kinetics, metabolism, potential effects and target organs of the sweeteners acesulfame, aspartame, cyclamate, and saccharin do not give any indications for combination effects of toxicological concern. For quantities as ingested by consumers no combination effects must be anticipated, as even under unfavourable circumstances ingested single doses remain far below a potential threshold for toxicological effects, and interactions of the individual sweeteners due to different kinetics must not be anticipated.

This view is apparently shared by committees and governments worldwide. Use of sweetener blends is therefore possible in virtually all countries approving intense sweeteners without any additional restrictions. Occasionally proportional clauses apply, allowing sweeteners not exceeding certain proportions to be used, although this seems rather unnecessary. In sweetener blends, whenever there is synergism, reductions in quantity can be achieved, and the self-limiting effects of too strong a sweetness will also come into action.

14.6 Intake considerations

With normal food habits the intake of intense sweeteners is fairly low. Consumption studies demonstrate that sweetener intake is normally well within the limits of the acceptable daily intake values allocated by the WHO and the FAO and the Scientific Committee for Foods of the EUC. Even very high consumers of certain foods or beverages containing intense sweeteners almost never exceed acceptable levels (MAFF, 1993; Hinson and Nicol, 1992). It should be noted that temporary intake exceeding the ADI should not bring about any sutstantial health concern.

Consumption of single sweeteners will be lowered by making a variety of sweeteners available for combined use. Employing a variety of sweeteners in different products will definitely be a move towards lowering the consumption of any one sweetener. The use of blends will reduce sweetener intake further as only part of the sweetness is provided by each individual sweetener, utilising the property of synergism.

References

Beyts, P.K. and Latymer, Z. (1982) EP 64 361, assigned to Tate & Lyle.
Frank, R.A., Mize, S.J.S. and Carter, R. (1989) *Chemical Senses* **14**, 621–632.
German Federal State Council (1989) Circular No. 411/89, 43.
Helgren, F.J. and Kirchmeyer, F.J. (1955) *Journal of the American Pharmaceutical Association* **44**, 442–446.
Hinson, A.L. and Nicol, W.M. (1992) *Food Additives and Contaminants* **9**, 669–681.
Hoppe, K. (1991) *Lebensmittelindustrie* **38**, 13–14.
Lotz, A. and Meyer, E. (1995) *International Food Marketing and Technology* **8**(2), 4–6, 8, 9.
Ministry of Agriculture, Fisheries & Food (1993) *Dietary Intake of Food Additives in the UK*, London.
Paulus, K. and Braun, M. (1988) *Ernährungs-Umschau* **35**, 384–391.
von Rymon Lipinski, G.-W. (1990) *International Food Marketing and Technology* **4**(5), 22–25.
von Rymon Lipinski, G.-W. (1991) *International Food Marketing and Technology* **5**(2), 5, 8–11.
von Rymon Lipinski, G.-W. and Lück, E. (1978) DE 2 628 294, assigned to Hoechst AG.
Scott, D. (1971) GB 1 256 995, assigned to G.D. Searle.
Stephens, C.R. and Torres, A. (1984) EP 139 430 (1984), assigned to Pfizer.

15 Cultural and legislative influences on the consumption of high intensity sweeteners in Europe

I.R. GORDON*

The cultural and legislative influences on the consumption of high intensity sweeteners in Europe were originally treated in a 500-page study published in 1994 (Giract, 1994a) which was the third of a series. Some of the main aspects are discussed below.

15.1 Objectives

The study had as its primary objectives:

(i) to determine the present penetration of, and thus demand for, high intensity sweeteners in Western Europe in three pertinent sectors:
- soft drinks
- table-top and culinary use
- solid foods and pharmaceutical/OTC formulations

(ii) to examine legislative and cultural influences in order to develop meaningful forecasts for future demand;

(iii) to examine the pattern of supply and probable future developments by the main producers.

15.2 Scope

In order to cope with such a vast subject, simplification into three key aspects is necessary:

(i) basic cultural divisions of Western Europe:

(ii) legislative influences from the past and changes taking place with the implementation of the EU Sweetener Directive, and

(iii) an examination of these effects on the major growth sector for high intensity sweeteners (HIS) – that of soft drinks.

Nine high-intensity sweeteners were considered. The main focus was on:

- aspartame
- acesulfame
- saccharin
- cyclamate

Forecasts were also made for:

- sucralose
- stevia
- alitame
- neohesperidene
- thaumatin

The full study looked at many markets, including the USA and Switzerland, but this chapter is limited to the three main European food cultures defined below.

15.3 Background

To set the scene, the market for high intensity sweeteners in Western Europe is of the order of a hundred million ECU, of which aspartame has an over 60% value share, although in volume terms its share is significantly less.

The concept of sucrose equivalence has been employed for comparing the sweeteners. This means that different sweeteners may be compared in terms of 'tons of sucrose equivalent'. This is fully valid when only the sweetening power of the substances is being compared. Thus in a large number of cases sucrose will fulfil other supplementary roles. In particular, sucrose and bulk sweeteners such as sorbitol and other polyols (Giract, 1994b) often fulfil major 'mouthfeel' bulking and carrier functions. Levels of potentiation have also been applied to allow for synergistic effects between sweeteners.

Figure 15.1 shows the ruling prices in 1993–94 and demonstrates how economically competitive the high intensity sweeteners are in comparison to full price sucrose, provided of course that all that is being sought in a formulation is sweetness. Once bulking effects other than those that may be provided by air or water are needed then completely different relationships exist.

The three basic European food cultures have been defined (Giract, 1994c) as:

(a) *Latin/Mediterranean*: characterised by:
  - a high proportion of fresh fruit, vegetable, pasta and rice in the diet
  - use of olive oil

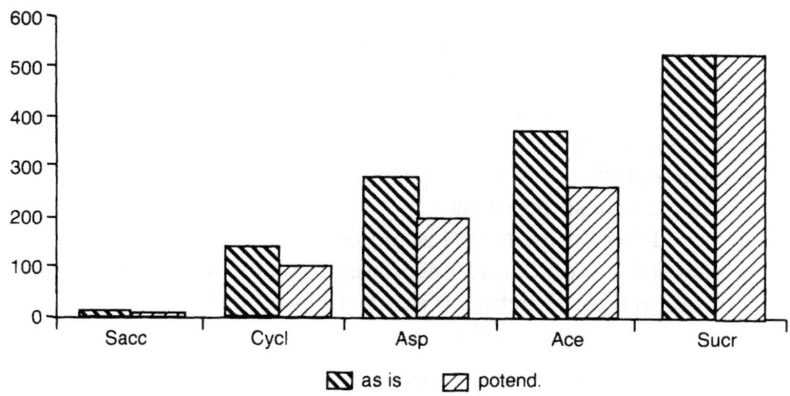

Figure 15.1 High intensity sweetener *vs.* sucrose prices (ECU/tse). tse, tons sugar equivalent; sacc, saccharin; cycl, cyclamate; asp, aspartame; ace, acesulfame-K; sucr. sucrose; potend, potentiated.

- the consumption of wine and water, including large quantities of still and carbonated mineral waters
- coffee usually drunk strong and black
- shopping still associated with the market and the house-wife preparing meals for her family, even though in many modern cities this is no more than a fond memory

For this culture France, Italy and Spain are used as examples.

(b) *Germanic*: covering a large swathe of Central and Northern Europe including The Netherlands, Belgium and the North-east of France. Here the traditional accent is on a higher calorific value of foods:

- large quantities of bread, potatoes, and meat in the diet
- beverages consumed are largely beer, though there are pockets of wine consumption where wine is locally grown
- for soft drinks waters are still important, but there is a growing interest in fruit-based drinks and colas, especially in Germany
- instead of olive oil, the focus is on dairy products and animal fats
- as a hot drink, medium strength coffee is all-pervasive
- we have now moved a long way from Kinder, Kirche and Kuche, and both partners generally work outside the household. Convenience shopping is the norm, with a high penetration of frozen goods and products stored and sold at ambient temperatures, and a relatively small use of chilled goods.

For this culture Germany serves as an example.

(c) *Anglo-Saxon/Scandinavian fringe*: to the north and west of Europe. A combination of special values and a closer overlap with the food cultures of North America give this culture:

- a base of potato and bread as carbohydrate
- a rapidly developing interest in a wide range of other ethnic foods, including pasta and rice-based ones
- a strong meat and dairy product orientation
- a basic consumption of beer
- carbonated soft drinks, especially colas, but also the typically British orange squash products
- a predeliction for tea, Nescafé or a relatively weak coffee
- many of the food preferences are driven by the very highly organised retail market structure (Giract, 1994d), and the best chill distribution chain in the world

For this culture the UK will be used as an example.

15.4 Legislation

In the last ten years, major changes have taken place in European legislation, mainly driven by the development of the Single European Market. This has occurred at a time when a number of new products have come onto the market to meet the latent demand for high intensity sweeteners. Such changes have included:

- the freeing of French legislation, permitting the use of high intensity sweeteners in foodstuffs and the marketing of table-top sweeteners in the retail grocery circuits
- similar relaxation of legislation in Italy and Spain, especially with regard to table-top sweeteners
- modification of the regulations affecting formulations in several European countries
- the rapid change in the pricing of aspartame with the expiry of the Monsanto/Nutrasweet patents, including anti-dumping levies in the EU whilst new capacity was being built
- soft drink manufacturers reformulating blends for sweetening (especially Coca Cola)
- the approval of the Sweetener Directive with its implication of complete harmonisation of national sweetener legislation during the 1990s

With the implemention of the directive, Table 15.1 shows typical maximum use levels for the primary high intensity sweeteners.

The legislation has a number of consequences including:

- the reintroduction of cyclamate into both UK and French markets
- a restriction on the use of saccharin, especially in soft drinks, since high intensity sweeteners in general, and saccharin in particular, fulfil roles not only in 'calorie-reduced' or 'Lite' soft drinks but also in 'regular' drinks for price reduction purposes.

Table 15.1 Legislation: directive, sample maximum use levels

Food	ace[a]	asp[b]	cycl[c]	sacch[d]	NHDC[e]
Soft drinks (general)	350	600	400	80	30
Soft drinks (gaseous)	350	600	400	100	50
Ices	800	800	250	100	50
Chewing gum	2000	5500	1500	1200	400
Sugar confectionery	500	1000	500	500	100
Bakery	1000	1700	1600	170	150

[a] Acesulfame-K. [b] Aspartame. [c] Cyclamate. [d] Saccharin. [e] Neohesperidine DC. Copyright © Giract Sarl.

A number of these formulations are expected to change to fall in line with the new legislation. This will have particular impact on Anglo-Saxon and Germanic cultures.

As discussed in earlier chapters, in addition to those high intensity sweeteners already approved within the EU Directive, a number of others are waiting in the wings, including sucralose and alitame. These are treated in this chapter as being 'aspartame/acesulfame equivalents' in that either or both of them may satisfy part of the demand for high intensity sweeteners. However, the uncertainty and timing of legislative approval is such that separate forecasts of volumes are not provided.

15.5 Soft drinks

The soft drinks sector is highly suitable as an example of the market for sweeteners since:

- it is the major growth sector of use, either first or second in all the countries studied
- the changes induced by legislation are particularly significant and have led or are leading to product reformulations

It is thus a market sector in rapid change, and since it is the biggest user of high intensity sweeteners world-wide, a comparison of consumption figures is of interest.

Figure 15.2 shows the total volumes of soft drinks for the 5 countries representing the three cultures, and their forecast development. Clearly the UK (Anglo-Saxon) and German markets outperform the other countries.

When the low-calorie sector is considered (Figure 15.3), the volume is still relatively small though growth has been substantial since 1986. The

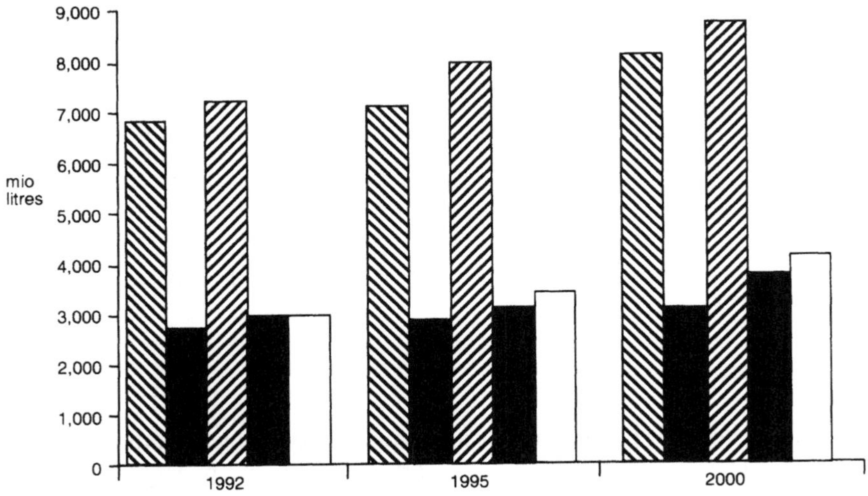

Figure 15.2 Five country comparision of all soft drinks sales (1986–2000). ▧, UK; ■, France; ▨, Germany; ■, Italy; □, Spain. mio, million.

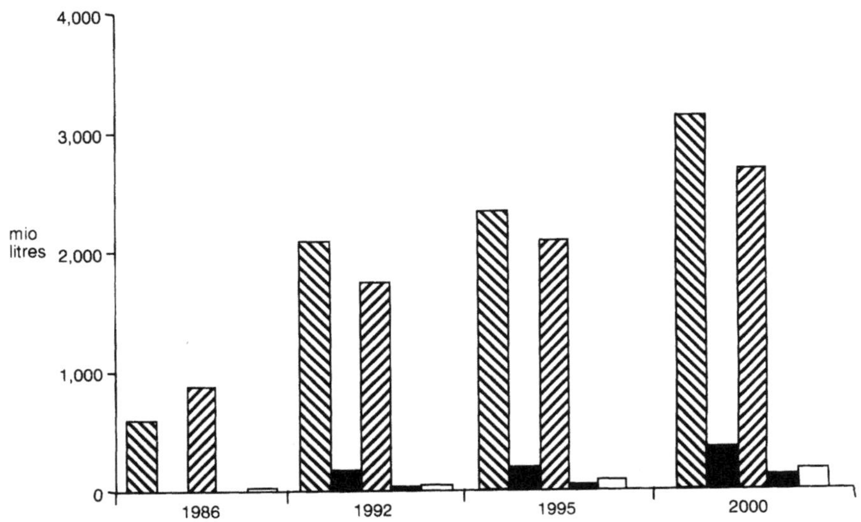

Figure 15.3 Five country comparison of all low-calorie soft drinks sales (1986–2000). ▧, UK; ■; France; ▨, Germany; ■, Italy; □, Spain. mio, million.

forecast for the year 2000 is that this sector will continue to grow at some 5% per annum, but that its proportion as a total of the entire sweetening market will increase less rapidly than total consumption due to the continued expansion of the market place. The reasoning behind this is explained below.

15.5.1 Latin/Mediterranean culture

Taking France as an example, Figure 15.4 shows the distribution of sweetened soft drink consumption. The low calorie sector is very small (Figure 15.5) having risen from zero since legislation was altered to allow the production and sale of such products in the late 1980s.

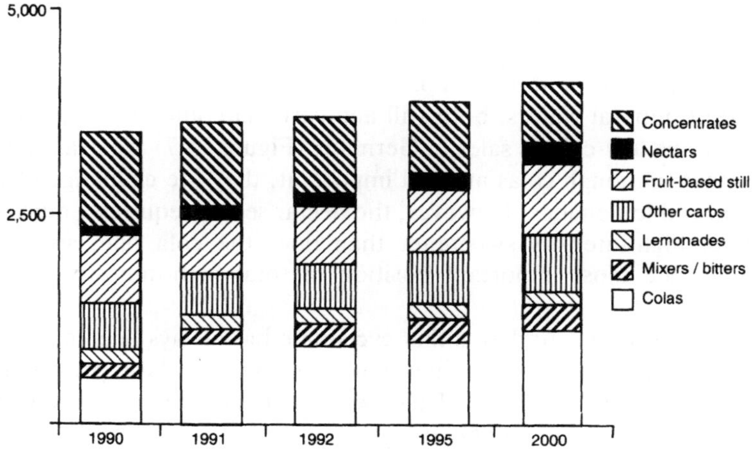

Figure 15.4 Soft drink volumes by type: France (1990–2000).

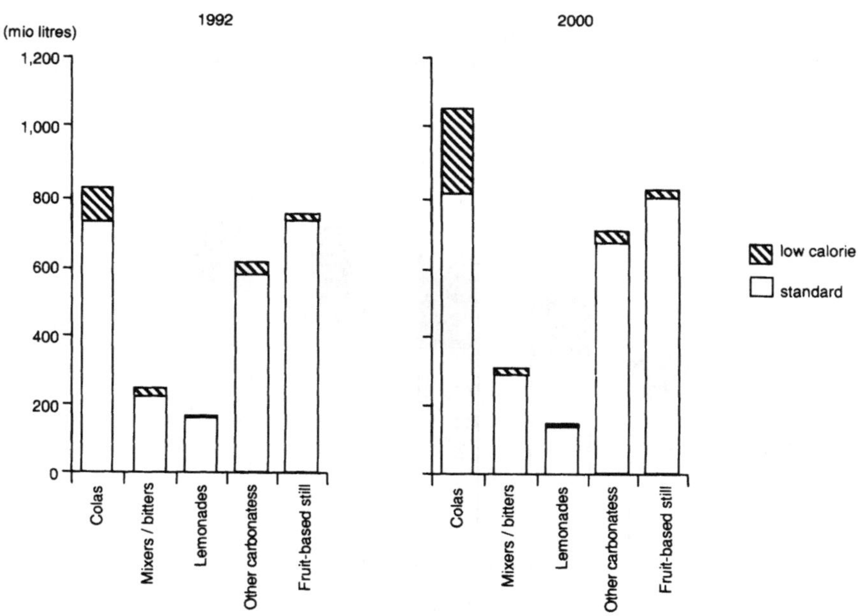

Figure 15.5 Standard and low-calorie soft drinks markets in France (1992 and 2000).

This volume is expected to more than double in the coming decade, but nevertheless this sector will lag well behind the more developed markets of Germany and the UK. As in many markets, low-calorie drink development will expand alongside the cola sector.

15.5.2 Germanic culture

In the Germanic culture a similar picture for total soft drinks (Figure 15.6) indicates the relative importance of the nectar sector, though both here and in France a very important part of total alcohol-free drink consumption is that of mineral waters, both still and carbonated/naturally sparkling.

When the low-calorie sales in Germany (Figure 15.7) are examined, it is clear that, although colas are still important, they are challenged by both nectars and lemonades. However, the nectar sector requires a much lower level of high-intensity sweetener than does the cola one, so cola still occupies the most important position in total high-intensity sweetener consumption.

A feature of this market is that cyclamate has always been allowed and saccharin/cyclamate mixtures have a well established place, providing cost-effective good taste profiles. These will be under some pressure from the legislative changes noted above.

15.5.3 Angle-Saxon culture

Finally, for the Anglo-Saxon culture, Figure 15.8 shows the total volumes of soft drinks in the very complex market in the UK. Concentrates

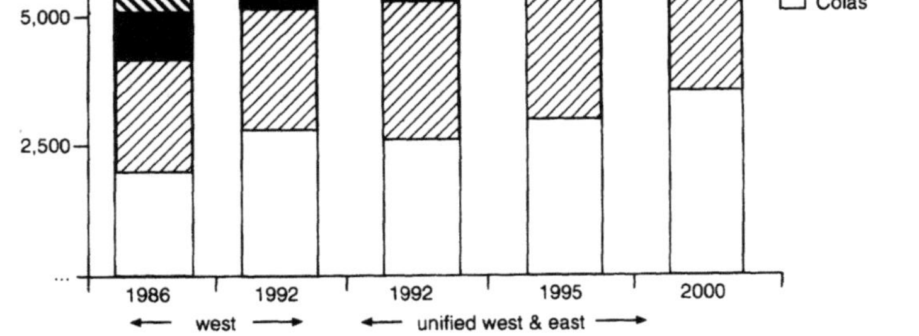

Figure 15.6 Soft drink volume by type: Germany (1986–2000).

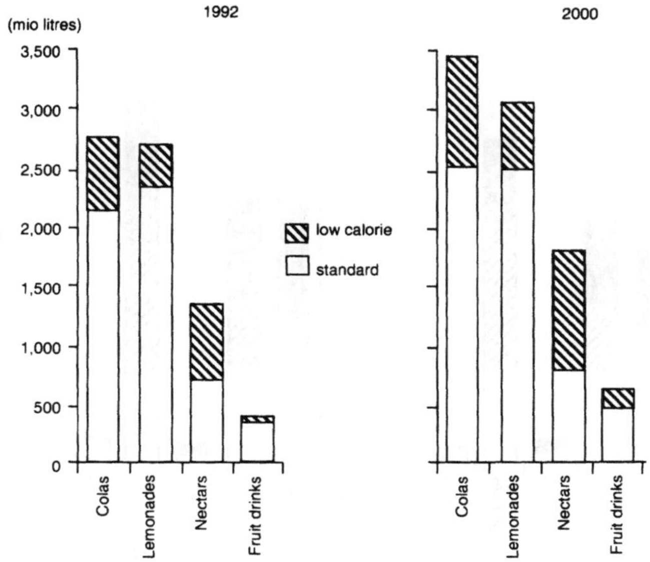

Figure 15.7 Standard and low-calorie soft drinks markets in Germany (1992 and 2000).

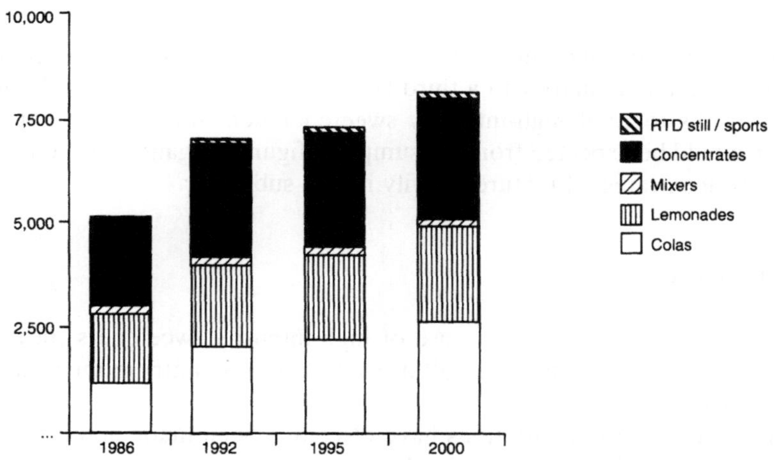

Figure 15.8 Soft drink volumes by type: UK (1986–2000). RTD, ready-to-drink.

(squashes) have been converted to ready-to-drink (RTD) equivalent using a multiplication factor of 5. The market is very large, with colas, lemonades and concentrates forming the major parts.

It is necessary, however, also to consider the use of high-intensity sweeteners in drinks not marketed as diet or low-calorie products. This volume is very large, more than double that of the low-calorie sector. However, since these products contain sugar together with high-intensity

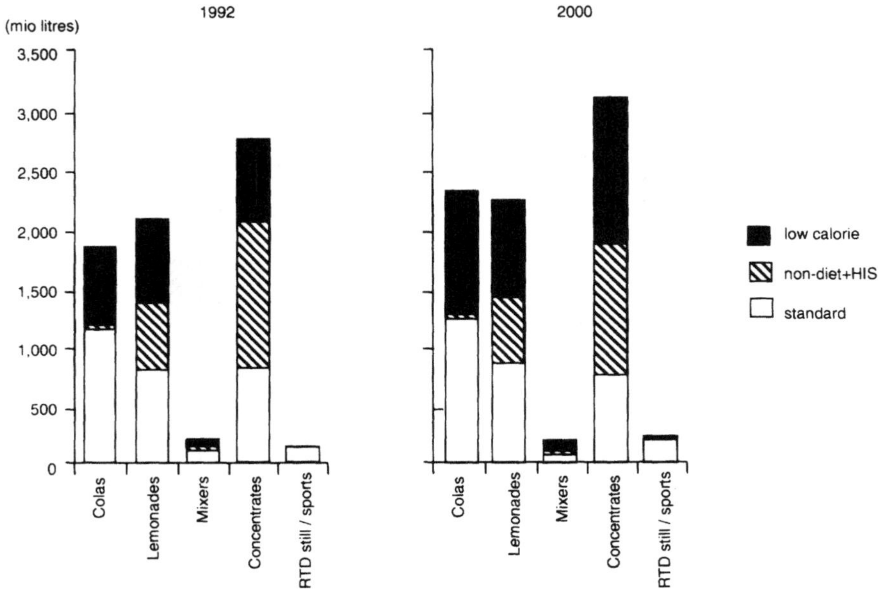

Figure 15.9 Standard and low-calorie soft drinks markets in the UK (1992 and 2000).

sweeteners, the high-intensity element of the total sweetening power is approximately a quarter to a third (Figure 15.9). Thus the contribution of the drinks to total high-intensity sweetener demand is somewhat lower than would be expected from consumption figures. Again colas, lemonades and concentrates all feature heavily in this subsector.

15.6 Formulations

Figure 15.10 summarises the use of high-intensity sweeteners (in tons of sugar equivalent) in the total soft drinks sector by country in the years 1992 and 2000.

In deriving this assumptions have of course been made about trends in formulations. The major single change here, apart from those forced by legislation, are moves away from pure aspartame because of its lack of stability both in low-pH drinks such as colas, and in environments where ambient storage temperatures are likely to be relatively high, such as in southern Europe.

The assumption has been made that the trends already observed in 1993, with Coca Cola reformulating its products' sweetening agents in Germany, France and Spain to blends varying by country but usually based on aspartame and acesulfame together (in the case of Germany also with cyclamate), will continue and spread to other parts of the community.

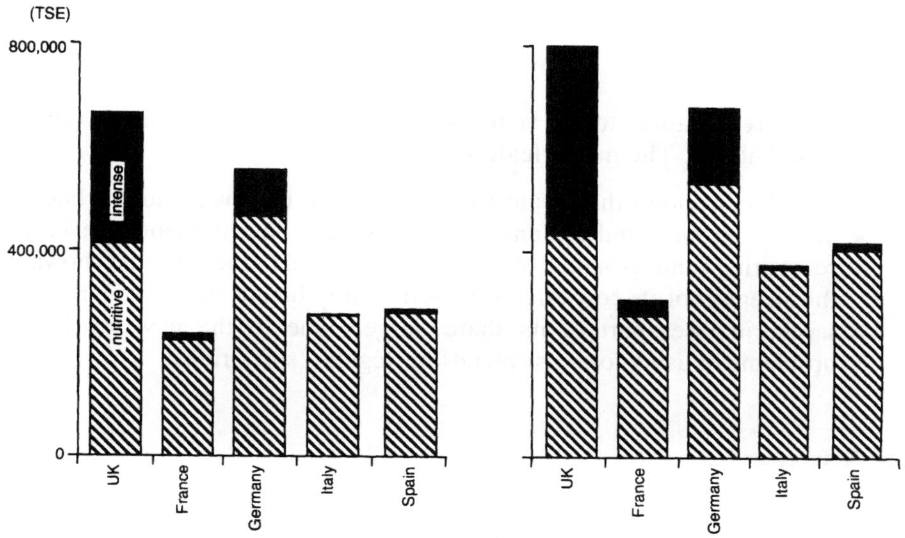

Figure 15.10 5 country use of high intensity and nutritive sweeteners (1992 and 2000).

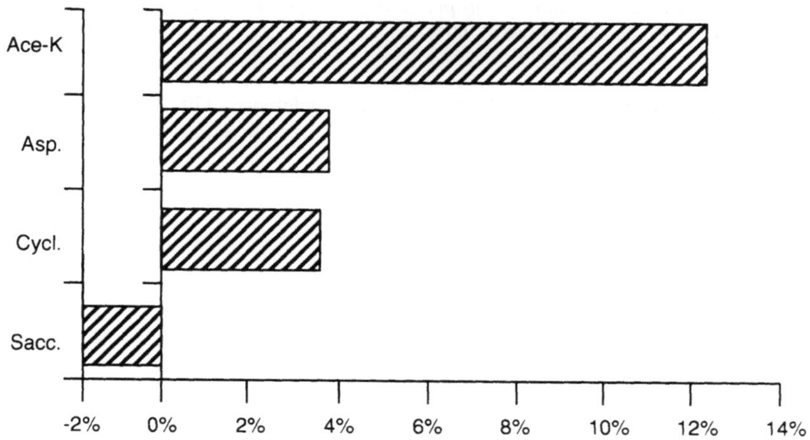

Figure 15.11 High intensity sweeteners: winners and losers. Volume growth p.a. (1992–2000). p.a., per annum.

These predictions have been incorporated in the forecast of disappearance of individual sweeteners in the year 2000 in the soft drinks sector.

15.7 Conclusions

Figure 15.11 provides an overview of the relative gains and losses of high-intensity sweeteners in all sectors:

- soft drinks
- table-top
- solid foods

in the three cultures to the year 2000, based on analyses similar to those outlined above. The major features are:

- acesulfame shows the greatest growth, from a relatively modest base
- both aspartame and cyclamates show steady development, the latter recovering some ground lost when it was banned in the UK, and showing the strength of its cost-effective good quality blends with saccharin
- saccharin loses part of its share, largely due to the new legislative problems in using low-cost blends in 'regular' soft drinks

References*

Giract (1994a) *High Intensity Sweeteners Update – Europe 1980–1992–2000*, Giract, Geneva.
Giract (1994b) *EU Market for Starch, Hydrolysates and Polyols 1995–2000*, Giract, Geneva.
Giract (1994c) *Food EEC 1992+ – The Food System of Europe 2000 under the Impact of the Single European Market – 1989–1993–2000*, Giract, Geneva.
Giract (1994d) *Retailer Brands – Europe: Strategic Opportunities in Food & Drink Products – 1987–1993–1995–2000*, Giract, Geneva.

* If difficulty is experienced in obtaining the above references, Giract may be contacted on telefax number (+41) 22 779 0505.

Index

MIX
Papier aus verantwortungsvollen Quellen
Paper from responsible sources
FSC® C105338
FSC
www.fsc.org

If you have any concerns about our products,
you can contact us on
ProductSafety@springernature.com

In case Publisher is established outside the EU,
the EU authorized representative is:
Springer Nature Customer Service Center GmbH
Europaplatz 3, 69115 Heidelberg, Germany

Printed by Libri Plureos GmbH
in Hamburg, Germany